Remediation Measures for Radioactively Contaminated Areas

Dharmendra K. Gupta • Anna Voronina
Editors

Remediation Measures for Radioactively Contaminated Areas

Springer

Editors
Dharmendra K. Gupta
Gottfried Wilhelm Leibniz Universität Hannover
Institut für Radioökologie und Strahlenschutz (IRS)
Hannover, Germany

Anna Voronina
Radiochemistry and Applied Ecology Department
Ural Federal University
Physical Technology Institute
Ekaterinburg, Russia

ISBN 978-3-030-10370-5 ISBN 978-3-319-73398-2 (eBook)
https://doi.org/10.1007/978-3-319-73398-2

Printed on acid-free paper

This Springer imprint is published by the registered company Springer International Publishing AG part of Springer Nature.
The registered company address is: Gewerbestrasse 11, 6330 Cham, Switzerland

Foreword

It is my great pleasure and honour to introduce this timely new multiauthored monograph *Remediation Measures of Radioactively Contaminated Areas*, edited by Drs. D. K. Gupta and A. Voronina.

The book *Remediation Measures for Radioactively Contaminated Area*s deals with some specific topical issues related to the recent developments and requirements concerning the safety and protection of the population against radiation exposure due to radionuclides released from areas that were contaminated during the past due to uranium mining and processing or other nuclear activities. It is well known that radiation protection standards and criteria have been constantly geared toward minimizing the impact of radioactively contaminated areas on the human environment. These aims have led to some stricter requirements to ensure adequate protection of the population in terms of their radiation exposure in order to satisfy the present dose limits and reference levels. To meet those requirements, some remediation actions have to be carried out in excessively radioactively contaminated areas, taking into account the latest approaches and recommendations of such expert bodies as the International Commission on Radiological Protection (ICRP) or the International Atomic Energy Agency (IAEA).

The ICRP has recommended the relevant derived levels, which should be applied within different exposure situations. In the past, when the standards for protection were not as strict as they are today, in those areas where nuclear activities such as uranium mining and processing were carried out or where nuclear accidents and other nuclear activities were undertaken, different potential pathways of radionuclides that could cause exposure to people must be taken into account. The situation obviously calls for some remediation actions, which can bring the exposure to the population affected in line with the present safety standards. This book demonstrates some procedures aimed at the reduction of this contamination and their consequences in terms of the potential exposure of humans.

The recommended remediation measures and the assessment of their efficiency rely on the latest general ICRP recommendations (ICRP 2007), which define existing exposure situations, such as situations resulting from sources that already existed when a decision to control them was taken. These recommendations have been recently followed and upgraded in the new International Basic Safety Standards compiled by the IAEA, which defines the situations of exposure due to residual radioactive material associated with past practices that were not subject to regulatory control or that remain after an emergency exposure situation. The latter circumstances are commonly referred to as legacies. As a result of evolving standards and/or because new information can come to light about past activities, new forms of legacy sites continue to be recognized, including sites associated with the processing and the use of naturally occurring radioactive material (NORM).

The IAEA, an autonomous organization under the UN, possesses world-class expertise in the safe and peaceful use of nuclear technologies; its safety standards provide a system of fundamental principles and reflect an international consensus on what constitutes a high level of safety for protecting people and the environment from the harmful effects of ionizing radiation (Copplestone et al. 2016). The IAEA safety standards have been developed for all types of nuclear facilities and activities, including radioactive waste management, the decommissioning of nuclear facilities, and environmental remediation. These standards are aimed at reducing radiation exposure from contaminated soil and groundwater and surface water, which may result from activities such as the mining and processing of uranium or the release of radioactive substances to the environment after a nuclear or radiological emergency. The generation of radioactive materials may also arise from some non-nuclear industries, such as oil and gas production, in which exploration and mining activities can increase the potential for exposure due to an excess in naturally occurring radioactive material.

In accordance with the IAEA (Amano 2016), there are four major elements that need to be considered in environmental remediation:

1. Determining the levels of radiation exposure to people that result from the contamination.
2. Reducing radiation doses and risks, making the best use of the available financial, technical, and labor resources.
3. Returning a site to the conditions that existed before the event that caused the contamination. This may not be necessary and is often not easily achievable anyway.
4. In many cases, the main driver for remediation is the public perception of the risks and benefits of undertaking the cleanup activity. In such situations, the overall well-being of the local community is an important factor in determining the planned final state of the site.

The 14 chapters of the book address and discuss various aspects relevant to remediation actions, including the use of some specific technological procedures whose goal is to identify and reduce the impact of radioactively contaminated areas on the exposed population by the decommissioning and environmental remediation measures implemented following the current international standards and best practice.

It is hoped that the book will serve its purpose and will be a good source of relevant information about the recent developments related to methods and techniques being used to reliably treat radioactively contaminated areas.

Lecturer and Sr. Researcher, Department of Crises Management
PACR, Prague, Czech Republic

Lecturer and Sr. Researcher, Czech Technical University, Prague, Czech Republic

Former Radiation Protection Manager, IAEA, Vienna, Austria

Visiting Professor, Institute of Nuclear Sciences, NTHU, Hsinchu, Taiwan

Senior Nuclear Safety Consultant, Pebble Bed Modular (Nuclear) Reactors, Ltd, Pelinbaba, South Africa

Jozef Sabol

References

Amano Y (2016) Decommissioning and remediation: enhancing safety of the public and the environment. IAEA Bulletin, April 2016. www.iaea.org/bulletin

Copplestone D, Larsson CM, Strand P, Sneve MK (2016) Protection of the environment in existing exposure situations. Annals ICRP 45:91–105

ICRP (2007) The 2007 Recommendations of the International Commission on Radiological Protection. ICRP Publication no. 103. Annals ICRP 37:2–4

Preface

Environmental cleanup or remediation denotes to reduce environmental contamination including decontamination of radioactive soil or water. The purpose of remediation is not just to clean radiation sources but also to protect humans, animals, and the environment against any probable harmful effects from ionizing radiation (such as decrease of doses of external and internal irradiation due to consuming radioactive contaminated drinking water and food) as well as to return radioactive contaminated lands to farming use (IAEA).

Radioactive pollution may be defined as "the land in which the radioactivity levels are above the ubiquitous natural and artificial background that is typical of the area in which the land is located." The universal artificial background is taken to include radioactivity from atmospheric testing of nuclear weapons in the 1950s and 1960s, fallout from the Chernobyl and Fukushima accident, and radioactivity resulting from effluent discharges from distant nuclear amenities. As places contaminated by artificial and natural radionuclides or even exposures of natural origin may give upsurge to the requirement for environmental remediation, remediation can only start once a harmony on the necessity to reduce existing or future exposures to ionizing radiation (IAEA). In all cases, the actual work, i.e., adopting certain environmental remediation actions, is always a case specific decision.

Atmosphere, hydrosphere, and soils are the media of initial contamination by radionuclides. Among them, the atmosphere may be a medium of radionuclides transfer to hydrosphere and soil. Depending on the conditions of radionuclides release and transfer, these radionuclides may present in the environmental objects in various species; this will condition the method of rehabilitation measures.

The elementary routes controlling movement of radionuclides (and other trace components) in soil comprise convective conveyance by flowing water, spreading caused by spatial dissimilarities of convection velocities, diffusive crusade within the fluid, and sometimes physicochemical exchanges with soil matrix (Walther and Gupta 2015). After deposition of radionuclides on the soil surface, relocation and movement of these radionuclide in soil depends basically on the soil properties, i.e., pH, texture, interchangeable calcium and potassium, and also on organic matter

(Gupta and Walther 2017), and later these fused radionuclides mixed soil ultimately enter into the soil edifice and taken up by plants via food chain and finally relocating to animals and also to humans (Gupta and Walther 2014). In most cases, mobility of radionuclides for relocation from adulterated land to humans and other organisms is primarily via plant root uptake and transfer to groundwater, which is fundamentally determined by the physicochemical aspects inducing the circulation of radionuclides between the solid and solution phases of soil. The uptake of most components by plant roots occurs mainly through soil solution. The important exchanges of any chemical species in solution, which can influence its mobility in soils and ultimate root uptake, include charge interactions, complexation, and precipitation responses with other element species. Till now radionuclides that have been recognized as relevant to assessments of polluted land are: ^{3}H; ^{14}C; ^{55}Fe; ^{60}Co; ^{63}Ni; ^{90}Sr; ^{99}Tc; ^{129}I; $^{134,137}Cs$; ^{226}Ra; ^{232}Th; $^{235,238}U$; ^{241}Am; and $^{238,239,240,241,242}Pu$.

There are numerous approaches present nowadays, which are useful in remediation of radioactively contaminated areas, and henceforth commonly used methods include deep plowing, addition of fertilizes, ameliorants, and sorption-active materials as well as some environmental biotechnology methods, i.e., transgenic plants and bacteria. After Chernobyl (1986) and Fukushima (2011) accidents, various native bioremediation (phytoremediation) approaches have been employed on terrestrial plants to reduce radionuclides from soil environment. Terrestrial plants commonly accumulate radioactivity by two foremost mechanisms: (1) direct deposition from the atmosphere and (2) root uptake. The transfer factor (TF) is significant expressive parameters that measures the uptake of radionuclides from soil to plants and are valuable devices toward assessment of quantity to the population through ingestion (Gupta and Walther 2016). The total absorption of the radionuclide per unit dry mass in the plants (Bq kg^{-1}) resulting by the one in the soil (also given in Bq kg^{-1}). When any radioactively contaminated land is being remediated with the aim of returning for agricultural use, procedures for decreasing radionuclide transference from soil to vegetation are obligatory (Voronina et al. 2015).

At present, plenty of remediation tools are in existence but officials often tend to value proven expertise; in some cases, the existing expertise are not suitable to accomplish the anticipated goals and further developments are required. For the sustainability of nuclear energy, modern nuclear facilities and operations are designed in a way that also take into account the end of the operation life cycle. In this way, the requirement for extensive environmental remediation activities are minimized (IAEA).

This book focuses mainly on the broad overview of reviews on a number of original publications on remediation strategies mainly after Chernobyl and Fukushima nuclear power plant accidents with some case studies showing the latest technological developments and future trends for affected area decontamination. The key features of this book are related to the radionuclide toxicity in soil and its possible remediation technologies and strategies.

Chapter 1 presents a sound review of the current approaches following the decommissioning of nuclear installations, including a summary of different remediation methods recommended for dealing with radioactive pollutants in water and soil media carried out after decommissioning nuclear facilities worldwide.

The following chapter (Chap. 2) deals with the physicochemical methods used to treat radioactively contaminated sites, with special emphasis on the use of soil washing, soil flushing, and electrokinetic approaches.

The sorption method as a prospective technique for the rehabilitation of radioactively contaminated lands, including the decontamination of natural waters and soils from radionuclides and the decrease of radionuclide transfer into agricultural vegetation, is discussed in Chap. 3.

Chapter 4 focuses attention on presenting the theory of concurrent sorption as an instrument for predicting the conditions for desorption of radionuclides, with special emphasis on removing strontium from contaminated solid-water systems.

Applications of biological, chemical, and nanosorption methods in remediation of metal wastes are discussed in Chap. 5. The method described presents a recent technique, where nanoparticles, nanocomposites, core/shell nanoparticles, as well as nanotubes are employed as adsorbents for the removal, transformation, sorption, and detection of all types of pollutants, including noxious radioactive wastes from soil, air, and water.

The potential of biochar as a measure for decreasing the bioavailability of Cs-137 in soil is a topic discussed in Chap. 6. The chapter focuses on the assessment of the possibility of the use of biochar (the biomass which has undergone pyrolysis processing) and soil-improving additives and their potential for decreasing the transfer of cesium-137 to crop production.

Chapter 7 presents a discussion of the large-scale contamination of agricultural lands that resulted from global fallout and accidents in the Southern Urals (PA "Mayak," 1957) and Chernobyl NPP (1986), including the development, testing, and application of remediation measures proposed to prevent the entrance of Sr-90 into the human diet through soil.

The use of bioremediation/phytoremediation techniques for the rehabilitation of radioactively contaminated soil is described in Chap. 8. Available studies demonstrate that these techniques, combined with other chemical or physically based strategies, have the potential to be used inexpensively and effectively in the restoration of contaminated environments.

Chapter 9 describes biological methods for the decontamination of radioactively contaminated areas. Attention is paid especially to phytoremediation of the environment using plants. These processes can potentially be used for the bioremediation of radioactively contaminated areas.

Chapter 10 presents some prospective uses of modified sorbents based on various supports (nonwoven filtering fabrics, aluminosilicates) for the decontamination of natural water with various salt content being contaminated by natural and anthropogenic radionuclides such as cesium, strontium, uranium, radium, lead, bismuth, and thorium.

The problems related to the treatment of radioactive waste after the rehabilitation of contaminated areas are the topics presented in Chap. 11, which also describes some modern approaches to the decontamination and deactivation solutions of radioactively contaminated soils, the treatment of radioactive biological materials and spent sorption materials, as well as the disposal of radioactive waste after rehabilitation.

The next chapter (Chap. 12) concentrates on modeling the effect of the mechanical remediation of dose rates above the surface of contaminated soil by radiocesium. Some examples of applying this strategy are illustrated as well.

Chap. 13 reflects specific practical experience in applying remediation methods at the Semipalatinsk test site (in Kazakhstan). In addition, some results of remedial efficiency in radioactively contaminated territory by plowing soil are documented.

The final Chap. 14 introduces the basic elements of radiation protection requirements needed for the assessment of the impact of remediation measures in radioactively contaminated areas. Relevant quantities and units for the quantification of radiation exposure of persons due to the presence of radioactive contamination are also outlined.

In whole, the information collected in this book will bring in-depth understanding and expansion of knowledge in the field of radionuclide toxicity and their possible remediation. Dr. Dharmendra K. Gupta and Dr. Anna Voronina individually acknowledge the authors for contributing with their precious time, knowledge, and interest to bring this book into the present shape.

Hannover, Germany — Dharmendra K. Gupta
Ekaterinburg, Russia — Anna Voronina

References

Gupta DK, Walther C (2014) Radionuclide contamination and remediation through plants. Springer, Cham

Gupta DK, Walther C (2016) Impact of cesium on plants and the environment. Springer, Cham

Gupta DK, Walther C (2017) Behaviour of strontium in plants and the environment. Springer, Cham

IAEA, iaea.org/OurWork/NE/NEFW/_nefw-documents/Environmental_Remediation.pdf. Downloaded on 23.10.2017

Voronina AV, Blinova MO, Semenischev VS, Gupta DK (2015) Returning lands, contaminated as a result of radiation accidents, to farming use. J Environ Radioact 144:103–112

Walther C, Gupta DK (2015) Radionuclides in the environment: influence of chemical speciation and plant uptake on radionuclide migration. Springer, Cham

Contents

Review of Remediation Approaches Implemented in Radioactively Contaminated Areas 1
Mirjana Ćujić, Jelena Petrović, and Snežana Dragović

Physicochemical Methods for the Remediation of Radionuclide Contaminated Sites 31
Claudio Cameselle and Susana Gouveia

Sorption Methods in Rehabilitation of Radioactively Contaminated Lands Using Surface-Modified Sorbents Based on Natural Aluminosilicates 51
Anna V. Voronina, Marina O. Blinova, Vladimir S. Semenishchev, and Dharmendra K. Gupta

Competitive Adsorption as a Physicochemical Ground for Self-Sufficient Decontamination Areas from Radioactive Pollutants 65
Evgeny V. Polyakov, Aleksey A. Ioshin, and Ilya V. Volkov

Biological, Chemical and Nanosorption Approaches in Remediation of Metal Wastes 93
Santa Ram Joshi and Debajit Kalita

Potential of Biochar as a Measure for Decreasing Bioavailability of ^{137}Cs in Soil 113
Aleksander Nikolaevich Nikitin, Olga Aleksandrovna Shurankova, Olga Il'inichna Popova, Ihar Anatol'evich Cheshyk, and Ruslan Kovsarovich Spirov

Remediation of Agricultural Lands Contaminated by ^{90}Sr 139
Natalia I. Sanzharova, Irina V. Geshel, Rena A. Mikailova, Dmitry V. Krylenkin, and Ekaterina V. Gordienko

Rehabilitation of Radioactively Contaminated Soil: Use of Bioremediation/Phytoremediation Techniques 163
Joana Lourenço, Sonia Mendo, and Ruth Pereira

Bioremediation and Phytoremediation: Best Approach for Rehabilitation of Soils for Future Use 201
Sergiy Dubchak and Olexander Bondar

Water Decontamination at Radioactively Contaminated Lands 223
Anna V. Voronina, Nickolay D. Betenekov, and Vladimir S. Semenishchev

Treatment of Radioactive Waste After Rehabilitation of Contaminated Areas 245
Anna V. Voronina, Vladimir S. Semenishchev, and Dharmendra K. Gupta

Modelling the Effect of Mechanical Remediation on Dose Rates Above Radiocesium Contaminated Land 259
Alex Malins, Hiroshi Kurikami, Akihiro Kitamura, and Masahiko Machida

Practical Experience in Applying Remediation Methods at the Semipalatinsk Test Site (Kazakhstan) 273
Andrey V. Panitskiy, Sergey N. Lukashenko, Erlan G. Batyrbekov, Sergey A. Berezin, Murat A. Umarov, Yuriy G. Strilchuk, and Vitaliy V. Yakovlev

Basic Radiation Protection Related to the Assessment of Remediation Measures in Radioactively Contaminated Areas 291
Jozef Sabol

Index 315

Contributors

Ihar Anatol'evich Cheshyk State Scientific Institution, Institute of Radiobiology of the National Academy of Sciences of Belarus, Gomel, Republic of Belarus

Erlan G. Batyrbekov National Nuclear Center of Kazakhstan, Kurchatov, Kazakhstan

Sergey A. Berezin National Nuclear Center of Kazakhstan, Kurchatov, Kazakhstan

Nickolay D. Betenekov Radiochemistry and Applied Ecology Department, Physical Technology Institute, Ural Federal University, Ekaterinburg, Russia

Marina O. Blinova Radiochemistry and Applied Ecology Department, Physical Technology Institute, Ural Federal University, Ekaterinburg, Russia

Olexander Bondar State Ecological Academy of Postgraduate Education and Management, Kyiv, Ukraine

Claudio Cameselle Department of Chemical Engineering, University of Vigo, Vigo, Spain

Mirjana Ćujić Vinča Institute of Nuclear Sciences, University of Belgrade, Belgrade, Serbia

Snežana Dragović Vinča Institute of Nuclear Sciences, University of Belgrade, Belgrade, Serbia

Sergiy Dubchak State Ecological Academy of Postgraduate Education and Management, Kyiv, Ukraine

Irina V. Geshel Russian Institute of Radiology and Agroecology, Obninsk, Russian Federation

Ekaterina V. Gordienko Russian Institute of Radiology and Agroecology, Obninsk, Russian Federation

Susana Gouveia Department of Chemical Engineering, University of Vigo, Vigo, Spain

Dharmendra K. Gupta Institut für Radioökologie und Strahlenschutz (IRS), Gottfried Wilhelm Leibniz Universität Hannover, Hannover, Germany

Olga Il'inichna Popova State Scientific Institution, Institute of Radiobiology of the National Academy of Sciences of Belarus, Gomel, Republic of Belarus

Aleksey A. Ioshin Laboratory for Physicochemical Analytical Methods, Institute of Solid State Chemistry, Ural Branch Russian Academy of Sciences, Ekaterinburg, Russia

Santa Ram Joshi Microbiology Laboratory, Department of Biotechnology and Bioinformatics, North-Eastern Hill University, Shillong, India

Debajit Kalita Microbiology Laboratory, Department of Biotechnology and Bioinformatics, North-Eastern Hill University, Shillong, India

Akihiro Kitamura Center for Computational Science and e-Systems, Japan Atomic Energy Agency, Kashiwa, Chiba, Japan

Fukushima Environmental Safety Center, Japan Atomic Energy Agency, Tamura-gun, Fukushima, Japan

Dmitry V. Krylenkin Russian Institute of Radiology and Agroecology, Obninsk, Russian Federation

Hiroshi Kurikami Center for Computational Science and e-Systems, Japan Atomic Energy Agency, Kashiwa, Chiba, Japan

Fukushima Environmental Safety Center, Japan Atomic Energy Agency, Tamura-gun, Fukushima, Japan

Joana Lourenço Department of Biology & CESAM, University of Aveiro, Aveiro, Portugal

Sergey N. Lukashenko National Nuclear Center of Kazakhstan, Kurchatov, Kazakhstan

Masahiko Machida Center for Computational Science and e-Systems, Japan Atomic Energy Agency, Kashiwa, Chiba, Japan

Alex Malins Center for Computational Science and e-Systems, Japan Atomic Energy Agency, Kashiwa, Chiba, Japan

Sonia Mendo Department of Biology & CESAM, University of Aveiro, Aveiro, Portugal

Rena A. Mikailova Russian Institute of Radiology and Agroecology, Obninsk, Russian Federation

Aleksander Nikolaevich Nikitin State Scientific Institution, Institute of Radiobiology of the National Academy of Sciences of Belarus, Gomel, Republic of Belarus

Andrey V. Panitskiy National Nuclear Center of Kazakhstan, Kurchatov, Kazakhstan

Ruth Pereira Faculty of Sciences, Department of Biology & GreenUPorto, University of Porto, Porto, Portugal

Jelena Petrović Vinča Institute of Nuclear Sciences, University of Belgrade, Belgrade, Serbia

Evgeny V. Polyakov Laboratory for Physicochemical Analytical Methods, Institute of Solid State Chemistry, Ural Branch Russian Academy of Sciences, Ekaterinburg, Russia

Jozef Sabol Department of Crises Management, PACR, Prague, Czech Republic

Czech Technical University, Prague, Czech Republic

IAEA, Vienna, Austria

Institute of Nuclear Sciences, NTHU, Hsinchu, Taiwan

Pebble Bed Modular (Nuclear) Reactors, Ltd, Pelinbaba, South Africa

Natalia I. Sanzharova Russian Institute of Radiology and Agroecology, Obninsk, Russian Federation

Vladimir S. Semenishchev Radiochemistry and Applied Ecology Department, Physical Technology Institute, Ural Federal University, Ekaterinburg, Russia

Olga Aleksandrovna Shurankova State Scientific Institution, Institute of Radiobiology of the National Academy of Sciences of Belarus, Gomel, Republic of Belarus

Ruslan Kovsarovich Spirov State Scientific Institution, Institute of Radiobiology of the National Academy of Sciences of Belarus, Gomel, Republic of Belarus

Yuriy G. Strilchuk National Nuclear Center of Kazakhstan, Kurchatov, Kazakhstan

Murat A. Umarov National Nuclear Center of Kazakhstan, Kurchatov, Kazakhstan

Ilya V. Volkov Laboratory for Physicochemical Analytical Methods, Institute of Solid State Chemistry, Ural Branch Russian Academy of Sciences, Ekaterinburg, Russia

Anna V. Voronina Radiochemistry and Applied Ecology Department, Physical Technology Institute, Ural Federal University, Ekaterinburg, Russia

Vitaliy V. Yakovlev National Nuclear Center of Kazakhstan, Kurchatov, Kazakhstan

About the Editors

Dharmendra K. Gupta is Sr. Scientist of environmental biotechnology/radioecology and already published more than 85 refereed research papers/review articles in peer-reviewed journals and edited twelve books. His field of research includes abiotic stress by radionuclides/heavy metals and xenobiotics in plants, antioxidative system in plants, and environmental pollution (radionuclides/heavy metals) remediation through plants (phytoremediation).

Anna Voronina is Chair of the Radiochemistry and Applied Ecology Department at the Ural Federal University, Physical Technology Institute, Ekaterinburg, Russia. She published more than 60 papers in peer-reviewed journals and 9 textbooks and tutorials for students. Her field of research includes development and study of inorganic sorbents, decontamination of radioactively contaminated soils and natural/drinking waters, liquid radioactive waste treatment, and study of radionuclides migration in the environment.

Review of Remediation Approaches Implemented in Radioactively Contaminated Areas

Mirjana Ćujić, Jelena Petrović, and Snežana Dragović

1 Introduction

Closing of the nuclear facilities worldwide is a regulated process which includes their decommissioning or dismantling, followed by appropriate remedial actions to enable that remedial areas can be newly used for the different purposes with the safe levels of the residual radioactivity. Complexity of nuclear site remediation activities involves following steps: problem formulation, investigation before remediation, planning of remediation, actions of remediation, close out and post remedial control of the site. One of the main remediation goals at many nuclear sites is preventing uncontrolled dispersion and transport of radionuclides through the environment. Remediation projects in countries with significant legacy clean-up challenges usually considered an approach of remediation which is based on the risk and considered two main issues: sustainability and costs. International Atomic Energy Agency (IAEA) in International Basic Safety Standards set up a string of requirements to which attention must be paid in the time of the performance of remediation actions (IAEA 1996, 2011). The United States Environmental Protection Agency (US EPA) published technical guidance documents with the purpose to improve the effectiveness of decontamination technologies. In the document 'Technology Reference Guide for Radioactively Contaminated Media' different technologies suitable for treatment of radioactive contamination in different media, liquid (groundwater, surface and/or waste water) or solid (soil, sediment and/ or solid waste) are summarized (USEPA 2007). This Guide provides information on 21 technologies that could be evaluated and mutually compared for site-specific applications. Some additional

M. Ćujić · J. Petrović · S. Dragović (✉)
Vinča Institute of Nuclear Sciences, University of Belgrade, Belgrade, Serbia
e-mail: sdragovic@vinca.rs

D. K. Gupta, A. Voronina (eds.), *Remediation Measures for Radioactively Contaminated Areas*, https://doi.org/10.1007/978-3-319-73398-2_1

information regarding the technologies which are not included in the EPA documents (ITRC 2008) could be found in the literature 'Decontamination and Decommissioning of Radiologically Contaminated Facilities'.

Radioactivity of naturally occurring radionuclides arises mostly from uranium and thorium decays. It is generally low-level and is not considered as contamination. In the area with industrial facilities, e.g. coal fired power plant, there is an increased probability for environmental media to become enriched with natural radionuclides (Dai et al. 2007; Charro et al. 2013; Lu et al. 2012; Ćujić et al. 2015). On the other side, anthropogenic levels of radioactivity in the environment originating from testing of nuclear weapons, nuclear disasters (e.g. Fukushima and Chernobyl) and routine authorised releases to the environment, are contamination, but it is either not under the control of the nuclear facilities operator or authorised by the national regulatory body. Major deposition on surface soils of anthropogenic radionuclide ^{137}Cs throughout Europe arises from nuclear accident in Chernobyl (Raffery et al. 2000; Petrović et al. 2013, 2016a, b; Dragović et al. 2015). Common radionuclides in radioactive wastes are: ^{137}Cs (physical half-life 30.5 year), ^{90}Sr (physical half-life 28.8 year),^{3}H (physical half-life 12.3 year), ^{14}C (physical half-life 5730 year), ^{60}Co (physical half-life 5.3 year), ^{99}Tc (physical half-life 2.1 $\times$ 10^{5} year), ^{129}I (physical half-life 1.6 $\times$ 107 year), ^{226}Ra (physical half-life 1601 year), uranium nuclides (physical half-life 2.5 $\times$ 10^{5} year for ^{234}U, 4.7 $\times$ 10^{9} year for ^{238}U) and transuranic nuclides (longest physical half-life 2.4 $\times$ 10^{4} year for ^{239}Pu). Some elements, such as C, Sr, Tc and I, show high mobility in the environment, and because of that they are of long-term environmental concern and their varying biogeochemical behaviour makes their co-treatment by traditional remediation technique challenging. Monitoring of remediated nuclear sites is an important issue as sometimes contamination from the sites is being spread beyond controlled areas and could potentially affect groundwater users. The selection of appropriate remediation technique is of special importance and it depends on several factors, e.g. existing problems, selection of the contaminants for removing, the soil type, the geology on the local level and goals which want to be reached. It sometimes happens that outside pressure and participation of stakeholders and regulators bring to the wrong finding. This was the case at the Hanford Site where a pump-and-treat system was installed and operated to remediate a ^{90}Sr plume adjacent into the Columbia River. In spite of scientists' warnings to regulators that the system would not be enough effective, it was operated almost 8 years and only less than 1% (67 GBq) of ^{90}Sr was removed. During the same period the activity of 11,840 GBq was removed by radioactive decay. At the end, the costly remedial system was replaced with the permeable reactive barrier to decrease flux to the river and the ^{90}Sr has been left to natural attenuation. Monitored natural attenuation could be an option of clean-up and IAEA recommended its applicability to media contaminated with different radionuclides (IAEA 2006a). The migration velocity of radionuclides which are not attenuated easily ought to significantly reduced and appropriate techniques for that purpose should be developed. *In situ* waste treatment can be an option as it avoids problems originating from 'dilute and disperse' or the 'redistribution' of exposures to settlers in the area surrounding the disposal site (Nisbet and Woodman 2000; Nisbet 2002;

Oughton 2013). The groundwater clean-up systems which are widely used are expensive and need to be applied in practice during long time period. One of the key roles in controlling the speciation and mobility of uranium and other redox sensitive radionuclides (e.g. Tc, Np, Pu) occurred via biogeochemical interactions through direct metabolic processes such as microbial respiration, or indirectly by changing ambient redox/pH conditions, producing ligands or new bio minerals, or altering mineral surfaces. Control of radionuclide mobility via natural attenuation can be stimulated by biogeochemical processes to accelerate clean-up of contaminated environments through bioremediation. Instead of using traditional methods in remediation of contaminated groundwater or surface water, could be used more efficient and cost-effective biological treatment methods. In the recent years, the remediation research is focused on development of new clean-up systems and improvement of the efficiency of the existing ones.

2 Remediation of Water: Case Studies

2.1 Pump and Treat System at Hanford Site (USA)

The US Department of Energy's Hanford Site is located in south-eastern Washington State along the Columbia River. At Hanford, the groundwater pump and treat system is projected to grab and treat contaminated groundwater plume emanating from a plutonium separations plant. It was estimated that approximately 180 billion kilolitres of contaminated water was discharged from the Site to the soil and that some of the contaminants have migrated to groundwater under the site. Contaminated groundwater with the radioactivity levels above groundwater protection standards spread over area of 207 km^2. The major constituents of concern (COC) for the Site are carbon tetrachloride and ^{99}Tc. The other COCs are total chromium, nitrate, trichloroethylene (TCE), ^{129}I, ^{3}H. With the desire to rich clean-up levels contaminants in 125 years, components are projected and applied in practice in combination with monitored natural attenuation. Concentrations of carbon tetrachloride in the groundwater over 100 mg L^{-1} roughly corresponds to 95% of the mass of carbon tetrachloride currently exists in the aquifer. To decrease the mass of carbon tetrachloride by 95%, the estimated pumping rate for this purpose should be 9500 L min^{-1} in the expected timeframe. After extraction, the COCs in groundwater is treated to reach the clean-up grade and returned to the aquifer out of injection wells. Natural attenuation processes are utilized and relied on to decrease concentrations to below the clean-up grade and parts of these processes include abiotic degradation, dispersion, sorption and, radioactive decay for tritium. Flow-path control is reached after the treated groundwater was syringed into the aquifer of the groundwater contamination.

The construction of the treatment facility included six process buildings: two extraction transfer buildings, two injection transfer buildings and the radiological and biological buildings. In the treatment system for the groundwater both radioactive ion exchange system and a biological system in airless conditions to remove the non-radioactive contamination from the groundwater were implemented. Groundwater from extraction wells that feed extraction transfer building contained concentrations of ^{99}Tc higher than 33 Bq L^{-1} and in the radiological building using ion exchange resins the concentrations of radionuclides were reduced. Cartridge filters removed fine particulate matter from incoming groundwater. The ion exchange effluent flow through bag filters which serve as a resin trap, and for a next treatment it was directed to the biological building. When the ion exchange resin reached its loading limit, it was removed from the vessel and was treated for clean-up and returned to plant. After treatment the stripping water was pumped to the biological building for treatment. The resin was sluiced with finished water and placed in a container to allow drainage and the dewatered resin to be transported to the landfill for disposal (US DOE 2010; US DOE, US EPA, Ecology 2009).

2.2 In situ *Permeable Treatment Wall System at West Valley (USA)*

In West Valley (in western New York, USA) at a former commercial nuclear fuel recycling and vitrification test site *in situ* permeable treatment wall (PTW) was implemented with the aim to remove ^{90}Sr from groundwater. The potential of exploiting a permeable barrier was evaluated and found to be reproducible and a pump-and-treat system that was not satisfactory treating ^{90}Sr was replaced. Moor et al. (2000) provided a general overview of the pilot PTW project at the West Valley. Rabideau et al. (2005) reported the results of laboratory and numerical analysis carried out to characterize the potential performance of the West Valley Demonstration Project treatment wall. In the groundwater at the site a plume of ^{90}Sr with radioactivity in the range 37–3700 Bq kg^{-1} was detected. Groundwater flows to the north and east and appears at depths in the range from 3 m below ground level up to the near ground surface. The main remedial actions objectives were to reduce or eliminate presence of ^{90}Sr in groundwater leakage, leaving activity concentrations on the site at the minimum level as practically achievable; minimise the possibilities for expansion of the ^{90}Sr plume beyond its current limits in the future; ensure that a selected technology does not excludes any plans for addressing the plume during site decommissioning. Based on ion exchange reaction ^{90}Sr was removed *in situ* from progressing plume in groundwater. For that purpose thick zone of zeolite (granular formation made of approximately 85% of the mineral clinoptilolite) dimensions 260 m length by 1 m thick was constructed. In wastewater treatment systems zeolite materials have been used extensively as ion exchange media. Robinson et al. (1991) demonstrated that certain natural zeolites possessed a strong selectivity for the caesium radioisotopes and ^{90}Sr. Modelling approach in correlation with the treatment media assessment showed that a 1 m zeolite thickness could satisfactory the desired design-life of 20 years. As the best installation solution for the 1 m thick PTW, single-pass trenching was applied. This technique includes a trencher moving along the alignment and bringing trench spoils to the surface using a chain-saw like cutting boom at the rear of the trencher. The trench was sustained and instantly backfilled behind the cutting boom using a delivery system. Some of the gains compared to other familiar methods are increased efficiency of installation, simply for construction and cost effectiveness.

The PTW performance monitoring system was installed on the PTW platform. Estimated hydraulic performance and ^{90}Sr removal from groundwater was in compliance with experimental results, according to obtained monitoring data. Activity concentrations of analysed ^{90}Sr within the PTW were mainly below 3.7 Bq L^{-1} and often below detectable level. It should be stated that activity concentrations of analysed ^{90}Sr immediately after installation of the PTW were higher than 370 Bq L^{-1}.

2.3 *Monitored Natural Attenuation at the Monticello Mill Tailing Site (USA)*

The Monticello Mill produced uranium and vanadium for military purposes in the period 1942–1960. Mill tailings placed of in piles along a small stream which flows through the site Montezuma Creek were the main source of contamination. Leaching of uranium from the tailings led to groundwater uranium plume within the underlying alluvial aquifer. Approximately 1.5 km down gradient of the former mill site reached the groundwater plume area where maximum contaminant levels were exceeded (US DOE 2009).

After the removal of mill tailings and an additional time of rating, an assessment of plume stability was carried out. The assessment covered the flow directions characterization of groundwater and analysis of contaminant concentrations in groundwater and aquifer solids (US DOE 2004). The removal of tailings piles has led to a significant decreasing of uranium concentration in groundwater of the investigated area. Applying model based on an equilibrium partitioning process between groundwater and aquifer materials, uranium attenuation was simulated and predicted activity concentrations of uranium were likened to observed trends at selected site locations to assess consistency. The solute transport model, with the lower K_d value of 2.8 kg L^{-1} foresee remediation by monitored natural attenuation and results pointed out uranium groundwater concentrations below maximum contaminant level values for approximately 42 years. Davis et al. (2004) reported that K_d values for uranium can vary several orders of magnitude over the pH range 6–9.

2.4 Enhanced Attenuation Technologies at Savannah River Site (USA)

Nuclear processing facilities operating in the period 1955–1988 produced seven billion litres of acidic aqueous waste storage in the F-Area Seepage Basins in Savannah River site. At the beginning it was assumed that under the basin head on connection between radionuclides and soil particles would be occurred. This assumption was correct for many disposed radionuclides, e.g. plutonium isotopes and ^{137}Cs, but for ^{90}Sr, ^{129}I, ^{99}Tc, ^{3}H, uranium isotopes migration into the groundwater in the direction of stream in its nearby has occurred. Heterogeneous geology exists at the site and the aquifer sediments underlying the basins were under the acidic solutions influence more than 40 years period, which contributed to modification in mineralogy of the sediments. The uppermost mechanism of attenuation for all radionuclide contaminants, excluding tritium, is adsorption, and attenuation at particular locations occurred due to dilution.

To measure radionuclide concentrations in the site, monitoring of wells was implemented. Radionuclide activity concentrations in groundwater from neighbouring wells and downgradient of the constrained basins indicated a major flux of radionuclides in the direction from the vadose zone below the basins to the saturated zone. Because of this, monitored natural attenuation was not a solution of the problem. One of the recommended solutions for contamination problem, the pump-and-treat system, was not acceptable because of relatively long period to meet regulatory goals for stream concentrations (several decades) and potentially radioactive solid waste in large quantities will be produced during that time. Because of that decision was made to apply *in situ* treatment technologies that would lead to the low pH and uranium, ^{90}Sr and ^{129}I plumes. Intensive research which included laboratory column studies followed by a field study, pointed out that relieved stabilisation will occurred after application of alkaline solutions with high buffering capacity. Implementation of the enhancement technology involves the assembly of a modified funnel-and-gate system. Monitoring downgradient of the treatment zone showed that adsorption of uranium and ^{90}Sr has occurred. This system showed satisfactory results regarding stabilization of both ^{90}Sr and uranium. The barrier wall was decreasing ^{3}H concentrations in the stream, while ^{129}I remained untreated. One of proposed solution for ^{129}I treatment in this site was interaction with silver chloride particles. Therefore, as ^{129}I in the groundwater interacts with these particles silver iodide is formed as precipitate, reacting with the silver chloride and releasing chloride ion. This access will lead to stabilization of the ^{129}I for a long-term period (PNNL 2009).

3 Microbial Bioreduction as the Potential Co-treatment of the Groundwater Radionuclide Contaminants

The biostimulation of anaerobic microbial communities was investigated as a method for removal of redox-sensitive radionuclides such as U(VI) from contaminated groundwater at nuclear sites. Stimulated uranium bioreduction has been achieved by adding an electron donor to promote enzymatic reduction of aqueous U (VI) to insoluble U(IV). This has been demonstrated in laboratory experiments (Wilkins et al. 2007; Begg et al. 2011; Law et al. 2011) and *in situ* (Istok et al. 2004; Wu et al. 2007; Williams et al. 2011). The direct enzymatic reduction under ambient environmental conditions is the dominant mechanism which mediates the reduction of U(VI) (Williams et al. 2012; Bargar et al. 2013). Biosorption is the passive uptake of contaminants (e.g. radionuclides) to the surface of microbial cells. Suzuki and Banfield (1999) reported that microbial biosorption capacity of uranium uptake in bacteria ranged from 45 to 615 mg g^{-1} cell dry weight. It has been shown that prokaryotes could enzymatically reduce U(VI) (Williams et al. 2012). U (VI) reduction occurred in the sulfate-reducers *Desulfovibrio desulfuricans* and *Desulfovibrio vulgaris*. These species produced the U(IV) mineral uraninite using c-type cytochrome activity (Lovley and Phillips 1992a, b). Enzymatically induced bioreduction of U(VI) involved a wide spectrum of the sulfate-reducer species: *Anaeromyxobacter species* (Sanford et al. 2007), *C. ferrireducens* (Khijniak et al. 2005), *Clostridium* species (Francis et al. 1994; Suzuki et al. 2003; Madden et al. 2007), *Cellulomonas* species (Sivaswamy et al. 2011), *Desulfosporosinus* (Suzuki et al. 2003), *Paenibacillus* (Ahmed et al. 2012). The bioreduction mechanisms of U (VI) were not fully explained. Renshaw et al. (2005) used *Geobacter sulfurreducens* and concluded that U(VI) was decreased to the unstable intermediate U(V), which became disproportionate to the end product U(IV). Gram-positive bacteria may use cytochromes to reduce Fe(III) which could be correlated to U(VI) and radionuclide reduction (Carlson et al. 2012).

Thorpe et al. (2014) investigated potential of microbial enrichment cultures obtained from Sellafield UK representative sediments to reductively precipitate an Fe(II)-bearing biomineral assemblage at both neutral and alkaline pH. This approach was used to investigate effects of microbially-mediated Fe(III) cutting down on the mobility of ^{99}Tc and Sr (stabile Sr^{2+} and radioactive $^{90}Sr^{2+}$). In alkaline conditions occurs microbially-mediated Fe(III) reduction which leads to ^{99}Tc reduction and as a result appear less mobile Tc(IV) phases, and ^{90}Sr via increased sorption and/or precipitation formed at higher pH values. The mobility of redox active ^{99}Tc can be attenuated by reduction from soluble Tc(VII) to poorly soluble Tc(IV) (Lloyd et al. 2000; Burke et al. 2005). The mobility of the $^{90}Sr^{2+}$ ion is mostly controlled by sorption and mineral precipitation reactions (Langley et al. 2009; Thorpe et al. 2012; Wallace et al. 2012, 2013). It is known that species of the *Clostridiales* could reduce Fe(III) in freshwater and marine environments over a wide pH range either via dissimilatory Fe(III) reduction (Dobbin et al. 1999). It was demonstrated that during fermentation process reduction will appear if Fe(III) acts as a minor electron acceptor (Lovley and Phillips 1988; Lehours et al. 2010). It was demonstrated that TcO^{-}_{4} added at 1.6×10^{-6} M was eliminated (more than80%) from solution concurrent with Fe(III) reduction in enrichment culture experiments at pH 7 and pH 9. Reduction of Tc(VII) to Tc(IV) was confirmed by X-ray absorption spectroscopy of the reduced bio minerals. Microbial active systems demonstrated enhanced removal of ^{90}Sr at pH 7 in ultra-trace experiments, compared to the sterile control because of sorption of $^{90}Sr^{2+}$ to the Fe(II)-bearing bio minerals that generated *in situ*. Similar removal of ^{90}Sr to the microbial active experiment shown at pH 9 in sterile control and was ascribed to Sr-sorption to mineral phases in the case of chemically precipitated in sterile controls, and also for the biologically precipitated in the microbial active systems. Throughout the different systems, the amount of the added ^{90}Sr was removed approximately 60–70% in bioreduction experiments. Different treatment approaches involving bioreduction and the promotion of Fe(III)-reducing conditions to remediate Tc(IV) are not coincide with treatment of groundwater ^{90}Sr contamination (Thorpe et al. 2014). The results obtained by Thorpe et al. (2014) pointed out the potential for removal of trace levels of ^{99}Tc and $^{90}Sr^{2+}$ from groundwater during stimulated bioreduction and pointed out that if there exists stable Sr^{2+}, than optimal removal for technetium and strontium is likely to occur under mildly alkaline, reducing conditions. Much greater quantities of Tc(VII) were removed by *Geobacter*, *Anaeromyxobacter* and *Shewanella* in the presence of ferrihydrite compared to experiments with just cells, indicate the importance of biogenic Fe(II) in Tc (VII) bioreduction (Plymale et al. 2011).

Experiments with enrichment cultures from the Oak Ridge site Nyman et al. (2007) found that the inhibition co-efficient for U(VI) was around 100 μM and at this concentration the effective yield and growth rate were reduced by 50%. Spain and Krumholz (2011) reported concentrations of U(VI) up to 11 μM in groundwater close to the S3 ponds at the Oak Ridge site, while Cho et al. (2012) reported value of 250 μM in well FW113–47.

The Old Rifle processing site was permitted to DOE for custody and long-term care and managed by the Office of Legacy Management. The site is under DOE jurisdiction from 1998 and requires routine inspection and maintenance, records-related activities and stakeholder support (USDOE 1999). Groundwater at this former uranium ore processing facility are contaminated with low levels of uranium, which leached from mill tailings into the unconfined aquifer, which is an alluvial deposit of the nearby Colorado River. Groundwater uranium concentrations inside of the test area ranged from 0.4 to 1.4 μM and were above the maximum contaminant limit of 0.18 μM (Anderson et al. 2003). It was also demonstrated that there is potential for *in situ* effective removing of uranium present in contaminated groundwater, by stimulating the activity of *Geobacter* species in the subsurface. When acetate was added to enhance the growth of *Geobacter* species, U(VI) was actively removed (Anderson et al. 2003). Zachara et al. (2013) pointed out that vertical migration of uranium-contaminated groundwater is limited by the Wasatch formation and that release of contaminant U(IV) from naturally reduced zones by oxidation, together with migration of up-gradient groundwater naturally high in U(VI), are responsible for the persistence of elevated concentrations of uranium in groundwater. Williams et al. (2011) carried out research at the Rifle site and concluded that concentrations of U decreased from 1–1.5 μM to 0.05–0.1 μM when acetate was supplied as an electron donor and U(VI) was in a complex uranyl-calcium-carbonate. These low concentrations were achieved by ensuring the concentration of acetate remained greater than the 10 mM present sulfate. Stable isotope probing and gene expression analysis confirmed *Geobacter* were active and oxidising acetate, even during sulfate reduction, and so are likely to be responsible for U(VI) reduction and maintaining low concentrations of U(VI) in groundwater. Bopp et al. (2010) reported that uranium isotope ratios, $^{238}U/^{235}U$, in groundwater decreased significantly during *in situ* bioreduction.

Areas where research should continue include determination of dominant mechanisms in bacterial electron transport, in both natural environment and bio stimulation trials. Newsome et al. (2014) pointed out in their review paper that further work in this area should be directed towards: determination of the exact mechanism(s) of electron transfer to U(VI) in circum neutral aquifer sediments; the reduction of U (VI) at alkaline pH, especially the role played by Gram-positive bacteria and determining the longevity of bioreduced U(IV).

4 Remediation of Soil: Excavation and Disposal

Common clean-up method of radioactive contaminated soil involves excavation/ removal, transportation and disposal of contaminated soil on site or off site (Environment Agency 2002; IAEA 1999, 2006b). During the radiological clean-up of Enewetak atoll of the Marshall Islands in the Pacific Ocean, former United States nuclear test site, conducted from 1972 to 1980, large amount of plutonium-contaminated soil from different islands was removed by excision and collected and transported to the Runit Island and deposited in the Cactus nuclear test crater (DNA 1981). Contaminated soil was mixed with cement to form a slurry and placed into the Cactus crater, formed in 1958 as a result of the nuclear explosion, and the rest of the contaminated material was mixed with concrete and put above ground over the crater in the shape of a dome, and finally the crater was sealed with concrete cap (DNA 1981; Paajanen and Lehto 1992; Lehto 1994; Noshkin and Robison 1997). After the nuclear accidents at Chernobyl NPP in Ukraine (1986) and Fukushima Daiichi NPP in Japan (2011) (Balonov 2007; Hardie and McKinley 2014) various environmental media (including soil) over large areas were contaminated with radionuclides and the different decontamination methods were used in order to reduce radiation exposure from contaminated soil. Vovk et al. (1993) gave an overview of mechanical and physical methods of decontamination of land after Chernobyl NPP accident. The clean-up of the Chernobyl NPP site included removal of highly radioactive soil surface layers (5–10 cm) and their transport to the solid waste storage vault of the fifth unit and also removal of the less radioactive soil and their transport to the disposal areas near to the site (IAEA 1989; Vovk et al. 1993; Rudy and Vovk 1996). A number of waste disposal pits and trenches were dug at different locations adjacent to the Chernobyl NPP (Vovk et al. 1993; Rudy and Vovk 1996). Removal of radioactive soil was commonly implemented as remediation measure in contaminated areas following the accident at the Fukushima Daiichi NPP (Wada et al. 2012; Hardie and McKinley 2014; EPA 2016). According to the results presented in the MOE (2015) report, the very high decontamination rate of surface contamination density (over 90%) was achieved after scraping topsoil by 5 cm from the playgrounds in a kindergarten facility and after scraping topsoil by

5 cm and ploughing of rice fields. Soil and waste generated from decontamination work can be stored in temporary storage site (no more than 3 years), interim storage facilities (less than 30 years) until its final disposal (disposal out of Fukushima) (MOE 2013, 2015; Hardie and McKinley 2014). Some temporary storage sites after the accident were established in Koori town, Yugawa village, Kawamata town, Ten-ei village, Miharu town, Tamakawa village etc. (MOE 2015). In 2012, during the decontamination of sports field, grassed and forest areas around the Yamakiya Elementary School, Kawamata town, located 40 km from the Fukushima Daiichi nuclear power plant, top soil from the sports field and grassed areas was removed and replaced with sand or fresh turf, and leaf litter from the forest areas was removed to a distance of 20 m from the forest edge (Cresswell et al. 2016). In 2014, test decontamination measure carried out at the lower part of the Kami-Oguni River, a tributary of the Abukuma River in Fukushima, located 55 km from the Fukushima Daiichi NPP, reduced the air dose rate by a factor of approximately two (Nishikiori and Suzuki 2017). During the decontamination test process all plants were removed from the flood channel and taken to a waste incineration plant, the upper 5 cm of soil was removed from the dike slopes using excavator and turf was put on the decontaminated slope, and sediments up to 15–35 cm in depth from the flood channel were removed. The removed soil and sediments were packed into flexible container bags, and transport to a temporary storage site for decontaminated soil (Nishikiori and Suzuki 2017). Before decontamination, the air dose rates at 1 and 100 cm above ground were 0.78 and 0.66 μSv h^{-1}, respectively, while after completion of decontamination air dose rates at 1 and 100 cm above ground were 0.34 μSv h^{-1} (Nishikiori and Suzuki 2017). Smaller nuclear accident in 1987 in Goiania, Brazil, compared to the accident at Chernobyl and Fukushima, resulted in the contamination of the environment and various remedial actions were undertaken (such as decontamination of property, collection of contaminated clothing, removal of contaminated soil etc.) in order to clean up the contamination (IAEA 1988). Topsoil was removed (about 60% of caesium total activity was retained in the top 1.5 cm of soil after the accident) and the surfaces were covered either with uncontaminated (clean) soil or concrete (Amaral et al. 1991). Radioactive waste generated during the decontamination was packed and over 200 lorry loads of waste were transported to the temporary disposal site (IAEA 1988). Contaminated waste arising from the clean-up of Salt Lake City, USA, processing site (from 1984 to 1989) in Port Hope, Canada (from 1976 to 1981) – has been transported over long distances to disposal sites (IAEA 1992; Paajanen and Lehto 1992; Lehto 1994). During the clean-up of – Salt Lake City processing site, large amount of contaminated waste (including tailings, contaminated soils and building debris) was transported by trains to the disposal site near Clive, Utah, 140 km away (IAEA 1992; Paajanen and Lehto 1992; Lehto 1994; DOE 2016). From the city of Port Hope more than 100,000 tons of radioactively contaminated soil was transported by trucks to a storage site at the Chalk River Nuclear Laboratories, 350 km away (IAEA 1989, 1992; Paajanen and Lehto 1992; Lehto 1994; Screening Report – Port Hope Project 2006).

5 Remediation of Soil: *In situ* Treatment Technologies

In situ remediation's technologies are used to treat contaminated soil on-site (IAEA 1999). Depending on whether *in situ* remediation technologies reduce mobility, toxicity and/or volume through the treatment, they can be categorized into following groups: (1) containment technologies; (2) stabilization technologies, and (3) treatment technologies (IAEA 1999; EPA 2007). *In situ* treatment technologies are source control technologies that apply various physical/chemical, biological or thermal processes in order to reduce the toxicity and/or volume of the contaminated soil by destroying or removing contaminants (IAEA 1999). According to EPA (2006, 2007) and IAEA (2006b) different *in situ* technologies (such as: electrokinetic remediation, *in situ* vitrification (ISV), phytoremediation etc.) are suitable for treatment of radioactively contaminated soil.

5.1 Electrokinetic Remediation

Separation and extraction of radionuclides from saturated or unsaturated soils by electrokinetic remediation process, requires the application of a direct electric current of low intensity (DC) across electrodes (anode – positive electrode and cathode – negative electrode) placed in the soil (EPA 1995, 2007; Cameselle et al. 2013a, b). Electromigration and electroosmosis are main mechanisms of contaminants' movement in an electric field involved in electrokinetic technology (Kim et al. 2010, 2016a; Cameselle et al. 2013a; Gill et al. 2014; Yang et al. 2014; Silva et al. 2017). Kim GN et al. (2008, 2010, 2011, 2012, 2015, 2016a); Kim SS et al. 2016, were studying electrokinetic-flushing, electrokinetic-electrodialytic and electrokinetic methods to decontaminate radioactive contaminated soil.

Laboratory-scale experiments of the removal of uranium from soil using electrokinetic method performed by Booher et al. (1997) demonstrated that this methods can be used for removal of uranium, in uranyl form (UO_2^{2+}), from unsaturated contaminated soil. Original and improved electrokinetic equipment was manufactured in order to remove uranium from soil (Kim et al. 2011). Original pilot-scale electrokinetic equipment of a 50 L size consisted of a reagent reservoir, an anode room (material for the anode electrode was titanium), an electrokinetic soil cell, a cathode room (material for the cathode electrode was DSA-Dimensional Stable Anode), an equipment support system, a power supply, pH controller. During the experiment with this equipment, lots of metal oxides were generated and covered the cathode plate. Therefore, original pilot-scale electrokinetic equipment was improved by manufacturing immersion-washing device, metal oxide separator and circulation system, and the remediation experiments were conducted again using improved pilot-scale electrokinetic remediation equipment (Kim et al. 2011). Remediation experiments show that when the initial uranium concentration of soil was 50, 75 and 100 Bq g^{-1} the electrokinetic remediation time required to remediate the uranium concentration below clearance concentration of 1.0 Bq g^{-1} was about

34, 42 and 49 days, respectively, and the created waste-solution was 3.8, 4.4 and 5.0 ml g^{-1}, respectively (Kim et al. 2011). The 50 L size electrokinetic remediation equipment (consisted of a electrokinetic soil cell, an anode/cathode rooms, a reagent reservoir, an equipment support system, a pH controller, a power supply) appropriate for the soil characteristics of a nuclear facility site in South Korea, was manufactured in order to remove ^{60}Co and ^{137}Cs from soil (Kim et al. 2010). Remediation experiments were carried out for soil contaminated with 0.01 M of Co^{2+} and Cs^{+} and radioactive soil from storage facility using manufactured electrokinetic remediation equipment. From contaminated soil the removal efficiencies of Co^{2+} was 98.4% and of Cs^{+} was 94.9%, after 15 days, and 3.4 ml g^{-1} of waste solution was generated (Kim et al. 2010). Experiment results for radioactive soil show that reagent type, soil radioactivity concentration, electric current in a soil cell and particle size of soil have influence on the removal efficiencies of cobalt-60 and cesium-137 from soil (Kim et al. 2010). Kim et al. (2010) found that electrokinetic remediation time required to reduce soil radioactivity concentration below clearance concentration level (100 Bq kg^{-1}) depends on initial radioactivity concentrations in soils. After an electrokinetic remediation experiment (conditions – remediation time: 55 days; electric current: 15 mA cm^{-2}) the total removal efficiency of cobalt-60 and cesium-137 from the radioactive soil of 2 kBq kg^{-1} was 95.8% (Kim et al. 2010). The study of the soil remediation using electrokinetic-flushing technology, show that electrokinetic-flushing remediation enhances the removal efficiencies of Co^{2+} and Cs^{+} from the contaminated soil when compared to the electrokinetic remediation (Kim et al. 2008). Electrokinetic method was used to remove cesium from the artificially contaminated Hanford sediment with nonradioactive ^{133}Cs (0.01 M $CsNO_3$) (Jang et al. 2015). After an electrokinetic remediation experiment for 68 days, the removal efficiency of Cs from the bulk sediment was 47% (312 mg kg^{-1}), and the higher removal efficiency of Cs was obtained from the silt-clay fraction (average 52%) than from the sand fraction (20–55%) (Jang et al. 2015).

5.2 In situ *Vitrification*

In situ vitrification (ISV) is a thermal treatment process that uses extremely high temperatures to melt and vitrify contaminated soil (Campbell and Koegler 1990; Spalding et al. 1992; IAEA 1999, 2006b; EPA 2007). Two methods can be used for producing heat for melting the contaminated soil, the older one uses electrodes and electrical resistance (traditional and planar *in situ* vitrification) while the second one uses plasma arc technology. *In situ* vitrification process is described in different studies (Campbell and Koegler 1990; Spalding et al. 1992; IAEA 2006b; EPA 2007).

In the 1980s, *in situ* vitrification process was developed by Pacific Northwest Laboratory (PNL) for the U.S. Department of Energy (DOE), primarily for treating soil contaminated with transuranic radionuclides (Dragun 1991). In 1987, joint ORNL (Oak Ridge National Laboratory) – PNL pilot-scale demonstration of *in situ* vitrification technology for the stabilization of radioactively contaminated soil was completed after 110 h of operation and consumption of about 29 MWh of power, which led to the formation of 25 × 10^3 kg vitrified soil approximately 4.9 m long × 1.2 m thick × 2.1 m wide (Jacobs et al. 1988). A large-scale demonstration of the *in situ* vitrification process, using existing DOE large-scale ISV equipment, was performed on soils contaminated with mixed wastes on the116-B-6A crib at Hanford (Luey et al. 1992; Luey 1993). Test began 11 April 1990 at 2:30 p.m. and ended 23 April 1990 at 2:55 p.m., after 288 h of operation and consumption of about 550 MWh of electrical energy, which resulted in 850 ton block of vitrified soil, the retention of chromium, and lead and cesium-137 in the vitrified block greater than 99.9% (Luey et al. 1992; Luey 1993). During the rehabilitation of the Maralinga, former British nuclear test sites, *in situ* vitrification had been identified by Technical Assessment Group – TAG (established by Australian Government to report on options, and associated costs, for the decontamination and rehabilitation of the former British nuclear test sites in Australia) as best options for stabilization a series of 21 burial pits containing soil and debris primarily contaminated with plutonium and uranium at Taranaki (MARTAC 2003). The GeoMelt-ISV project was developed through four phases – phase I testing and evaluation of Taranaki soils; phase II engineering and intermediate-scale ISV tests; phase III manufacture and testing of the full-scale ISV equipment; and phase IV operation of ISV on the Taranaki pits (MARTAC 2003). Prior to ISV operations the concrete pit caps were removed, the surface area and depth of pit was determined, and the refractory sand trenches and associated instrumentation as well as soil berm were installed. ISV equipment at Taranaki, 1999, is shown in Fig. 1 (MARTAC 2003). A total of 11 pits were treated

Fig. 1 ISV equipment at Taranaki, March 1999 (This figure has been obtained from the Commonwealth of Australia, Report titled "Rehabilitation of Former Nuclear Test Sites at Emu and Maralinga (Australia)", MARTAC 2003)

with the GeoMelt-ISV process until its suspension, the first melt was carried out on pit no. 19 from 21 May 1998 to 25 May 1998 and the last melt was carried out on pit no. 17 from 10 March 1999 to 21 March 1999 when the melt was terminated by the explosion (MARTAC 2003). The ISV materials were removed from Taranaki pits and disposed of either in ISV burial trench or debris burial trench (MARTAC 2003).

5.3 Phytoremediation

Phytoremediation is a bioremediation process that uses plants in order to extract, degrade, contain, or immobilize contaminants that are present in soil (IAEA 2004a, 2006b; Khan et al. 2004; EPA 2007). Among different mechanisms of phytoremediation, phytoextraction and phytostabilization can be used for remediation of radionuclides, since they cannot be biodegraded (EPA 2007). Phytoextraction or phytoaccumulation is a process that involves the uptake/absorption of radionuclides from soil through the plant root system and translocation/accumulation of radionuclides in the aboveground biomass, i.e. shoots and leaves (Dushenkov 2003; IAEA 2004a; EPA 2007). Phytoextraction removes radionuclides without destructive impact on soil structure and with a limited impact on fertility and it is especially suitable for treatment of large areas with a low level of radioactivity (Dushenkov 2003). Above-ground biomass of the plants loaded with radionuclides can be harvested and disposed of (IAEA 2004a; Saleh 2016; Sheoran et al. 2016). Identification of hyperaccumulator of radionuclides is a key step in successful application of phytoextraction. In 2009, the concentrations of U, Th, Ba, Ni, Sr and Pb in the tailings and dominant plant parts (stalk, shoot, root) from the uranium mill tailings repository in South China were determined (Li et al. 2011). Among studied plants, only *Cyperus iria* (U concentration in the shoot, root and tailings was 36.4, 2.43 and 6.03 μg g^{-1}, respectively) and *Parthenocissus quinquefolia* (Sr concentration in leaf, stalk and tailings was 190, 72.2 and 154 μg g^{-1}, respectively) satisfied the criteria for a hyperaccumulator of uranium and strontium (Li et al. 2011). Similar study was conducted in 2011, when the ^{226}Ra activity were determined in the tailings and dominant plant species (stalk, shoot, root) from the uranium mill tailings impoundment in South China (Hu et al. 2014). Among studied plants, only *Pteris multifida* (activities of ^{226}Ra and transfer factors (TFs) found in leaf, stalk and root were: ^{226}Ra – 150.6, 53.65 and 1.782 Bq g^{-1}, respectively, and TF – 9.131, 3.251 and 0.108, respectively), *Pteridium aquilinum* (activities of ^{226}Ra and transfer factors (TFs) found in leaf, stalk and root were: ^{226}Ra – 122.2, 42.36 and 13.14 Bq g^{-1}, respectively, and TF – 7.409, 2.568 and 0.797, respectively), and *Dryopteri sscottii* (activities of ^{226}Ra and transfer factors (TFs) found in leaf, stalk and root were: ^{226}Ra – 105.7, 29.685 and 5.699 Bq g^{-1}, respectively, and TF – 6.408, 1.800 and 0.346, respectively) satisfied the criteria for a hyperaccumulator for ^{226}Ra (Hu et al. 2014). According to Hu et al. (2014) and Li et al. (2011) these plants could be the candidates for phytoremediation of uranium, strontium and ^{226}Ra contaminated soils. Huang et al. (1998) found that citric acid was the most effective in enhancing uranium accumulation in *Brassica juncea* and *Brassica chinensis*. Soudek et al. (2007) investigated the ^{226}Ra activities in the leaves, flowers and seeds of three woody species alder (*Alnus glutinosa*), birch (*Betula pendula*) and elder (*Sambucus nigra*) growing at a mill tailing dump K1 at a previous uranium mill in Mydlovary, South Bohemia. Plant samples were collected in period May–October during 3 years and results showed that^{226}Ra activities in the leaves generally increased during the vegetation periods and the highest ^{226}Ra accumulation occurred in the birch and lowest in alder. Soudek et al. (2007) suggested that these pioneer woody species can be used as remediation alternative to the use of herbs. By analysing plants grown in uranium mining-impacted soils, –, Chen et al. (2005) reported that the root uptake of ^{238}U, ^{226}Ra and ^{232}Th from the soil is plant-specific. The highest TF values for ^{238}U, ^{226}Ra and ^{232}Th were found for lupine (*Lupinus albus*) shoot, white clover (*Trifolium pratense*) shoot, and ryegrass (*Lolium perenne*) shoot, respectively (Chen et al. 2005). *Tridax procumbens L.*, *Azadirachta indica A. Juss.* and *Euphorbia hirta L.* were reported to be effective in phytoremediation of ^{90}Sr contaminated sites, based on the study conducted in the vicinity of Tummalapalle uranium mining site Kadapa district, A.P. India (Ahammad et al. 2015). Dushenkov et al. (1999) found that *Amaranthus retroflexus cvs. L. cv. aureus* and *PT-95* (Table 1) shows highest total removal of ^{137}Cs from soil in the vicinity of Chernobyl NPP. Sugiura et al. (2016)

Table 1 Bioaccumulation coefficient and total ^{137}Cs removal from soil[a]

Species and cultivars	Bioaccumulation coefficient	Total removal (Bq m^{-2})
Amaranthus retroflexus L. cv. PT-95	1.50	3225
Amaranthus retroflexus L. cv. aureus	1.90	2440
Amaranthus retroflexus L. cv. belozernii	1.41	1392
Amaranthus cruentus L.	1.32	1251
Helianthus tuberosum L. x Helianthus annuus L.	0.49	1221
Amaranthus caudatus L.	2.03	1144
Amaranthus cruentus L. cv. myronivka	1.07	1053
Helianthus tuberosus L.	0.30	846
Amaranthus hybridus L.	0.60	719
Amaranthus retroflexus L. cv. Antey	1.07	641
Amaranthus bicolor L.	0.59	417
Amaranthus cruentus L. cv. paniculatus	0.53	412
Zea mays L.	0.28	409
Helianthus annuus L.	0.24	319
Pisum sativum L.	0.48	44
Brassica juncea (L.) Czern.	0.47	194

Adopted from Dushenkov et al. (1999)

[a]Plants were grown at the experimental plot at the Northwest border of Chernobyl, Ukraine, approximately 10 km south of the ChNPP fourth reactor that was damaged in 1986. Bioaccumulation coefficient was calculated as a ratio of ^{137}Cs specific activity in the plant versus ^{137}Cs specific activity in the soil

reported that *Chengiopanax sciadophylloides* has an ability to accumulate radiocesium. The concentrations of ^{137}Cs in leaves of *Chengiopanax sciadophylloides* (28.1 kBq kg^{-1}) were higher comparing to those in leaves of *Quercus serrata* (4.13 Bq kg^{-1}), *Acer crataegifolium* (9.67 Bq kg^{-1}), *Fraxinus sieboldiana* (3.34 Bq kg^{-1}), and *Ilex macropoda* (2.55 Bq kg^{-1}) growing at same site near the town of Kawamata, Fukushima prefecture, located approximately 37 km northwest of F1 NPP.

6 Mathematical Modelling Approaches for Remediation of Contaminated Environments

In order to obtain the most efficient remediation strategy environmental decision aiding technologies are coupled with models for radionuclide transfer in different environmental compartments, models for radiation risk assessment and information on available regulatory and management strategies. A number of software based models (ERICA Tool, RESRAD Family of Codes, NORMALYSA, SYMBIOSE, CROM), internet based calculators (PRG, DCC) and excel based models (UK RCLEA, NJRaSoRs) have been developed so far for radiological risk and dose assessment (ANL 2001; US EPA 2002, 2004; Bureau of Environmental Radiation, NJ 2003; UK DEFRA 2003; Brown et al. 2008; CIEMAT 2011; IAEA 2015; IRSN n.d.). Following Chernobyl accident, the development of a wide range of programs has been started on the international level with the purpose to support the methods and models for assessing environmental impacts of radionuclides and other contaminants were designed, such as VAMP (Validation and Model Predictors), BIOMOVS (BIOsphericMOdel Validation Study) and BIOMASS (BIOsphere Modelling and ASSessments Methods). Within BIOMASS program the particular emphasis is directed towards the improvement of modelling techniques and accuracy of predicted results (IAEA 2004b). Among other important radioecological issues, the program addressed environmental assessment modelling in remediation of contaminated environment due to nuclear accidents, unlimited releases, even unsatisfactory management practices. To estimate the success of the remediation steps environmental transfer models were tested within two scenarios based on the results from the remediation program carried out on a radium extraction site in Olen, Belgium. Olen scenario Type A was aimed at testing the accuracy of predictions of different models used (CLRP, DOSDIM, OLENRAD-A, RISKOLEN and TAMDYN) based on data when remedial actions are performed. It was related to the influence of these actions on the ^{226}Ra concentrations in food chain. Within Olen scenario Type B the following remedial actions were evaluated: removal of the most contaminated soil layer (1 m depth) and capped with a uncontaminated soil layer (0.5 m depth), and their effectiveness assessed in terms of individual doses to an adult farmer arise from external irradiation, inhalation of resuspended particles and radon, consumption of soil particles, drinking water and food products. Six different models (CLRP-RAD, DOSDIM, OLENRAD-B, RESRAD-OFFSITE, RESRAD-ONSITE and TAMDYN-UV) were tested within this type of Olen scenario. The analysis of the model prediction results pointed out that the different interpretations of the scenario contributed more to the differences among the model predictions than the differences in modelling approaches (IAEA 2004b).

As rural environment can remain polluted for long time of period after nuclear accident, the effective recovering strategies are highly needed to protect the population. Within STRATEGY project the model was used to help decision makers in recognising optimal agricultural countermeasure actions under user-defined requirements (Cox and Crout 2003). The application to the model to a hypothetical nuclear power plant accident scenario in Cumbria (UK) showed the significance of relevant information on local level useful for preparation of effective remediation (Cox et al. 2005). This model enabled decision makers to point out the optimal combination of ploughing, administration of ammonium-ferric-cyano-ferrate, clean feeding and food restrictions, taking into account local conditions and implementation costs.

After years of application of countermeasures in radioactively contaminated areas it was recognized that they are not quite effective for ecosystems and society in general. The effectiveness of remediation of radioactively contaminated areas varied with implemented management options (Table 2).A reliable evaluation of cost-benefit balance through predictive models was highly needed in optimization of remediation strategies. Under the European Commission project MOIRA (A Model based computerised system for management support to identify optimal remedial strategies for restoring radionuclide contaminated aquatic ecosystems and drainage areas) the software model was developed for the cost-benefit analysis of applied countermeasures in affected areas and identification of optimal remediation strategies in freshwater ecosystems and complex catchments (Appelgren et al. 1996). Because of site-specific models in some cases could be costly and time consuming there was the demand if predictive models have generic strictures, than can be used for different environmental conditions. The model for assessing radionuclide transport and countermeasure effects in complex catchments (MARTE) established on summarized collective parameters was developed to predict sediment removal, water

Table 2 Summary of reduction factors for selected management options to decrease contamination of agricultural products implemented after Kyshtym and Chernobyl accidents

	Reduction factor		
	Kyshtym accident (crops)	Chernobyl accident	
Management option	^{90}Sr	^{137}Cs	^{90}Sr
Normal ploughing	10–16	2.0–4.0	n.d.
Liming	upto 3	1.5–3.0	1.5–2.6
Application of mineral fertilizers	1.1–1.4	1.5–3.0	1.0–2.0
Clean feeding	3.0–4.0	2.0–5.0	2.0–5.0
Administration of Cs binders to animals	n.d.	3.0–5.0	n.d.
Administration of clay minerals to animals	n.d.	2.0–3.0	n.d.

Adopted from IAEA (2012a)
Note: *n.d.* no data

diversion from sub-catchments and cleaning up of sub-catchments (Monte 2001). The model included sub-models for: (1) radionuclide migration, (2) the assessment of the water balance and the water fluxes and (3) predicting the effects of countermeasures. It was used to a variety of European water bodies in Sweden, Italy and Ukraine contaminated by ^{137}Cs and ^{90}Sr and the obtained outcomes suggested the effectiveness of some remediation strategies, e.g. the control of water from polluted catchments, in complex catchments of radioactively contaminated areas (Monte 2001).

Taking into consideration temporal and spatial variations of radionuclide activities in areas of former Soviet Union affected by Chernobyl accident due to hydrochemical conditions, an environmental decision support system (EDSS) based on geographical information systems (GIS) which integrates all these factors was developed in the frame of the RESTORE project to enable local authorities to manage the remediation of contaminated land (Van der Perk et al. 1998). Data on surface contamination (maps of soil contamination produced by geostatistical interpolation), soil type, land use, food production rates, the density and structure of population were used as input for the EDSS. The EDSS integrated radioecological data (data on transfer of radionuclides to agricultural and wild food products, external doses etc.) with geographical database through dynamic modelling toolkit which included geostatistical analysis, conditional simulation, and topological modelling. The EDSS enabled an insight into radionuclide fluxes in the environment, particularly in the food chain, and also identification of areas vulnerable due to the presence of population groups which received an increased radiation doses.

After the decade of remediation in affected areas in Belarus, Russia and Ukraine it was noticed that due to a large variety of factors governing the radionuclide transfer in the environment the identical countermeasures are not appropriate for each contaminated area and that there is a need for integrated remediation strategies adapted to site-specific conditions. In a due course an approach was developed based on wide radioecological data collected from 70 representative contaminated colonies in affected areas taking into account the distribution of different soil types in these colonies (Jacob et al. 2001). For eight remedial actions for large-scale application (radical improvement, Prussian blue, supply clean milk, unpolluted feed for pigs, mineral fertilizer for potatoes, limit mushroom consumption, food monitors and removal of soil), were consider basic data on reduction factors, time periods of effectiveness and costs. Taking into account radioecological data for each settlement (^{137}Cs activity per unit area, transfer factors for foodstuffs) and data on fraction of the foodstuffs produced locally and on their annual consumption in affected areas, a dose model was established and representative remediation strategies were derived. These data were used to evaluate remediation strategies for different radioecological classes in which affected area was subdivided and results were then generalised to the total affected population (Jacob et al. 2001).

The experiences in implementation of decision aiding technologies in areas affected by Chernobyl accident pointed out that these technologies, to be effectively used, have to be supported by models for radionuclide transfer in the environment, radiation risk models and information on available management and decision making options (IAEA 2012b). Two decades after Chernobyl accident there were a few hundreds of settlements in Belarus, Russia and Ukraine in which population suffered annual effective doses due to ionizing radiation caused by accident which exceeded 1 mSv. Therefore, the IAEA initiated the project Radiological support for the rehabilitation of the areas affected by the Chernobyl nuclear power plant accident. For the purpose of this project, a software tool ReSCA–Remediation Strategies after the Chernobyl Accident has been established (Ulanovsky et al. 2011) to facilitate optimization of remediation strategies in the rural settlements of the affected areas. The model used settlement-specific data on ground contamination and activity contents in foodstuffs (Jacob et al. 2001). Only exposures due to ^{137}Cs as only relevant radionuclide after decades from the accident were considered in the model. Based on radioecological data collected for rural settlements in which the annual effective dose exceeds 1 mSv annual effective doses were calculated with ReSCA and validated with data from whole body counter measurements (Jacob et al. 2009). The implementation of the model has shown that a substantial collective dose can be averted by quite cost-effective remediation strategies (Jacob et al. 2009). The FORESTLAND EDSS applied in Bryansk region of Russian Federation for justification of remediation strategies of forests affected by Chernobyl accident. The results indicated the ingestion of milk from cows grazing forest meadows as the main pathway of contamination and that forest management options are the most effective way to decrease the long term impact of radiocaesium contamination (IAEA 2012b).

With the aim to evaluate dose rates to marine biota after Fukushima accident Vives i Batlle et al. (2016) report an inter-comparison of eight models designed to predict the radiological exposure of radionuclides. In this work model predictions of radionuclide uptake and turnover using kinetic calculations based on biological half-life ($T_{b1/2}$) and/or more complex metabolic modelling approaches were used to predict activity concentrations and dose rates of ^{90}Sr, ^{131}I and ^{137}Cs to fish, crustaceans, macroalgae and molluscs under circumstances where the water concentrations are changing with time.

The modelling approaches were also successfully applied to resolve some important issues in optimization of bioremediation strategies. It has been widely recognized that dissimilatory metal reducing bacteria grown on acetate could be applied as a technique for remediation purpose to immobilize hexavalent uranium [U(VI)] as precipitated U(IV). However, Anderson et al. (2003) identified the inefficiency of U (VI) removal from groundwater when acetate-oxidizing sulfate reducers became dominant as an important consideration in the optimization of a bioremediation strategy. The multicomponent reactive transport model based on acetate-oxidizing iron- and sulfate reducing microbial populations was developed to picture the bioreduction of a continuous influx of soluble U(VI) in groundwater to insoluble and immobile U(IV). The model was applied for identification of the dominant transport and biological processes controlling uranium mobility throughout acetate biostimulation in the unconfined aquifer of the Old Rifle UMTRA field site in western Colorado (Yabusaki et al. 2007). It was further developed to involve abiotic reactions along with the biologically mediated reaction network in the multicomponent biogeochemical reactive transport simulator (Fang et al. 2009).

In planning, guiding and evaluating remediation, locally varying conditions such as temporal variation of radionuclide activity concentrations in different environmental compartments, need to be taken into account. The linking of radioecological models with GIS enables optimization of remediation strategies according to local variations of radionuclide activity concentrations. Following Chernobyl accident, a number of EDSSs such as FORCON (Fesenko et al. 1996), SAVE (Howard and Wright 1999), RESTORE (Van der Perk et al. 1998) and ReSCA (Ulanovsky et al. 2011; Fesenko et al. 2012), were developed to facilitate optimization of remediation strategies in contaminated areas.

The understanding of physical, chemical and biological processes governing behaviour of contaminants in aqueous environments is necessary for assessment of efficiency of terrestrial environmental remediation strategies. Therefore, a number of mechanistically-based reactive chemical transport models have been developed in the decades after Chernobyl accident (Yeh and Tripathi 1990; Pruess 1991; Zyvoloski et al. 1994; Fang et al. 2003; Zhang et al. 2007). Yeh et al. (2013) reviewed the development of mechanistically-based numerical models and demonstrated their flexibility and range of capabilities through the application to environmental remediation problems such as waste disposals and uranium mill tailings.

Mathematical modelling methods were successfully used for solving complex decision-making problems such as the selection of alternative remediation options. Jimenez-Martin et al. (2016) proposed an approach based on multi-criteria and fuzzy logic which takes into account environmental, radiological, social and economic impact as the main criteria for the appropriate management of the site (Perez-Sanchez et al. 2015). This approach was applied for evaluating different remediation options for the Zapadnoe uranium mill tailings in Ukraine (Jimenez-Martin et al. 2016).

One of the most challenging issues in remediation of closed uranium mining sites is the treatment of mine water. To ensure the sufficient removal of pollutants from these waters it is necessary to understand underlying reaction processes which could be interpreted by hydrogeochemical modelling. This approach was used at Wismut mine sites to investigate reaction processes during the treatment of mine water, i.e. for hydrogeochemical understanding of precipitation processes in a CO_2 stripping column with a focus on the influence of pH value and for characterization of uranium speciation of mine water (Lietsch et al. 2015). The modelling derived the prognosis of necessary pH values in order to prevent oversaturation inside the stripping column. The model outputs were tested at mine water treatment plant where lower pH values showed significant improvements. The application of hydrogeochemical modelling for identification of typical uranium species and complexes in mine water revealed the dominance of calcium uranyl carbonates in mine water at alkaline and neutral conditions and uranyl sulfates in acidic conditions (Lietsch et al. 2015). Authors emphasized the need for further investigation of thermodynamic parameters of uranyl species.

The widespread contamination in subsurface environments makes the remediation process economically challenging and in such cases there is a need for alternative approaches. In the case of uranium contamination such an approach implies changes in physicochemical conditions of the environment to promote *in situ* immobilization of uranium, for example by addition of phosphate amendments (Wellman et al. 2008; Mehta et al. 2014). The understanding of processes governing U(VI) transport under the influence of phosphate amendments, can result in identification of conditions for reduction of its mobility. Mehta et al. (2015) used a reactive transport model to evaluate a transport of U(VI) through contaminated sediments. The application of the model in batch and column experiments with sediments from a Rifle site, Colorado, indicated that addition of phosphate could result in uranium retention in sediments to some extent.

Up to now, much effort has been devoted to explore both new technologies and new types of materials for remediation of contaminated wastewater. Modelling is often used in characterization of capacity and mechanisms of sorption of radionuclides on materials proposed to be used for environmental remediation. Zhang et al. (2015) investigated interaction between U(VI) and carbonaceous nanofibers by applying double layer sorption model. They found out that when co-existing with montmorillonite, carbonaceous nanofibers could extract the sorbed uranium onto their surface by a pseudo-second order kinetic process. The modelling results indicate carbonaceous nanofibers as a promising material for wastewater remediation. Different kinetic and equilibrium models are used to understand the mechanism of uranium adsorption from contaminated water. Naeem et al. (2017) investigated mechanism of uranium adsorption on green mung (*Vigna radiata*) by applying Freundlich isotherm and pseudo-second order kinetic model and proved its efficiency for uranium removal from aqueous solutions.

Within EMRAS (Environmental Modelling for Radiation Safety) Programme several mathematical models (EXPURT, METRO-K, CPHR, RESRAD-RDD) used for assessment related to the remediation of urban areas contaminated with radionuclides were tested (IAEA 2012a). The EXPURT model calculates surface activity densities and external gamma dose rates as a function of time in built environments and can represent the implementation of countermeasures such as decontamination and soil mixing. The EXPURT embedded within CONDO software tool can assist decision makers by providing estimations of normal living doses, doses from inhalation of resuspended radionuclides and waste activity concentrations (IAEA 2012a). The METRO-K model with a simple mathematical structure gives concentrations of radioactive materials and associate exposure doses as a function of time for different environments. Different countermeasures such as cutting and removal of grass, removal of trees, leaves, soil and washing of different surfaces could be considered within this model (IAEA 2012a). The compartment model CPHR is based on Ecolego tool[1] and it is used to assess time dependence of radioactive contamination and associated dose rates due to environmental processes (IAEA 2012a). The RESRAD-RDD compartment model is used to derive external exposure rates at receptor locations from different surfaces, radionuclide surface concentrations, annual and cumulative external doses for receptors

[1]http://www.facilia.se/products/ecolego.asp

(IAEA 2012a). It also considers different countermeasures such as cutting and removal of grass, removal of soil, washing and relocation. The models were tested on three hypothetical scenarios (point source, area source and area source + river). By comparing the model outputs a number of issues were derived where model capabilities could be improved and uncertainties reduced (information on initial distribution of contaminants and their transport in urban conditions, environmental data and data on radionuclides other than ^{137}Cs).

7 Conclusions

In the past few decades a much effort has been made to clean up the surrounding environment after decommissioning of nuclear installations. It is important to point out that there is no universal remediation plan how to remediate some media (water or soil). Because of plenty of factors that influence the outcome of the media rehabilitation process, it is important to made selection of optimal solution for each considered case study. International organisations (e.g. IAEA, EPA) made a lot of effort to provide guides and recommendation for different remediation approach based on research studies and the practical experience. More mobile radionuclides seek for more effective methods of remediation and that should be an area for coordinated international research in the future. It is of major importance in groundwater remediation strategy to develop new clean-up systems or to improve the efficiency of the existing ones with the goal to minimise the generation of secondary wastes. Monitored natural attenuation may be an alternative to clean-up radiologically contaminated media. Regarding soil remediation, developing remediation techniques for land underneath operating facilities is of major interest. Instead the most common conventional *ex situ* (soil removal, chemical extraction) remediation technique used for soil, *in situ* technologies (e.g. bioremediation, phytoremediation) is becoming increasingly used. Different approaches based on mathematical modelling methods have been proven to be successful in a case of complex decision-making problems, e.g. the selection of alternative remediation options.

Acknowledgments The authors gratefully acknowledge the financial support from the Ministry of Education, Science and Technological Development of the Republic of Serbia (Project No. III43009).

References

Ahammad JS, Sumithra S, Yamuna RT (2015) Screening for strontium accumulator plants in the vicinity of Tummalapalle uranium mining site Kadapa District, A.P. India. Int J Environ Sci 6:37–45

Ahmed B, Cao B, McLean JS, Ica T, Dohnalkova A, Istanbullu O, Paksoy A, Fredrickson JK, Beyenal H (2012) Fe(III) Reduction and U(VI) immobilization by *Paenibacillus sp.* strain

300A, isolated from Hanford 300A subsurface sediments. Appl Environ Microbiol 78:8001–8009

Amaral EC, Vianna ME, Godoy JM, Rochedo ER, Campos MJ, do Rio MA, Oliveira JP, Pereira JC, Reis WG (1991) Distribution of ^{137}Cs in soils due to the Goiânia accident and decisions for remedial action during the recovery phase. Health Phys 60:91–98

Anderson RT, Vrionis HA, Ortiz-Bernad I, Resch CT, Long PE, Dayvault R, Karp K, Marutzky S, Metzler DR, Peacock A, White DC, Lowe M, Lovley DR (2003) Stimulating the in situ activity of Geobacter species to remove uranium from the groundwater of a uranium-contaminated aquifer. Appl Environ Microbiol 69:5884–5891

ANL (2001) User's manual for RESRAD version 6. Environmental Assessment Division, Argonne National Laboratory, Argonne, https://web.evs.anl.gov/resrad/RESRAD_Family/

Appelgren A, Bergström U, Brittain J, Gallego Diaz E, Håkanson L, Heling R, Monte L (1996) An outline of a model-based expert system to identify optimal remedial strategies for restoring contaminated aquatic ecosystems: the project "MOIRA", ENEA, RT/AMB/96/17, Roma

Balonov MI (2007) The Chernobyl Forum: major findings and recommendations. J Environ Radioact 96:6–12

Bargar JR, Williams KH, Campbell KM, Long PE, Stubbs JE, Suvorova EI, Lezama-Pacheco JS, Alessi DS, Stylo M, Webb SM, Davis JA, Giammar DE, Blue LY, Bernier-Latmani R (2013) Uranium redox transition pathways in acetate-amended sediments. Proc Natl Acad Sci U S A 110:4506–4511

Begg JDC, Burke IT, Lloyd JR, Boothman C, Shaw S, Charnock JM, Morris K (2011) Bioreduction behavior of U(VI) sorbed to sediments. Geomicrobiol J 28:160–171

Booher WF, Lindgren ER, Brady PV (1997) Electrokinetic removal of uranium from contaminated, unsaturated soils. Prepared by Sandia National Laboratories for U.S. DOE. Sandia Report SAND97–0122

Bopp CJ, Lundstrom CC, Johnson TM, Sanford RA, Long PE, Williams KH (2010) Uranium 238U/235U isotope ratios as indicators of reduction: results from an in situ biostimulation experiment at Rifle, Colorado, USA. Environ Sci Technol 44:5927–5933

Brown JE, Alfonso B, Avila R, Beresford NA, Copplestone D, Pröhl G, Ulanovsky A (2008) The ERICA tool. J Environ Radioact 99:1371–1383

Bureau of Environmental Radiation, NJ (2003) Radioactive Soil Remediation Standards (RaSoRS). http://www.nj.gov/dep/rpp/rms/rad_cleanups.htm

Burke IT, Boothman C, Lloyd JR, Mortimer RJM, Livens FR, Morris K (2005) Effects of progressive anoxia on the solubility of technetium in sediments. Environ Sci Technol 39:4109–4116

Cameselle C, Chirakkara RA, Reddy KR (2013a) Electrokinetic-enhanced phytoremediation of soils: status and opportunities. Chemosphere 93:626–636

Cameselle C, Gouveia S, Akretche DE, Belhadj B (2013b) Advances in electrokinetic remediation for the removal of organic contaminants in soils. In: Rashed MN (ed) Organic pollutants – monitoring, risk and treatment. Intech Open Publisher, pp 209–229

Campbell BE, Koegler SS (1990) In situ vitrification of a mixed radioactive and hazardous waste site. Presented at the American Institure of Chemical Engineer 1990. Annual Meeting November 11–16, Chicago, Illinois

Carlson HK, Iavarone AT, Gorur A, Yeo BS, Tran R, Melnyk RA, Mathies RA, Aue M, Coates JD (2012) Surface multiheme c-type cytochromes from *Thermincola potens* and implications for respiratory metal reduction by Gram-positive bacteria. Proc Natl Acad Sci U S A 109:1702–1707

Charro E, Pardo R, Peña V (2013) Chemometric interpretation of vertical profiles of radionuclides in soil near a Spanish coal-fired power plant. Chemosphere 90:488–496

Chen SB, Zhu YG, Hu QH (2005) Soil to plant transfer of ^{238}U, ^{226}Ra and ^{232}Th on a uranium mining-impacted soil from southeastern China. J Environ Radioact 82:223–236

Cho K, Zholi A, Frabutt D, Flood M, Floyd D, Tiquia SM (2012) Linking bacterial diversity and geochemistry of uranium-contaminated groundwater. Environ Technol 33:1629–1640

CIEMAT (2011) Screening model for environmental assessment (CROM). ftp://ftp.ciemat.es/pub/CROM/CROM_8/

Cox GM, Crout NMJ (2003) A methodology for the selection and optimisation of countermeasure strategies, report for the STRATEGY project (CEC Contract No: FIKR-CT 2000-00018), University of Nottingham, UK

Cox G, Beresford NA, Alvarez-Farizo B, Oughton D, Kis Z, Eged K, Thørring H, Hunt J, Wright S, Barnett CL, Gil JM, Howard BJ, Crout NMJ (2005) Identifying optimal agricultural countermeasure strategies for a hypothetical contamination scenario using the strategy model. J Environ Radioact 83:383–397

Cresswell AJ, Kato H, Onda Y, Nanba K (2016) Evaluation of forest decontamination using radiometric measurements. J Environ Radioact 164:133–144

Ćujić M, Dragović S, Đorđević M, Dragović R, Gajić B, Miljanić Š (2015) Radionuclides in the soil around the largest coal-fired power plant in Serbia: radiological hazard, relationship with soil characteristics and spatial distribution. Environ Sci Pollut Res 22:10317–10330

Dai L, Wei H, Wang L (2007) Spatial distribution and risk assessment of radionuclides in soils around a coal-fired power plant: a case study from the city of Baoji, China. Environ Res 104:201–208

Davis JA, Meece DE, Kohler M, Curtis GP (2004) Approaches to surface complexation modelling of uranium (VI) adsorption on aquifer sediments. Geochim Cosmochim Acta 68:3621–3641

DNA (1981) The radiological cleanup of Enewetak Atoll. Defense Nuclear Agency, Washington, DC

Dobbin PS, Carter JP, San Juan CGS, von Hobe M, Powell AK, Richardson DJ (1999) Dissimilatory Fe(III) reduction by *Clostridium beijenrickii* isolated from freshwater sediment using Fe (III) maltol enrichment. FEMS Microbiol Lett 176:131–138

DOE (2016) Salt Lake City, Utah, processing and disposal sites –fact sheet, U.S. Department of Energy Office of Legacy Management

Dragović S, Petrović J, Dragović R, Đorđević M, Đokić M, Gajić B (2015) The influence of edaphic factors on spatial and vertical distribution of radionuclides in soil. In: Walther C, Gupta DK (eds) Radionuclides in the environment: influence of chemical speciation and plant uptake on radionuclide migration. Springer, Cham, pp 61–80

Dragun J (1991) Geochemistry and soil chemistry reactions occurring during in situ vitrification. J Hazard Mater 26:343–364

Dushenkov S (2003) Trends in phytoremediation of radionuclides. Plant Soil 249:167–175

Dushenkov S, Mikheev A, Prokhnevsky A, Ruchko M, Sorochinsky B (1999) Phytoremediation of radiocesium-contaminated soil in the vicinity of Chernobyl, Ukraine. Environ Sci Technol 33:469–475

Environment Agency (2002) Guidance on the characterisation and remediation of radioactively contaminated land

EPA (1995) In situ remediation technology: electrokinetics. EPA542-K-94-007. U.S. Environmental Protection Agency Office of Solid Waste and Emergency Response Technology Innovation Office Washington, D.C.

EPA (2006) In situ treatment technologies for contaminated soil. EPA 542/F-06/013. U.S. Environmental Protection Agency

EPA (2007) Technology reference guide for radioactively contaminated media. EPA 402-R-07-004. U.S. Environmental Protection Agency Office of Air and Radiation Office of Radiation and Indoor Air Radiation Protection Division Center for Radiation Site Cleanup

EPA (2016) Current and emerging post-Fukushima technologies, and techniques, and practices for wide area radiological survey, remediation, and waste management. EPA/600/R-16-140. U.S. Environmental Protection Agency

Fang Y, Yeh GT, Burgos WD (2003) A generic paradigm to model reaction-based biogeochemical processes in batch systems. Water Resour Res 33:1083–1118

Fang Y, Yabusaki SB, Morisson SJ, Amonette JP, Long PE (2009) Multicomponent reactive transport modeling of uranium bioremediation field experiments. Geochim Cosmochim Acta 73:6029–6051

Fesenko SV, Sanzharova NI, Wilkins BT, Nisbet AF (1996) FORCON: local decision support system for the provision of advice in agriculture–methodology and experience of practical implementation. Radiat Prot Dosim 64:157–164

Fesenko S, Jacob P, Ulanovsky A, Chupov A, Bogdevich I, Sanzharova N (2012) Justification of remediation strategies in the long term after the Chernobyl accident. J Environ Radioact 119:39–47

Francis AJ, Dodge CJ, Lu F, Halada GP, Clayton CR (1994) XPS and XANES studies of uranium reduction by Clostridium sp. Environ Sci Technol 28:636–639

Gill RT, Harbottle MJ, Smith JWN, Thornton SF (2014) Electrokinetic-enhanced bioremediation of organic contaminants: a review of processes and environmental applications. Chemosphere 107:31–42

Hardie SML, McKinley IG (2014) Fukushima remediation: status and overview of future plans. J Environ Radioact 133:75–85

Howard BJ, Wright SM (1999) Spatial analysis of vulnerable ecosystems in Europe: spatial and dynamic prediction of Radiocaesium fluxes into European foods (SAVE). Final Report, Commission of European Communities, Brussels

Hu N, Ding D, Li G, Zheng J, Li L, Zhao W, Wang Y (2014) Vegetation composition and ^{226}Ra uptake by native plant species at a uranium mill tailings impoundment in South China. J Environ Radioact 129:100–106

Huang JW, Blaylock MJ, Kapulnik Y, Ensley BD (1998) Phytoremediation of uranium-contaminated soils: role of organic acids in triggering uranium hyperaccumulation in plants. Environ Sci Technol 32:2004–2008

IAEA (1988) The radiological accident in Goiânia.International Atomic Energy Agency (IAEA). Vienna

IAEA (1989) Cleanup of large areas contaminated as a result of a nuclear accident. Technical Reports Series No. 300. International Atomic Energy Agency (IAEA), Vienna

IAEA (1992) Disposal of waste from the cleanup of large areas contaminated as a result of a nuclear accident. Technical Reports Series No. 330. International Atomic Energy Agency (IAEA), Vienna

IAEA (1996) International basic safety standards for protection against ionising radiation and for the safety of radiation sources. Safety Series No. 115. International Atomic Energy Agency (IAEA), Vienna

IAEA (1999) Technologies for remediation of radioactively contaminated sites, IAEA TECDOC-1086. International Atomic Energy Agency (IAEA), Vienna

IAEA (2004a) Remediation of sites with dispersed radioactive contamination. Technical Reports Series No. 424. International Atomic Energy Agency (IAEA),Vienna

IAEA (2004b) Testing of environmental transfer models using data from the remediation of a radium extraction site report of the remediation assessment working group of BIOMASS theme 2, IAEA-BIOMASS-7, Vienna

IAEA (2006a) Applicability of monitored natural attenuation at radioactively contaminated sites. Technical Reports Series No. 445. International Atomic Energy Agency (IAEA), Vienna

IAEA (2006b) Remediation of sites with mixed contamination of radioactive and other hazardous substances. Technical Reports Series No. 442. International Atomic Energy Agency (IAEA), Venna

IAEA (2011) Radiation protection and safety of radiation sources: international basic safety standards for protecting people and the environment INTERIM EDITION No. GSR Part 3 (Interim) General Safety Requirements. International Atomic Energy Agency (IAEA), Vienna

IAEA (2012a) Environmental modelling for radiation safety (EMRAS). A summary report of the results of the EMRAS Programme (2003–2007), Vienna

IAEA (2012b) Guidelines for remediation strategies to reduce the radiological consequences of environmental contamination. Technical Report Series No. 475. International Atomic Energy Agency (IAEA), Vienna

IAEA (2015) NORM and Legacy Site Assessment (NORMALYSA). http://project.facilia.se/normalysa/software.html

IRSN (n.d.) SYMBIOSE. https://gforge.irsn.fr/gf/project/symbiose/

Istok JD, Senko JM, Krumholz LR, Watson D, Bogle MA, Peacock A, Chang YJ, White DC (2004) In situ bioreduction of technetium and uranium in a nitrate contaminated aquifer. Environ Sci Technol 38:468–475

ITRC (2008) Decontamination and decommissioning of radiologically contaminated facilities. Interstate Technology and Regulatory Council (ITRC), Washington, DC

Jacob P, Fesenko S, Kirsakova SK, Likhtareva IA, Schotola C, Alexakhin RM, Zhuchenko YM, Kovgan L, Sanzharova NI, Ageyets V (2001) Remediation strategies for rural territories contaminated by the Chernobyl accident. J Environ Radioact 56:51–76

Jacob P, Fesenko S, Bogdevitch I, Kashparov V, Sanzharova N, Grebenshikova N, Isamov N, Lazarev N, Panov A, Ulanovsky A, Zhuchenko Y, Zhurba M (2009) Rural areas affected by the Chernobyl accident: radiation exposure and remediation strategies. Sci Total Environ 408:14–25

Jacobs GK, Spalding BP, Carter JG, Koegler SS (1988) In situ vitrification demonstration for the stabilization of buried wastes at the Oak Ridge National Laboratory. Nuc Chem Waste Manag 8:249–259

Jang HB, Yang JS, Um W (2015) Bench-scale electrokinetic remediation for cesium-contaminated sediment at the Hanford Site, USA. J Radioanal Nucl Chem 304:615–625

Jimenez-Martin A, Martin M, Perez-Sanchez D, Mateos A, Dvorzhak A (2016) Zapadnoe uranium mill tailings: a fuzzy approach. MVLSC 26:529–540

Khan FI, Husain T, Hejazi R (2004) An overview and analysis of site remediation technologies. J Environ Manag 71:95–122

Khijniak TV, Slobodkin AI, Coker V, Renshaw JC, Liven FR, Bonch-Osmolovskaya EA, Birkeland NK, Medvedeva-Lyalikova NN, Lloyd JR (2005) Reduction of uranium (VI) phosphate during growth of the thermophilic bacterium *Thermoterra bacterium ferri reducens*. Appl Environ Microbiol 71:6423–6426

Kim GN, Jung YH, Lee JJ, Moon JK, Jung CH (2008) Development of electrokinetic-flushing technology for the remediation of contaminated soil around nuclear facilities. J Indust Eng Chem 14:732–738

Kim GN, Lee SS, Shon DB, Lee KW, Chung US (2010) Development of pilot-scale electrokinetic remediation technology to remove ^{60}Co and ^{137}Cs from soil. J Indust Eng Chem 16:986–991

Kim GN, Shon DB, Park HM, Lee KW, Chung US (2011) Development of pilot-scale electrokinetic remediation technology for uranium removal. Sep Purif Technol 80:67–72

Kim GN, Kim SS, Park HM, Kim WS, Moon JK, Hyeon JH (2012) Development of complex electrokinetic decontamination method for soil contaminated with uranium. Electrochim Acta 86:49–56

Kim GN, Kim SS, Park UR, Moon JK (2015) Decontamination of soil contaminated with cesium using electrokinetic-electrodialytic method. Electrochim Acta 181:233–237

Kim GN, Kim I, Kim SS, Choi JW (2016) Removal of uranium from contaminated soil using indoor electrokinetic decontamination. J Radioanal Nucl Chem 309:1175–1181

Kim SS, Han GS, Kim GN, Koo DS, Kim IG, Choi JW (2016) Advanced remediation of uranium-contaminated soil. J Environ Radioact 164:239–244

Langley S, Gault AG, Ibrahim A, Takahashi Y, Renaud R, Fortin D, Clark ID, Ferris FG (2009) Strontium desorption from bacteriogenic iron oxides (BIOS) subjected to microbial Fe(III) reduction. Chem Geol 262:217–228

Law GTW, Geissler A, Burke IT, Livens FR, Lloyd JR, McBeth JM, Morris K (2011) Uranium redox cycling in sediment and biomineral systems. Geomicrobiol J 28:497–506

Lehours A, Rabiet M, Morel-Desrosiers N, Morel J, Jouve L, Arbeile B, Mailhot G, Fonty G (2010) Ferric iron reduction by fermentative strain BS2 isolated from an iron-rich anoxic environment (Lake Pavin, France). Geomicrobiol J 27:714–722

Lehto J (1994) Cleanup of large radioactive –contaminated areas and disposal of generated waste. Final report of the KAN2 project. Copenhagen Tema Nord 1994:567

Li GY, Hu N, Ding DX, Zheng JF, Liu YL, Wang YD, Nie XQ (2011) Screening of plant species for phytoremediation of uranium, thorium, barium, nickel, strontium and lead contaminated soils from a uranium mill tailings repository in South China. Bull Environ Contam Toxicol 86:646–652

Lietsch C, Hoth N, Kassahun A (2015) Investigation of phenomena in uranium mine waters using hydrogeochemical modeling – a case study, In: Merkel BJ, Arab A (eds) Uranium – Past and Future Challenges. Proceedings of the 7th international conference of uranium mining and hydrogeology, Springer

Lloyd JR, Sole VA, van Praagh CVG, Lovley DR (2000) Direct and Fe(II)-mediated reduction of technetium by Fe(II) reducing bacteria. Appl Environ Microbiol 66:3743–3749

Lovley DR, Phillips EJP (1988) Novel mode of microbial energy metabolism: organic carbon oxidation coupled to dissimilatory reduction of iron or manganese. Appl Environ Microbiol 54:1472–1480

Lovley DR, Phillips EJ (1992a) Reduction of uranium by *Desulfovibrio desulfuricans*. Appl Environ Microbiol 58:850–856

Lovley DR, Phillips EJP (1992b) Bioremediation of uranium contamination with enzymatic uranium reduction. Environ Sci Technol 26:2228–2234

Lu X, Li X, Yun P, Luo D, Wang L, Ren C, Chen C (2012) Measurement of natural radioactivity and assessment of associated hazards in soil around Baoji second coal-fired thermal power plant, China. Radiat Prot Dosim 148:219–226

Luey J (1993) In situ vitrification: demonstrated capabilities and potential applications. Presented at the Office of Technology Development 2nd semi-annual information meeting, January 25–27. Houston, Texas

Luey J, Koegler SS, Kuhn WL, Lowery PS, Winkelman RG (1992) In Situ Vitrification of a Mixed-Waste Contaminated Soil Site: The 116-B-6A Crib at Hanford. PNl-8281. Prepared for the U.S. Department of Energy under Contract DE-AC06-76RLO 1830

Madden AS, Smith AC, Balkwill DL, Fagan LA, Phelps TJ (2007) Microbial uranium immobilization independent of nitrate reduction. Environ Microbiol 9:2321–2330

MARTAC (2003) Rehabilitation of former nuclear test sites at Emu and Maralinga (Australlia). Report by the Maralinga Rehabilitation Technical Advisory Committee

Mehta VS, Maillot F, Wang Z, Catalano JG, Giammar DE (2014) Effect of co-solutes on the products and solubility of uranium(VI) precipitated with phosphate. Chem Geol 364:66–75

Mehta VS, Maillot F, Wang Z, Catalano JG, Giammar DE (2015) Transport of U(VI) through sediments amended with phosphate to induce *in situ* uranium immobilization. Water Res 69:307–317

MOE (2013) Decontamination guidelines. Ministry of the Environment. Retrieved from http://josen.env.go.jp/en/framework/pdf/decontamination_guidelines_2nd.pdf

MOE (2015) FY2014 decontamination report. Ministry of the Environment. Retrieved from http://josen.env.go.jp/en/cooperation/pdf/decontamination_report1503_01.pdf

Monte L (2001) A generic model for assessing the effects of countermeasures to reduce the radionuclide contamination levels in abiotic components of fresh water systems and complex catchments. Env Model Soft 16:669–690

Moore HR, Steiner RE, Fallon E, Repp CL, Hemann MR, Rabideau AJ (2000) Permeable treatment wall pilot project at the West Valley demonstration project. Proceedings of waste management 2000, 27 February–2 March, 2000. Tucson, AZ

Naeem H, Bhatti HN, Sadaf S, Iqbal M (2017) Uranium remediation using modified *Vigna radiata* waste biomass. Appl Radiat Isot 123:94–101

Newsome L, Morris K, Lloyd JR (2014) The biogeochemistry and bioremediation of uranium and other priority radionuclides. Chem Geol 363:164–184

Nisbet AF (2002) Management options for food production systems contaminated as a result of a nuclear accident. Radioprot Colloq 37:115–120

Nisbet AF, Woodman RFM (2000) Options for the management of Chernobyl restricted areas in England and Wales. J Environ Radioact 51:239–254

Nishikiori T, Suzuki S (2017) Radiocesium decontamination of a riverside in Fukushima, Japan. J Environ Radioact 177:58–64

Noshkin V, Robison W (1997) Assessment of a radioactive waste disposal site at Enewetak Atoll. Health Phys 73:234–247

Nyman JL, Wu HI, Gentile ME, Kitanidis PK, Criddle CS (2007) Inhibition of a U(VI)- and sulfate-reducing consortia by U(VI). Environ Sci Technol 41:6528–6533

Oughton DH (2013) Social and ethical issues in environmental remediation projects. J Environ Radioact 119:21–25

Paajanen A, Lehto J (1992) Disposal of Radioactive Wastes from the Cleanup of Large Areas Contaminated in Nuclear Accidents, A Literature Survay. Nordic Nuclear Safety Research Program 1990–93. Projekt KAN-2

Perez-Sanchez D, Jimenez A, Mateos A, Dvorzhak A (2015) Fuzzy MCDA for remediation of a uranium tailing. In: Merkel BJ, Arab A (eds) Uranium–past and future challenges. Proceedings of the 7th international conference of uranium mining and hydrogeology, Springer, Cham

Petrović J, Ćujić M, Đorđević M, Dragovic R, Gajic B, Miljanic S, Dragovic S (2013) Spatial distribution and vertical migration of ^{137}Cs in soils of Belgrade (Serbia) 25 years after the Chernobyl accident. Environ Sci Proces Imp 15:1279–1289

Petrović J, Dragović S, Dragović R, Đorđević M, Đokić M, Ćujić M (2016a) Spatial and vertical distribution of ^{137}Cs in soils in the erosive area of south eastern Serbia (Pčinja and South Morava River Basins). J Soil Sed 16:1168–1175

Petrović J, Dragović S, Dragović R, Đorđević M, Đokić M, Zlatković B, Walling D (2016b) Using ^{137}Cs measurements to estimate soil erosion rates in the Pčinja and South Morava River Basins, southeastern Serbia. J Environ Radioact 158–159:71–80

Plymale AE, Fredrickson JK, Zachara JM, Dohnalkova AC, Heald SM, Moore DA, Kennedy DW, Marshall MJ, Wang CM, Resch CT, Nachimuthu P (2011) Competitive reduction of pertechnetate ((TcO_4^-)-Tc-99) by dissimilatory metal reducing bacteria and biogenic Fe(II). Environ Sci Technol 45:951–957

PNNL (2009) Scientific opportunities to reduce risk in groundwater and soil remediation. PNNL 18516. Pacific Northwest National Laboratory (PNNL)

Pruess K (1991) TOUGH2: a general numerical simulator for multiple phase fluid and heat flow, report LBL-29400. Lawrence Berkeley Laboratory, Berkeley

Rabideau AJ, van Benschoten J, Patel A, Bandilla K (2005) Performance assessment of a zeolite treatment wall for removing Sr-90 from groundwater. J Contam Hydro 79:1–24

Rafferty B, Brennan M, Dawson D, Dowding D (2000) Mechanisms of ^{137}Cs migration in coniferous forest soils. J Environ Radioact 48:131–143

Renshaw JC, Butchins LJC, Livens FR, May I, Charnock JM, Lloyd JR (2005) Bioreduction of uranium: environmental implications of a pentavalent intermediate. Environ Sci Technol 39:5657–5660

Robinson SM, Arnold WD, Byers CH (1991) Multicomponent ion exchange equilibria in zeolite. In: Tedder DW, Pohland FG (eds) Emerging Technologies in Hazardous Waste Treatment. ACS Symposium Series No. 486. Kluwer Academic Publishers, New York

Rudy CG, Vovk IF (1996) Radioactive waste management issues related to the conversion of the Chernobyl Sarcophagus into an ecologically safe system. One decade after Chernobyl: summing up the consequences of the accident. IAEA-TECDOC-964. Vienna, Austria

Saleh HM (2016) Biological remediation of hazardous pollutants using water hyacinth – a review. J Biotech Res 2:80–91

Sanford RA, Wu Q, Sung Y, Thomas SH, Amos BK, Prince EK, Löffler FE (2007) Hexavalent uranium supports growth of *Anaeromyxobacter dehalogenans* and *Geobacter spp.* with lower than predicted biomass yields. Environ Microbiol 9:2885–2893

Screening Report –Port Hope Project (2006) The port hope long-term low-level radioactive waste management project. Natural Resources Canada, the Canadian Nuclear Safety Commission and Fisheries and Oceans Canada, Ottawa

Sheoran V, Sheoran AS, Poonia P (2016) Factors affecting phytoextraction: a review. Pedosphere 26:148–166

Silva EBS, Fernandes NS, Costa ECTA, Silva DR, de Araújo DM, Martínez-Huitle CA (2017) Scale-up on electrokinetic treatment of polluted soil with petroleum: effect of operating conditions. Int J Electrochem Sci 12:4001–4015

Sivaswamy V, Boyanov MI, Peyton BM, Viamajala S, Gerlach R, Apel WA, Sani RK, Dohnalkova A, Kemner KM, Borch T (2011) Multiple mechanisms of uranium immobilization by *Cellulomonas sp.* strain ES6. Biotechnol Bioeng 108:264–276

Soudek P, Petrova S, Benesova D, Tykva R, Vankova R, Vanek T (2007) Comparison of ^{226}Ra nuclide from soil by three woody species *Betula pendula*, *Sambucus nigra* and *Alnus glutinosa* during the vegetation period. J Environ Radioact 97:76–82

Spain AM, Krumholz LR (2011) Nitrate-reducing bacteria at the nitrate and radionuclide contaminated Oak ridge integrated field research challenge site: a review. Geomicrobiol J 28:418–429

Spalding BP, Jacobs GK, Dunbar NW, Naney MT, Tixier JS, Powell TD (1992) Tracer-level radioactive pilot-scale test of in situ vitrification for the stabilization of contaminated soil sites at ORNL. Prepared by the Oak Ridge National Laboratory, Oak Ridge

Sugiura Y, Kanasashi T, Ogata Y, Ozawa H, Takenaka C (2016) Radiocesium accumulation properties of Chengiopanax sciadophylloides. J Environ Radio 151:250–257

Suzuki Y, Banfield JF (1999) Geomicrobiology of uranium. Rev Mineral Geochem 38:393–432

Suzuki Y, Kelly SD, Kemner KA, Banfield JF (2003) Microbial populations stimulated for hexavalent uranium reduction in uranium mine sediment. Appl Environ Microbiol 69:1337–1346

Thorpe CL, Lloyd JR, Law GTW, Burke IT, Shaw S, Bryan ND, Morris K (2012) Strontium sorption and precipitation behaviour during bioreduction in nitrate impacted sediments. Chem Geol 306:114–122

Thorpe CL, Boothman C, Lloyd JR, Law GTW, Bryan ND, Atherton N, Livens FR, Morris K (2014) The interactions of strontium and technetium with Fe(II) bearing biominerals: implications for bioremediation of radioactively contaminated land. Appl Geochem 40:135–143

UK DEFRA (2003) The Radioactively Contaminated Land Exposure Assessment Methodology (RCLEA). https://www.gov.uk/government/publications/rclea-software-application

Ulanovsky A, Jacob P, Fesenko S, Bogdevitch I, Kashparov V, Sanzharova N (2011) ReSCA: decision support tool for remediation planning after the Chernobyl accident. Radiat Environ Biophys 50:67–83

US DOE (1999) Final site observational work plan for the UMTRA project Old Rifle site GJO-99–88-TAR. Colorado, United States Department of Energy (US DOE)/Grand Junction

US DOE (2004) Monticello mill tailings site, operable unit III–remedial investigation addendum/ focused feasibility study. GJO-2003-413-TAC.United States Department of Energy

US DOE (2009) Explanation of significant difference for the Monticello mill tailings (USDOE) site operable unit III, surface water and ground water, Monticello, Utah.Doc. No S0429400. United States Department of Energy

US DOE (2010) 200 west area groundwater pump-and-treat remedial design report, DOE/RL-2010-13, rev. 1.United States Department of Energy, Richland Operations Office. Richland, Washington, DC

US DOE, US EPA, Ecology (2009) Explanation of significant differences for the 300-FF-2 operable unit interim action record of decision Hanford site Benton County, Washington. United States Department of Energy, United States Environmental Protection Agency, Washington State Department of Ecology. Richland, Washington, DC

US EPA (2002) Preliminary Remediation Goal (PRG). https://epa-prgs.ornl.gov/cgi-bin/radionuclides/rprg_search

US EPA (2004) Dose Compliance Concentration (DCC). https://epa-dccs.ornl.gov/cgibin/dose_search

US EPA (2007) The technology reference guide for radioactively contaminated media EPA-402-R-07-004. United States Environmental Protection Agency (US EPA). Washington, DC

Van der Perk M, Burrough PA, Voigt G (1998) GIS based modelling to identify regions of Ukraine, Belarus, and Russia affected by residues of the Chernobyl nuclear power plant accident. J Hazard Mater 61:85–90

Vives i Batlle J, Beresford NA, Beaugelin-Seiller K, Bezhenar R, Brown J, Cheng JJ, Ćujić M, Dragović S, Duffa C, Fiévet B, Hosseini A, Jung KT, Komboj S, Keum DK, Kryshev A, LePoire D, Maderich V, Min BJ, Periáñez R, Sazykina T, Suh KS, Yu C, Wang C, Heling R (2016) Inter-comparison of dynamic models for radionuclide transfer to marine biota in a Fukushima accident scenario. J Environ Radioact 153:31–50

Vovk IF, Blagoyev VV, Lyashenko AN, Kovalev IS (1993) Technical approaches to decontamination of terrestrial environments in the CIS (former USSR). Sci Total Environ 137:49–63

Wada K, Yoshikawa T, Murata M (2012) Decontamination work in the area surrounding Fukushima Dai-ichi Nuclear Power Plant: another occupational health challenge of the nuclear disaster. Arch Environ Occ Health 67:128–132

Wallace SH, Shaw S, Morris K, Small JS, Fuller AJ, Burke IT (2012) Effect of groundwater pH and ionic strength on strontium sorption in aquifer sediments: implications for $^{90}Sr^{2+}$ mobility at contaminated nuclear sites. Appl Geochem 27:482–1491

Wallace SH, Shaw S, Morris K, Small JS, Burke IT (2013) Alteration of sediments by hyperalkaline k-rich cement leachate: implications for strontium adsorption and incorporation. Environ Sci Technol 47:3694–3700

Wellman DM, Glovack JN, Parker K, Richards EL, Pierce EM (2008) Sequestration and retention of uranium(VI) in the presence of hydroxylapatite under dynamic geochemical conditions. Environ Chem 5:40–50

Wilkins MJ, Livens FR, Vaughan DJ, Beadle I, Lloyd JR (2007) The influence of microbial redox cycling on radionuclide mobility in the subsurface at a low-level radioactive waste storage site. Geobiology 5:293–301

Williams KH, Long PE, Davis JA, Wilkins MJ, N'Guessan AL, Steefel CI, Yang L, Newcomer D, Spane FA, Kerkhof LJ, McGuinness L, Dayvault R, Lovley DR (2011) Acetate availability and its influence on sustainable bioremediation of uranium-contaminated groundwater. Geomicrobiol J 28:519–539

Williams KH, Bargar JR, Lloyd JR, Lovley DR (2012) Bioremediation of uranium contaminated groundwater: a systems approach to subsurface biogeochemistry. Curr Opin Biotechnol 24:489–497

Wu WM, Carley J, Luo J, Ginder-Vogel MA, Cardenas E, Leigh MB, Hwang C, Kelly SD, Ruan C, Wu L, Van Nostrand J, Gentry T, Lowe K, Carroll S, Luo W, Fields MW, Gu B, Watson D, Kemner KM, Marsh T, Tiedje J, Zhou J, Fendorf S, Kitanidis PK, Jardine PM, Criddle CS (2007) In situ bioreduction of uranium (VI) to submicromolar levels and reoxidation by dissolved oxygen. Environ Sci Technol 41:5716–5723

Yabusaki SB, Fang Y, Long PE, Resch CT, Peacock AD, Komlos J, Jaffe PR, Morrison SJ, Dayvault RD, White DC, Anderson RT (2007) Uranium removal from groundwater via in situ biostimulation: field-scale modelling of transport and biological processes. J Contam Hydrol 93:216–235

Yang JS, Kwon MJ, Choi J, Baek K, O'Loughlin J (2014) The transport behavior of As, Cu, Pb, and Zn during electrokinetic remediation of a contaminated soil using electrolyte conditioning. Chemosphere 117:79–86

Yeh GT, Tripathi VS (1990) HYDROGEOCHEM: a coupled model of HYDROlogical transport and GEOCHEMical equilibrium of multi component systems, ORNL 6371, Oak Ridge National Laboratory. Oak Ridge, Tennessee

Yeh GT, Gwo JP, Siegel MD, Li M-H, Fang Y, Zhang F, Luo W, Yabusaki SB (2013) Innovative mathematical modeling in environmental remediation. J Environ Radioact 119:26–38

Zachara JM, Long PE, Bargar J, Davis JA, Fox P, Fredrickson JK, Freshley MD, Konopka AE, Liu C, McKinley JP, Rockhold ML, Williams KH, Yabusaki SB (2013) Persistence of uranium groundwater plumes: contrasting mechanisms at two DOE sites in the groundwater-river interaction zone. J Contam Hydrol 147:45–72

Zhang F, Yeh GT, Parker JC, Brooks SC, Pace MN, Kim YJ, Jardine PM, Watson DB (2007) A reaction-based paradigm to model reactive chemical transport in groundwater with general kinetic and equilibrium reactions. J Contam Hydrol 92:10–32

Zhang R, Chen C, Li J, Wang X (2015) Investigation of interaction between U(VI) and carbonaceous nanofibers by batch experiments and modeling study. J Collo Inter Sci 460:237–246

Zyvoloski GA, Robinson BA, Dash ZV, Trease LL (1994) Models and methods summary for the FEHM application, FEHM MMS SC-194, Los Alamos National Laboratory. Los Alamos, New Mexico

Physicochemical Methods for the Remediation of Radionuclide Contaminated Sites

Claudio Cameselle and Susana Gouveia

1 Introduction

The rapid industrial development during the twentieth century has resulted in a very serious environmental impact mainly due to poor management of the wastes. In the past, it was common the deposition or dispersion of the wastes in the environment without any proper treatment (Liu et al. 2017). As a result of this mismanagement, there are numerous contaminated sites all over the world, and this problem, instead of being reduced, is increasing, especially in developing countries (Guerrero et al. 2013). The society has become aware of the problems for the environment and public health of poor waste management, and public demand from the Administrations and the scientific community solutions to the problems associated to the environmental pollution (Cortner 2000). As a result of this social pressure, a great amount of resources have been devoted for the development of processes or technologies to reduce the impact of human activities on the environment, as well as recovering degraded or contaminated sites (Khan et al. 2004).

Uncontrolled discharges of solid or liquid wastes directly into the environment on soil or waterbodies had led to the contamination of soil and groundwater. Soil is an element with a significant buffer capacity and many of these pollutants are absorbed or precipitated onto soil particles (Li 2014). However, once the soil absorption capacity has been exceeded the pollutants begin to leach moving to deeper layers, contaminating groundwater and dispersing in the environment. Thus, contaminated sites act as a long-term source of contamination for the environment a. In addition, soils in these contaminated sites cannot be used for their normal uses such as agriculture, livestock grazing, residential uses, etc., because of the risk of exposure

C. Cameselle (✉) · S. Gouveia
Department of Chemical Engineering, University of Vigo, Vigo, Spain
e-mail: claudio@uvigo.es; gouveia@uvigo.es

D. K. Gupta, A. Voronina (eds.), *Remediation Measures for Radioactively Contaminated Areas*, https://doi.org/10.1007/978-3-319-73398-2_2

to pollutants and the possibility of entering the food chain, generating a serious public health problem (Lu et al. 2015).

Since the 1980s, various technologies have been developed for the remediation and restoration of contaminated soils (Sims 1990). Initially, the first approach was excavation and deposition of the contaminated soil in a security landfill. Excavation of the contaminated soil can solve the problem locally but the storage of the contaminated soil in a landfill is not the definitive solution; it is just a transfer of a problem from one site to another. The excavation and storage in a landfill of the contaminated soil is only appropriate for small areas affected by high concentration of pollutants with a high risk for ecosystems and population. When the amount of contaminated soil is very large, this process is economically unfeasible because of the transportation costs and large space required in the landfills to storage the contaminated soil. As an alternative, some off-site technologies were developed to treat ex-situ or off-site the excavated soil. Some examples of these technologies include soil washing and incineration. The advantages of these technologies are associated to the high contaminant removal efficiency mainly due to the close control of the operating variables when treating the soil in a specially designed device. However, the cost of energy and chemical reagents is often very high and therefore impracticable for large areas of contaminated soil with medium or low concentrations. These technologies are more appropriate for small areas contaminated with medium or high concentrations of pollutants (Kuppusamy et al. 2016). The on-site soil remediation technologies have arisen afterwards with great interest from scientist and practitioners since it is not necessary to excavate the contaminated soil as well as the lower cost of the remediation process. The lower cost is associated to the savings in excavation and transportation (Song et al. 2017).

The aim of this chapter is to review the main physicochemical technologies for the remediation of contaminated soil and groundwater with radionuclides.

2 Electrokinetic Remediation of Contaminated Soils

Electrokinetic remediation is a technology specifically designed for the decontamination of soils, sediments, sludge and other solid porous materials. This technology is also known in literature as electroreclamation, electrokinetics or electroremediation (Reddy and Cameselle 2009a). Electrokinetics has been extensively investigated in the last 20 years and this technology have shown several successes in the remediation of soils, sediments and sludges contaminated with both inorganic and organic contaminants (Reddy and Cameselle 2009b).

The principle of electrokinetic remediation relies upon the application of a low-intensity direct current through the soil between a couple of electrodes (cathode and anode). The electric field mobilizes charged species, causing ions and water to move toward the electrodes trough the soil (Pamukcu 2009). Electrokinetics can be applied *in situ* as depicted in Fig. 1. The electric current applied to the main electrodes forces the transportation of ions and water towards the electrodes. This

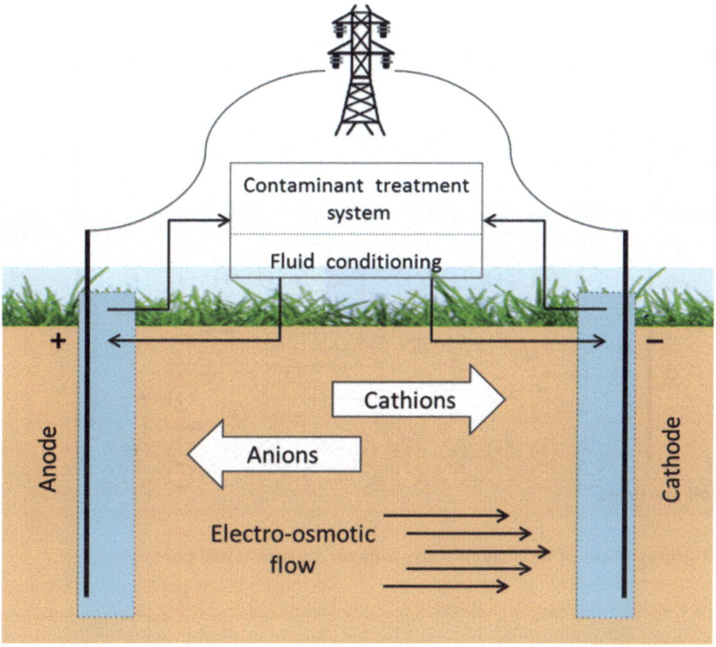

Fig. 1 Electrokinetic remediation of a contaminated site

movement removes the contaminants from the soil and accumulates them in the electrode wells. A circulation system extract the processing fluid (water or aqueous solutions) with the contaminants from the wells and after the proper treatment and conditioning the fluid is pumped back to the electrode wells. The system is operated until the contaminant concentration in soil and/or processing fluid is below the desired value (Reddy and Cameselle 2009b).

Advantages of electrokinetic remediation over other soil remediation technologies include low power consumption, close control over the direction of water and dissolved contaminants, and the confinement of pollutants in the electrode walls that makes easier the subsequent treatment of the removed pollutants (Virkutyte et al. 2002).

The main transport mechanisms in electrokinetics are electromigration and electroosmosis. Electromigration is defined as the transportation of ions by the electric field towards the electrode of opposite charge (Fig. 2). The electromigration is the responsible for the transportation of free ions such us H^+, OH^-, cations (heavy metals, ammonium, etc.) and anions (chloride, cyanide, fluoride, nitrate, etc.) but also ionizable organic compounds (Ricart et al. 2004). The migration rate mainly depends on the volume of the ion and its charge but also the interactions with the surface of the solid particles of soil that can greatly retard the movement (Pamukcu 2009).

Electroosmosis is defined as the net flux of water induced by the electric field in the porous structure of soil (Cameselle and Reddy 2012). Figure 3 shows the structure of the double electric layer for a soil particle electronegatively charged.

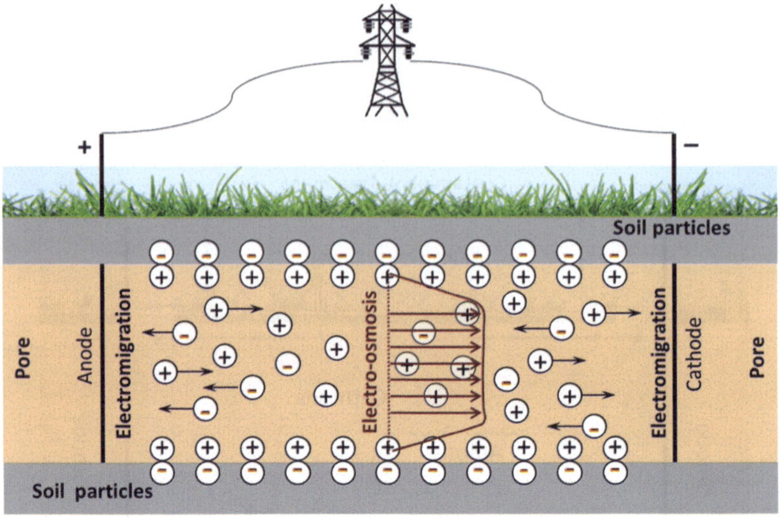

Fig. 2 Electromigration of ions and electro-osmosis flow in a soil pore

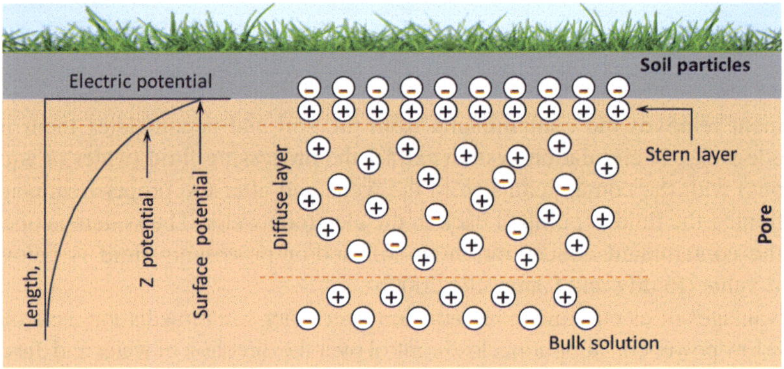

Fig. 3 Electric double layer induced in the interstitial fluid due to the surface charge of soil particles

Figure 4 shows the development of the electroosmotic flow in the porous structure of soil when the electric field induces the movement of the ions in the pore fluid close to the pore wall. In general, soil particles are electronegative, so the counter-ions will be positive; hence, the direction of the electro-osmotic will flow toward the cathode (Cameselle 2015a). Electro-osmosis is the dominant transport mechanism for inorganic and organic contaminants in solution o emulsified in the interstitial fluid. The electro-osmotic flow is an important transportation mechanism in the environmental applications of electrokinetics, since the flux of water removes all the compounds in solution (ionic or non-ionic) in the interstitial fluid. In general, the transportation by

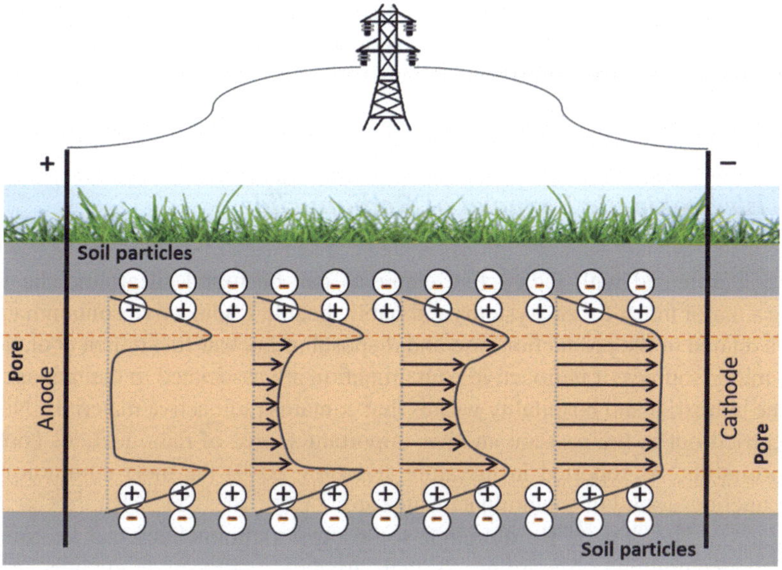

Fig. 4 Development of the electro-osmotic flow in a soil pore

electromigration is faster than electroosmosis (Pamukcu 2009). Electromigration is the main transport mechanism for ionic species (Ottosen et al. 2009) whereas electroosmosis is used for the removal of non-ionic organic contaminants (Yang and Lee 2009).

The application of an electric field to moisten porous matrix also induces chemical reactions into the soil and upon the electrodes. These reactions are critical for the efficiency of electrokinetics since they affect the speciation of contaminants, solubility, soil pH, etc. These chemical reactions include acid-base reactions, redox reactions, adsorption-desorption and dissolution-precipitation reactions. The main reaction in the electrochemical/electrokinetic system is the electrolysis of water at the electrodes (Eqs. 1 and 2). The electrolysis of water generates oxygen gas and hydrogen ions (H^+) in the anode and hydrogen gas and hydroxyl ions (OH^-) in the cathode.

$$2H_2O \rightarrow 4e^- + 4H^+_{(aq)} + O_{2(gas)} \tag{1}$$

$$4H_2O + 4e^- \rightarrow 2H_{2(gas)} + 4OH^-_{(aq)} \tag{2}$$

The H^+ and the OH^- ions electrogenerated upon the electrodes electromigrates into the soil modifying the soil pH and affecting the solubility and speciation of the contaminants (Ricart et al. 1999). The acid and alkaline fronts can be favored or depressed by the control addition of acid/base in the electrode chamber to induce the most appropriate pH in the soil for solubilization and transportation of the contaminants. Moreover, the addition to the electrode chambers of external facilitating agents (complexing agents, surfactants, co-solvents) can be introduced into the soil by electroosmosis and electromigration to enhance the solubilization and removal of contaminants (Cameselle and Pena 2016).

2.1 *Electrokinetic Removal of Radionuclides*

Soils contaminated with radioactive elements can be found all around the world posing a major threat for ecosystems and public health. Radioactive contamination is always related to the production, use and disposal of nuclear fuel (Chen et al. 2005). Other minor sources of radioactive contamination are associated to the management of some industrial and hospitality wastes that contains radioactive materials. Nuclear tests carried out in the past are another important source of radionuclides contamination in soils. As reported in literature (Korolev 2009) the most common radioactive nuclides found in contaminated sites are: ^{60}Co, ^{90}Sr, ^{90}Y, ^{106}Ru, ^{137}Cs, ^{144}Ce, ^{147}Pm, $^{238,\,239,\,240}Pu$, ^{226}Ra although some other radionuclides can be found in minor concentrations. According to their toxicity, radioactive nuclides can be classified in 5 groups (Table 1) from the group A, particularly high toxicity, to the group E, low toxicity (Korolev 2009).

The contamination by radionuclides in soil is mainly found in the upper layer of soil close to the surface adsorbed to the soil particles and organic matter. Thus, in most of the contaminated sites 95% of the radioactive contaminants are concentrated in the upper layer of soil forming various chemical species but always adsorbed to the soil components. These adsorptions retard, but not impede the transportation of the radionuclides deeper into the soil or leached to the groundwater (Yasunari et al. 2011). The actual speciation or fractionation of radionuclides in a contaminated site depends on the soil composition. In general, radionuclides can be found in the exchangeable form, 8–30%; in water soluble form, 2–10%; but most of the radionuclides are found bound to soil components, about 60–80%, which retard their transportation and dispersion in the environment (Nekrasova and Korolev 1997).

The electrokinetic technology has been tested for the removal of radionuclides from contaminated soils. The most common radionuclide that can be found in nuclear waste disposal sites was studied: ^{90}Sr, ^{88}Sr, ^{137}Cs, ^{133}Cs, etc., using as a matrix several types of soils: from loam to clay. ^{90}Sr and ^{137}Cs radionuclides are not

Table 1 Radionuclides classified by their toxicity

Group	Toxicity	Radionuclides
A	Particularly high toxicity	H
B	High toxicity	C, Cu, Zn, Ge
C	Medium toxicity	Na, P, S, Cl, Mn, Fe, Co, Br, Y, Zr, Nb, In, Sn, Cs, Tl
D	Small toxicity	Sr, Ru, Sb, I, Bi, Ce, Tm
E	Low toxicity	Pb, Po, Ra, Th, Pa, U, Np, Pu, Am, Cm

Adapted from Korolev (2009)

only two of the most common radionuclides in contaminated sites but the most toxic and dangerous for living organisms with a half-life of 28 and 30 years, respectively. ^{90}Sr and ^{137}Cs can be found in contaminated sites as free elements (dissolved in the interstitial fluid) and adsorbed to soil particles (Korolev 2009). During the electrokinetic treatment, ^{88}Sr isotope was observed to be transported towards the cathode. The soil sections closer to the anode showed lower concentration of strontium whereas the concentration of Sr increased towards the cathode. The Sr concentration profile proved the electromigration of this radionuclide towards the cathode as a cation where it was concentrated. At the end of the test, the free (dissolved) strontium was similar along the soil sample, whereas the adsorbed strontium showed and increasing profile along the soil sample. As a conclusion, the electrokinetic technology is able to transport the radionuclides but it is necessary to enhance the extraction and transportation of the radionuclides to achieve a significant removal.

A collaboration project in electrokinetics between the University of Vigo (Spain) and the University of Science and Technology Houari Boumediene (USTHB) in Algeria studied the removal of radioactive nuclides from contaminated soils. The soil used in this studied was mainly contaminated ^{137}Cs and ^{241}Am. The laboratory tests were carried out in a cylindrical cell (4 cm diameter, 20 cm long). Four tests were run at 10, 20, 30 and 40 mA/cm^2 for 7 days (Fig. 5). In order to favor the solubility and transportation of the radionuclides, the pH in the anode and cathode was kept below pH 2 with the addition of concentrated hydrochloric acid. The best results were found in the test at 40 mA/cm^2. Over 48% of ^{137}Cs and 61% of ^{241}Am were removed from the soil and collected in the anode. The use of ionic exchange

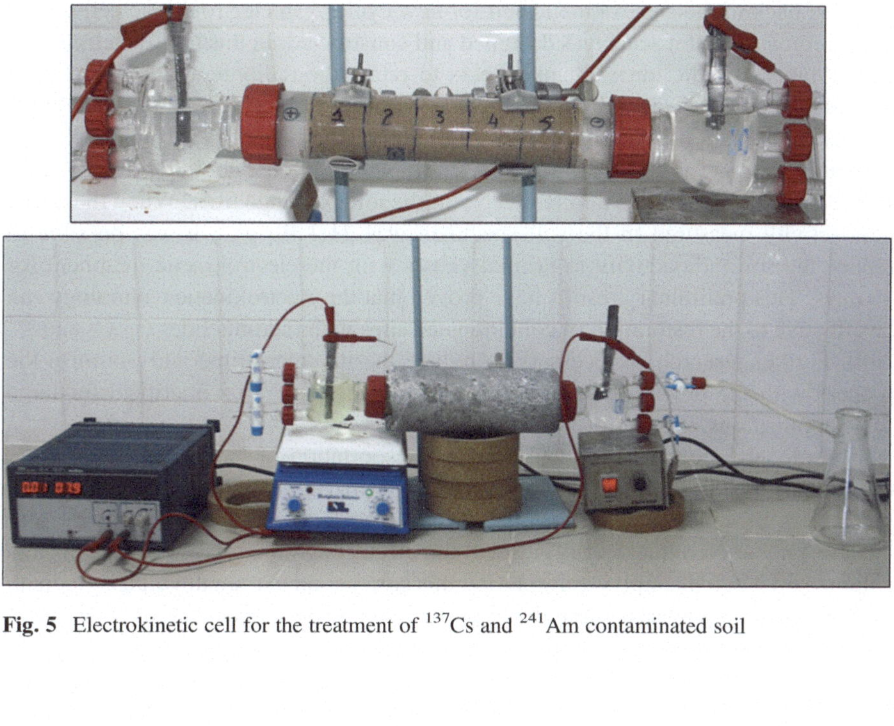

Fig. 5 Electrokinetic cell for the treatment of ^{137}Cs and ^{241}Am contaminated soil

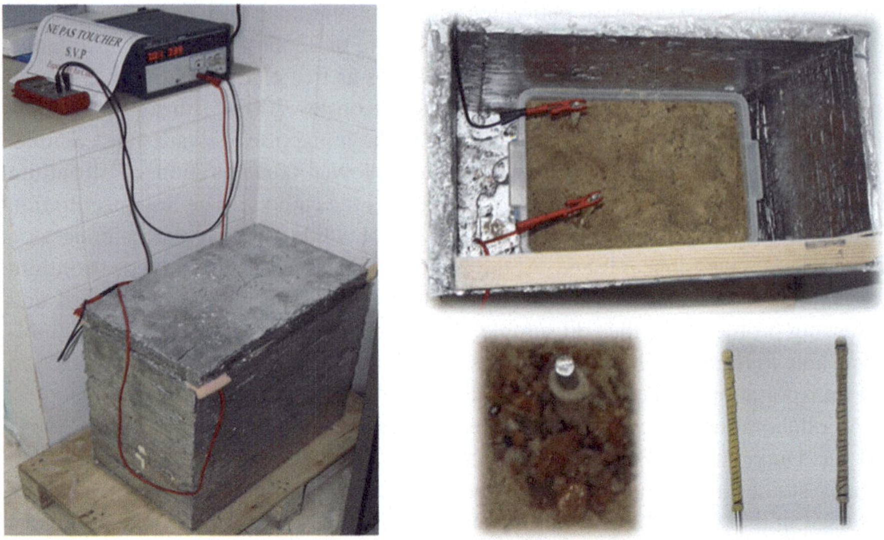

Fig. 6 Pilot cell for the direct electrokinetic removal of ^{137}Cs using ion exchange fibers in the electrodes

fibers was proposed to retain (adsorb) the radionuclides before they reach the cathode chamber (Belhadj et al. 2014). Thus, the removed radionuclides from the soil are concentrated in the fiber and the collection and disposal of the radionuclides is more effective. Once the radionuclides are desorbed from the fiber, it can be recycled back to the electrokinetic cell for new cycles. An electrokinetic cell for the ^{137}Cs removal at pilot scale was designed and constructed. In this test, the electrodes were wrapped in the ion exchange fibers to retain the radionuclides. This experimental device was designed with the idea of transfer this technology to the field for the decontamination in situ of Cs and Sr contaminated sites (Fig. 6). The electric current intensity was set at 30 mA/cm^2. In these conditions, about 27.5% of ^{137}Cs was removed from the soil and recovered in the ion exchange fiber. When the initial radioactivity measured in the soil specimen was 21.7 Bq g^{-1}, it was possible to reduce the soil radioactivity to normal values with the electrokinetic treatment for 40 days. This preliminary results have proved that the electrokinetic technology can be effective to the restoration of contaminated sites with radionuclides (^{137}Cs and ^{241}Am). Further research is necessary to reduce the treatment time and optimize the process with the objective of reducing operational costs. This is critical for large scale applications.

The modelization of the electrokinetic transportation of ^{137}Cs and ^{90}Sr form a contaminated soil (Yu and Neretnieks 1997) proved that the physicochemical parameters of soils and radionuclides have a decisive effect in the removal efficiency. Yu and Neretnieks (1997) tested the electromigration of ^{137}Cs and ^{90}Sr in a column soil 3 m long applying 100 DCV, this correspond to a small voltage gradient, 0.33 V cm^{-1}, compared with the typical value widely accepted of 1 V cm^{-1}

(Ricart et al. 1999). The prediction results from the model have shown that the efficiency of the electrokinetic treatment depends on the sorption and diffusion parameters. The treatment time will increase with increasing radionuclide sorption to the soil particles and with slower diffusion. Using bentonite as model soil, both ^{137}Cs and ^{90}Sr can be removed is the soil when flushed with saline water, because the saline environment favored desorption of the radionuclides from the soil matrix. Even in these conditions the complete cleaning of soil by electrokinetics will take a few months.

Frizon et al. (2003) have studied the transport of Cs through cementitious materials by electromigration. Two extreme cases were described in this study: the irreversible cesium binding to the solid matrix, and the reversible cesium sorption. It was observed that the main factor in Cs transportation and removal is the metal interactions with the solid phase. Their results suggest that electrokinetics appears to be an effective method for decontaminating cement-based materials (Frizon et al. 2005). No electro-osmosis was detected, so the transport of Cs was only due to electromigration. Electrokinetic decontamination extracted 96% of the initial Cs contamination after 3 weeks, when the material was initially contaminated with 100 mol/m^3 solution of cesium; 63–90% with the 1 mol/m^3 contaminating solution.

These studies have proved that it is possible the transportation and removal of radionuclides in various porous materials (soils and cement-based materials). However, the adsorption and interaction of typical radionuclides such as Cs, Am and Sr with the solid matrix retard and limit the removal efficiency. The enhanced transportation and removal of radionuclides from soil is discussed bellows. Different approaches have been proposed and tested to improve the removal of radionuclides from soil. They include the pH control in anode and/or cathode and the use of facilitating agents that enhance the solubility of radionuclides. Other studies have combined electrokinetics with other remediating technologies such as soil flushing or soil washing.

2.2 *Enhanced Electrokinetic Removal of Radionuclides*

The enhanced removal of radionuclides from contaminated soil implies the modification of the environmental conditions to increase their extractability and solubility. The typical approaches include pH control in anode and/or cathode, the use of facilitating agents (complexing agents etc.) and the combination of various remediating technologies to reach a synergistic effect.

Kim et al. (2003) have studied the contaminated soil around the TRIGA reactor in Korea. The soil was mainly contaminated with ^{137}Cs and ^{60}Co. Electrokinetic remediation was tested as a possible technology for the removal of those radionuclides from soil. They used NaCl solution as electrolyte to enhance the electric conductivity of the soil, but minor Cs and Co were removed because of the premature precipitation of the metals due to the alkaline environment on the cathode side. In order to enhance the metal removal, an acetate buffer solution was used in the cathode to avoid any increase of pH. Cs^{2+} and Co^{2+} ions were transferred by electromigration rather than electroosmosis during the initial remediation period, and no precipitate was formed in the soil column. 96% of the total amount of cesium in the soil column was removed after 5.9 days, while 94% of the total amount of cobalt was removed. However, it is important to analyze the chemical effects of any facilitating agent on the radionuclides. Kim et al. (2008) found a reduction of 10% in the removal efficiency of Cs and Co due to the negative complexes that these ions formed with the organic acid used to control de pH in the cathode. The negative complexes migrated towards the anode whereas the cationic Cs and Co migrated towards the cathode. This reduction in the removal was compensated with the combination of electrokinetics and soil flushing under hydraulic gradient.

NPO Radon organization (Moscow, Russia) has developed various methods and technologies for the nuclear waste management (Prozorov et al. 2000; Perera 2002). These technologies include cementation, bituminization and vitrification for the treatment of the liquid waste. The solid waste can be treated by compaction and plasma incineration. Electrokinetics was also studied as a decontamination method for solid materials containing radionuclides, as well as contaminated soils, sediments and sludges. Electrokinetics can be applied for the remediation of contaminated sites in situ; usually in sites were solid o liquid radioactive waste was dumped. The radionuclide ^{137}Cs is very common at those contaminated sites; this is why this element is the target radionuclide in many of the electrokinetic studies. The influence of the different parameters such as voltage gradient, electric current intensity and energy consumption were studied in the removal of ^{137}Cs contaminated soils. Preliminary tests have demonstrated that electrokinetic remediation showed promising results for the removal of radionuclides from loamy soil (Prozorov et al. 2000). Korolev et al. (2007) tested electrokinetics for the removal of ^{137}Cs in loamy soil. The origin of the contamination was an accidental spill of radioactive liquid waste that resulted in a radioactivity of 80 to 200 kBq kg^{-1}. The experimental setup for the electrokinetic tests is the typical cell with three compartments. One central compartment for the contaminated soil and two compartments on both sides for the electrodes. The electrode compartment was filled with a processing fluid. Stainless steel (cathodes) and platinum-coated titanium (anode) plates were used as electrodes. The unenhanced electrokinetic test resulted in minor Cs removal. In order to enhance the removal efficiency, several facilitating agents (NH_4NO_3, KNO_3, water) were added to the anode compartment at a concentration of 1 mol L^{-1}. The main objective of the addition of the facilitating agents was to improve the extraction solubilization of Cs. The alkaline environment in the cathode was neutralized by the controlled addition of concentrated nitric acid. Thus, the premature Cs precipitation in the soil was avoided. Test results showed that radionuclide ^{137}Cs removed from soil was accumulated in the catholyte. The catholyte was periodically refreshed with new solution to avoid high concentrations of Cs in the catholyte and the possible limitation in its removal due to concentration effects. About 50% of initial Cs was removed when the facilitating agent in the anode was ammonium nitrate. The use of potassium nitrate resulted in the removal of 30%, and the use of water resulted in the removal of only 5% of the initial Cs. These results confirm the important effect of the proper selection of the anolyte and catholyte conditioning.

Korolev (2009) run various tests with various facilitating agents that transform the Cs in soil in a more mobile species. These authors used a mixture 1:1 of nitric and phosphoric acids in the catholyte and their ammonium salts in the anolyte. This combination was the most effective in the removal of ^{137}Cs. Furthermore, the soil at the end of the test was not so acid than in previous tests. This was due to the use of the ammonium salts in the anolyte that partially neutralized the H^+ from the electrolysis of water. This result is very important for in situ large scale applications and for complete remediation and restoration of polluted sites.

The ammonium ion, besides of the neutralization of the acid front in the anode, also resulted to be very active in desorption of the Cs from soil. Korolev (2009) proved that ammonium was much more effective than potassium. The extraction of ^{137}Cs is 1.6 times more efficient with ammonium nitrate than potassium nitrate. Korolev et al. found that 50% of Cs was extracted with ammonium nitrate whereas only 30% of Cs was mobilized with potassium nitrate. When compared with water, the mobilization of Cs with ammonium nitrate is 5.4 times higher.

2.3 *Electrokinetic Pilot Scale and Field Applications*

There are limited studies about the use of electrokinetics in situ in contaminated sites with radionuclides. Kim et al. (2010) developed pilot-scale electrokinetic remediation equipment for the remediation of radioactive soil suitable to the geological characteristics of South Korean nuclear facility sites. The soil used in these tests was contaminated with ^{60}Co and ^{137}Cs and the total radioactivity of the soil specimen was about 2000 Bq kg^{-1}. A constant current intensity of 15 mA/cm^2 was applied for 55 days. The results were very promising for large scale application since 95.8% of initial radionuclides were removed. These authors concluded that the main factor affecting the radionuclide removal were the initial concentration, soil particle size and facilitating agents. Soils with larger particles are easier to remediate. The removal efficiency is also higher with the initial concentration. As per facilitating agents, nitric acid showed better removal efficiency than acetic acid, probably due to its capacity of mobilizing Cs. Furthermore, nitric acid had the advantage of reducing the electric power consumption.

The remediation of contaminated sites with radionuclides can be challenging not only for the dangerousness of dealing with radioactive materials but also for the difficulty of the operation in situ in large areas. Sometimes, instead of removing the contaminants, the solution could be the isolation of the contaminants in the site. Electrokinetics can be operated as a fence (electrokinetic fences) to prevent the radioactive material to migrate to the surrounding areas and groundwater. Figure 7a, b show schemes of such electrokinetic protective fences. Figure 7a shows a section of an electrokinetic protective fence around a site contaminated with radioactive nuclides. The fence of Fig. 7a is designed for a shallow contaminated site. The

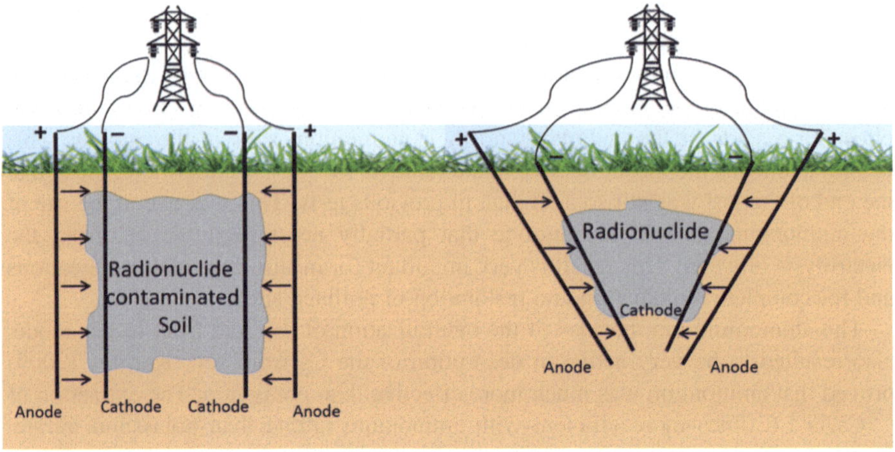

Fig. 7 Electrokinetic fences for radionuclide deposits or contaminated soils

electric field is applied around the site by two parallel rows of anodes and cathodes wells. Thus, the electric field attracts the radionuclides towards the center of the contaminated site, avoiding the migration of radionuclides far from the site. Thus, contaminant radionuclides are maintained in the limits of the contaminated area, avoiding any dispersion and the contamination of surrounding areas. Figure 7b shows another scheme of an electrokinetic protective fence. This is applicable to contaminated sites where the radionuclides are located in the surface of the ground, for instance, an accumulating pond of liquid radioactive waste. There is a risk in such ponds of migration of the radionuclides downwards contaminating soil and groundwater with the subsequent dispersion of the radionuclides. An inclined protective fence consisting of rows of anode and cathode wells should be constructed along the perimeter of the pond with the purpose of preventing the migrations of the radionuclides. As in the previous case, the electric field is applied to force the transportation of the radionuclides towards the center of the pond. Thus, the pollution of groundwater and surrounded soil will be prevented. The electrokinetic fences are applicable at emergency sites and in any other sites where radioactive pollution cannot be treated or removed. Additionally, several minerals with adsorptive capacity can be used to create an adsorption fence around the contaminated sites. This adsorptive fences can be used or not in combination with the electrokinetic fences. Usually clay grounds with high sorption capacity and containing minerals as smectite and illite are applied for constructing these fences. In addition, artificial protective fences made of other disperse grounds (from sand to loam) are applied; the sorption ability of such grounds is beforehand artificially enhanced with strengthening compounds, for instance, such as oxalic silica-alumina gel. These compounds decrease the filtration ability of the fence and raise its sorption characteristics.

3 In Situ Soil Flushing

In situ soil flushing is a remediation technology for contaminated soils with organic and/or inorganic contaminants (He et al. 2015; Trellu et al. 2016). Water or an aqueous solution of facilitating agents is pumped into the subsoil. The flow of water favors the dissolution of the contaminants and their transportation to wells where the mixture of water and contaminants is pumped out and treated. In the treatment process, the contaminants can be separated from water, so it can be reused in the soil flushing (Fig. 8). The effectiveness of the process relies in several factors such as soil permeability, stratification of soil, and contaminant extractability/solubility. Several facilitating agents can be selected based on the physicochemical characteristics of soil and the contaminants, Thus, plain water can be used for soluble inorganic salts (Zhou et al. 2009) whereas acid solutions, organic acids and other complexing agents have been used for the removal of heavy metals (Yun et al. 2015). The removal of hydrophobic organic pollutant requires the use of surfactant or biosurfactants to minimize the impact in the subsoil (Cheng et al. 2017).

Soil flushing has been tested as a remediation technology for soil contaminated with radionuclides (Cameselle 2015b). The remediation alternatives for a soil contaminated with ^{137}Cs was studied by Bea et al. (2004). Cesium was strongly adsorbed to clay aggregates where water flux is negligible due to the low permeability. The mobile fraction of the soil (macropores) retains little water and cesium. The mobilization of Cs was achieved by flushing of potassium (as KCl) through the exchange with K. These authors concluded that flushing is a viable alternative for the removal of Cs, but it is important a high rate periodic infiltration and draining, as

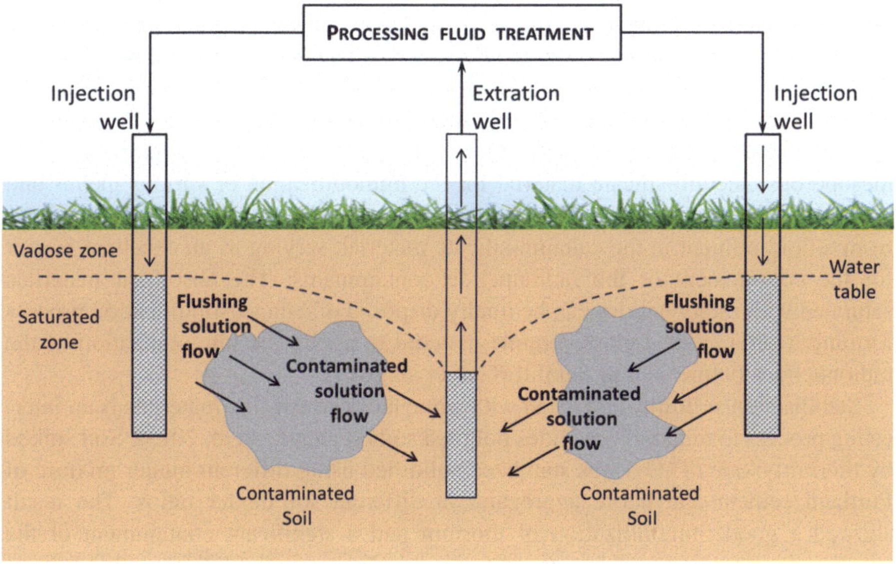

Fig. 8 In situ soil flushing

well as performing infiltration from independent cells to limit the effect of preferential flow paths, that would lead to non-uniform remediation. Similar to Cs, Ishchenko et al. (1990) found that the flow of water was able to transport uranium and thorium in irrigated soils. According to their results, about 68–72% of uranium is leached from the plow horizon whereas thorium removed about 36–41%. When uncontaminated water was flushed through the soil, uranium migrates over distances of up to 12 m but thorium only reached 6 m. These results confirm the different transportation behavior due to the interaction of radionuclides with the soil matrix. Sims et al. (1996) studied the adsorption of uranium in sandstone and the conditions of the reversibility uptake. The rapid adsorption of uranium implies a strong interaction with the rock matrix. The adsorption process was found to be highly concentration dependent and it is saturated within a relatively short time. Desorption of uranium was studied with the injection of cadmium and flushing with brine. Most of the uranium was readily recoverable but a small percentage is released very slowly on flushing, suggesting conversion to a more stable form.

The physicochemical interaction of radionuclides with soil is a limiting factor in the remediation of contaminated soils, but this information can be used to design adsorption systems to retain the radionuclides when the objective is the immobilization and confining of the radionuclides (Van et al. 2004).

4 Solidification and Stabilization

Remediation of contaminated soils by solidification and stabilization refers to a group of techniques that tend to immobilize contaminants in the original location instead of degrade, destroy or remove them. Although the contaminants remain in the same place, they do not show their negative effects in the environment because they are kept in the solid phase, with minor dissolution in interstitial water. Thus, the contaminants are immobilized in the solid phase and they are not bioavailable (Kogbara 2014). As an example, Qi et al. (2015) reported the preparation of mesoporous calcium-silicate material for the immobilization of various metals and radionuclides. These authors reported a complete removal of Co and radionuclides from a fluid effluent in the calcium-silicate material, serving as an excellent barrier for the confinement of the radionuclide contaminants. The absorbent materials saturated with radionuclides can be finally disposed of using a solidification process forming a ceramic or a glass-ceramic material to assure the immobilization of the radionuclides before sent to landfill (Guo et al. 2017).

Stabilisation/solidification (S/S) with cement or cement-like materials is an interesting process to treat radionuclides polluted soils (Falciglia et al. 2013). Soil spiked by thorium oxide (ThO_2) was stabilized/solidified using different binder mixture of Portland cement and barite aggregates at different soil:binder ratios. The result showed a good immobilization of thorium and a significant containment of the γ-radiation compared to the untreated soil. (de-Freitas and Al-Tabbaa 2013) have proposed the cementation of Intermediate Level Radioactive Waste (ILRW) using low pH cements. They claim that hyperalkaline pore waters compromises the effectiveness of the S/S with Portland cement. Moreover, the manufacture of Portland cement is responsible for significant CO_2 emissions. Low pH cements are environmentally more suitable and have emerged as a potential alternative for S/S of nuclear waste. These authors have reported that reactive magnesium oxide and magnesium phosphate cements feature important advantages in terms of reduced permeability, higher density and capacity for controlling acid-base reactions.

Falciglia et al. (2014) tested the enhanced cementation of thorium oxide contaminated soil in Portland cement with barite aggregates. The contaminated soil with thorium was successfully confined in the cement material and the best results were obtained with the higher ratios of binding material. The main benefits of this study are related to the use of barite aggregates that gave a significant containment of the γ radiation and an excellent reduction of contaminant leaching. The only drawback of the barite aggregates is the slightly reduction of mechanical resistance of the final cement material. These authors concluded that this is a good alternative remediation technology for ow level radionuclide polluted soil (Falciglia et al. 2015).

5 Vitrification

Vitrification is a remediation technology based in the transformation of contaminated soil into a glass-like material (Yuan et al. 2017). The contaminated soil is heated with the application of an electric current directly to the soil by a set of electrodes around the contaminated site. Temperatures above 1000 °C are necessary to reach the melting point of the soil components. The metals and other inorganic contaminants are trapped in the glass-like materials whereas the organic contaminants are incinerated or volatilized and collected in a hood over the treatment site (Fig. 9). Taylor-Pashow et al. (2017) studied the vitrification of low activity waste. The main radionuclide in the waste is technetium (isotope ^{99}Tc). These authors investigated Tc removal via reductive precipitation with stannous chloride from a condensate stream from the off-gas processes. The preliminary results showed a satisfactory separation and immobilization of Tc, although the authors said that more research is necessary before this technology can be applied at large scale.

Alternatively, Yuan et al. (2017) simulated the vitrification of radioactive contaminated soil by microwave heating (1200 °C). The analysis of the final material showed an amorphous structure with microwave sintering whereas crystalline silica structure was detected by conventional sintering technology. These authors concluded that microwave technology performs vitrification better than the conventional sintering method in vitrification of radioactive contaminated soil.

Overall, vitrification is a remediation technology that can be applied to polluted sites with both organic and inorganic contaminants and it is a good alternative treatment for soils and wastes with radionuclides. The main advantages are the application in situ, the short treatment time and the good stabilization of inorganic contaminants and the removal of organics. The main drawback is the high energy requirements.

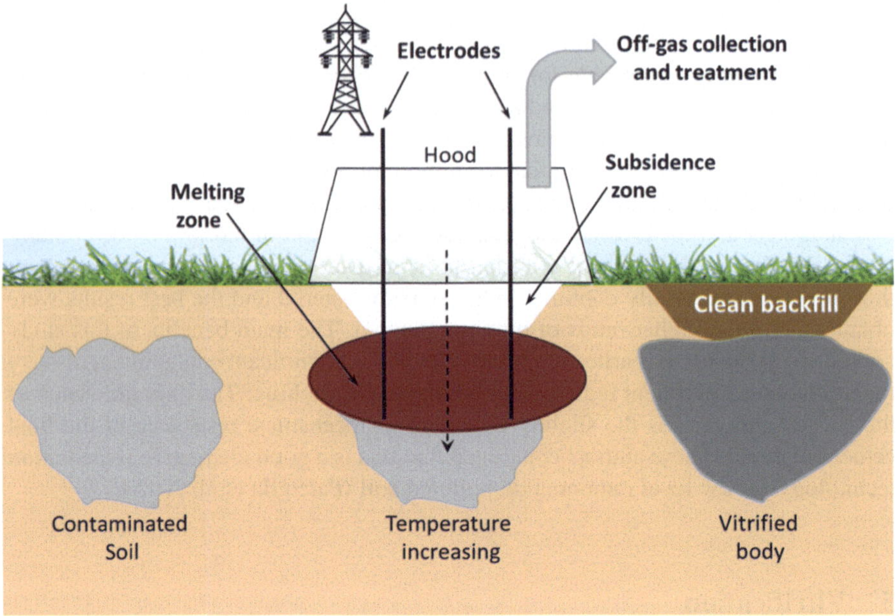

Fig. 9 Vitrification

6 Conclusions

Electrokinetic remediation is a proved technology for the removal of organic and inorganic contaminants in soils, sediments and other solid materials. Electrokinetics has been tested for the removal of radionuclides from soils and nuclear wastes reaching very promising results for large scale operation. The addition of facilitating agents and the pH control on the cathode to avoid alkaline environment, clearly increases the efficiency of the process. Considering the results from literature, it is necessary to test new facilitating agents that can be able to extract and keep in solution the radionuclides, so they can migrate under the effect of the electric field. The combination of various remediation technologies may enhance the radionuclide removal by a synergistic effect of the two (or more) coupled technologies. Thus, the combination of soil flushing and electrokinetics resulted in very promising results at lab and pilot scale tests. Soil flushing is able to remove the soluble and exchangeable fractions of radionuclides, whereas electrokinetics may mobilize and extract other radionuclide fractions more bonded to the soil. In situ soil flushing, solidification and stabilization and vitrification are alternative physicochemical technologies for the removal of radionuclides. It is widely recognized that they are mainly applicable to radionuclide contaminated soils with low/medium concentrations.

References

Bea SA, Carrera J, Soler JM, Ayora C, Saaltink M (2004) Simulation of remediation alternatives for a ^{137}Cs contaminated soil. Radiochim Acta De Gruyter 92:827–833

Belhadj B, Cameselle C, Akretche DE (2014) Physico-chemical effects of ion-exchange fibers on electrokinetic transportation of metal ions. Sep Purif Technol 135:72–79

Cameselle C (2015a) Enhancement of Electro-Osmotic flow during the electrokinetic treatment of a contaminated soil. Electrochim Acta 181:31–38

Cameselle C (2015b) Electrokinetic remediation and other physico-chemical remediation techniques for in situ treatment of soil from contaminated nuclear and NORM sites. In: Velzen LV (ed) Environmental remediation and restoration of contaminated nuclear and norm sites. Elsevier, Cambridge, pp 161–184

Cameselle C, Pena A (2016) Enhanced electromigration and electro-osmosis for the remediation of an agricultural soil contaminated with multiple heavy metals. Proc Saf Environ Prot 104:209–217

Cameselle C, Reddy KR (2012) Development and enhancement of electro-osmotic flow for the removal of contaminants from soils. Electrochim Acta 86:10–22

Chen SY, Moeller DW, Dornsife WP, Meyer HR, Lamastra A, Lubenau JO, Strom DJ, Yusko JG (2005) Toward the framework and implementation for clearance of materials from regulated facilities. Health Phys 89:115–126

Cheng M, Zeng G, Huang D, Yang C, Lai C, Zhang C, Liu Y (2017) Advantages and challenges of Tween 80 surfactant-enhanced technologies for the remediation of soils contaminated with hydrophobic organic compounds. Chem Eng J 314:98–113

Cortner HJ (2000) Making science relevant to environmental policy. Environ Sci Policy 3:21–30

de-Freitas RM, Al-Tabbaa A (2013) Development of low-pH cements for immobilisation of intermediate level radioactive waste: Achievements and challenges. In: Proceeding of ASME 2013, 15th international conference on environmental remediation and radioactive waste management, Brussels, Belgium.

Falciglia PP, Cannata S, Pace F, Romano S, Vagliasindi FGA (2013) Stabilisation/solidification of radionuclides polluted soils: a novel analytical approach for the assessment of the γ-radiation shielding capacity. Chem Eng Trans 32:223–228

Falciglia PP, Cannata S, Romano S, Vagliasindi FGA (2014) Stabilisation/solidification of radionuclide polluted soils, Part I: assessment of setting time, mechanical resistance, γ-radiation shielding and leachate γ-radiation. J Geochem Explor 142:104–111

Falciglia PP, Puccio V, Romano S, Vagliasindi FGA (2015) Performance study and influence of radiation emission energy and soil contamination level on γ-radiation shielding of stabilised/solidified radionuclide-polluted soils. J Environ Radioact 143:20–28

Frizon F, Lorente S, Ollivier JP, Thouvenot P (2003) Transport model for the nuclear decontamination of cementitious materials. Comput Mater Sci 27:507–516

Frizon F, Lorente S, Auzuech C (2005) Nuclear decontamination of cementitious materials by electrokinetics: an experimental study. Cem Concr Res 35:2018–2025

Guerrero LA, Maas G, Hogland W (2013) Solid waste management challenges for cities in developing countries. Waste Manag 33:220–232

Guo B, Sasaki K, Hirajima T (2017) Solidification of ettringite after uptaking selenate as a surrogate of radionuclide in glass-ceramics by using industrial by-products. J Mater Sci 52:12999–12913,011.

He F, Gao J, Pierce E, Strong PJ, Wang H, Liang L (2015) In situ remediation technologies for mercury-contaminated soil. Environ Sci Pollut Res 22:8124–8147

Ishchenko GS, Butnik AS, Ratopchina VA (1990) Migration of uranium and thorium resulting from soil leaching and irrigation. Sov Soil Sci 22:1–6

Khan FI, Husain T, Hejazi R (2004) An overview and analysis of site remediation technologies. J Environ Manage 71:95–122

Kim GN, Oh WZ, Won HJ, Choi WK (2003) Removal of cesium and cobalt from soil around TRIGA reactor using electrokinetic method. J Ind Eng Chem 9:306–313
Kim GN, Jung YH, Lee JJ, Moon JK, Jung CH (2008) An analysis of a flushing effect on the electrokinetic-flushing removal of cobalt and cesium from a soil around decommissioning site. Sep Purif Technol 63:116–121
Kim GN, Lee SS, Shon DB, Lee KW, Chung US (2010) Development of pilot-scale electrokinetic remediation technology to remove ^{60}Co and ^{137}Cs from soil. J Ind Eng Chem 16:986–991
Kogbara RB (2014) A review of the mechanical and leaching performance of stabilized/solidified contaminated soils. Environ Rev 22:66–86
Korolev VA (2009) Electrokinetic removal of radionuclides. In: Reddy KR, Cameselle C (eds) Electrochemical remediation technologies for polluted soils, sediments and groundwater. Wiley, Hoboken, pp 127–139
Korolev VA, Barkhatova Y, Shevtsova E (2007) Electrokinetic remediation of the radionuclide contaminated soils. In: Proceedings of 6th symposium on electrokinetic remediation. EREM 2007, Vigo, Spain, pp 169–170
Kuppusamy S, Palanisami T, Megharaj M, Venkateswarlu K, Naidu R (2016) Ex-situ remediation technologies for environmental pollutants: A critical perspective. Rev Environ Cont Toxicol 236:117–192
Li WC (2014) Occurrence, sources, and fate of pharmaceuticals in aquatic environment and soil. Environ Pollut 187:193–201
Liu Y, Kong F, Santibanez Gonzalez EDR (2017) Dumping, waste management and ecological security: Evidence from England. J Clean Prod 167:1425–1437
Lu Y, Song S, Wang R, Liu Z, Meng J, Sweetman AJ, Jenkins A, Ferrier EC, Li H, Luo W, Wang T (2015) Impacts of soil and water pollution on food safety and health risks in China. Environ Int 77:5–15
Nekrasova MA, Korolev VA (1997) Electrochemical cleaning of polluted soils. In: Proceedings of international symposium on engineering geology and the environment, organized by the Greek National Group of IAEG, Athens, Greece, vol 2, pp 2047–2052
Ottosen LM, Hansen HK, Jensen PE (2009) Electrokinetic removal of heavy metals. In: Reddy KR, Cameselle C (eds) Electrochemical remediation technologies for polluted soils, sediments and groundwater. Wiley, Hoboken, pp 95–126
Pamukcu S (2009) Electrochemical transport and transformations. In: Reddy KR, Cameselle C (eds) Electrochemical remediation technologies for polluted soils, sediments and groundwater. John, Hoboken, pp 29–64
Perera J (2002) Waste control and management in Moscow. Nucl Eng Int 47:39–41
Prozorov LB, Shcheglov MY, Nikolaevsky VB, Shevtsova EV, Korneva SA (2000) The influence of electric parameters on the dynamics of the electrokinetic decontamination of soils. J Radioanal Nucl Chem 246:571–574
Qi G, Lei X, Li L, Yuan C, Sun Y, Chen J, Chen J, Wang Y, Hao J (2015) Preparation and evaluation of a mesoporous calcium-silicate material (MCSM) from coal fly ash for removal of Co(II) from wastewater. Chem Eng J 279:777–787
Reddy KR, Cameselle C (2009a) Electrochemical remediation technologies for polluted soils, sediments and groundwater. Wiley, Hoboken
Reddy KR, Cameselle C (2009b) Overview of electrochemical remediation technologies. In: Reddy KR, Cameselle C (eds) Electrochemical remediation technologies for polluted soils, sediments and groundwater. Wiley, Hoboken, pp 1–28
Ricart MT, Cameselle C, Lucas T, Lema JM (1999) Manganese removal from spiked kaolinitic soil and sludge by electromigration. Sep Sci Technol 34:3227–3241
Ricart MT, Hansen HK, Cameselle C, Lema JM (2004) Electrochemical treatment of a polluted sludge: Different methods and conditions for manganese removal. Sep Sci Technol 39:3679–3689
Sims RC (1990) Soil remediation techniques at uncontrolled hazardous waste sites. J Air Waste Manag Assoc 40:704–732

Sims R, Lawless TA, Alexander JL, Bennett DG, Read D (1996) Uranium migration through intact sandstone: Effect of pollutant concentration and the reversibility of uptake. J Contam Hydrol 21:215–228
Song B, Zeng G, Gong J, Liang J, Xu P, Liu Z, Zhang Y, Zhang C, Cheng M, Liu Y, Ye Y, Yi H, Ren X (2017) Evaluation methods for assessing effectiveness of in situ remediation of soil and sediment contaminated with organic pollutants and heavy metals. Environ Int 105:43–55
Taylor-Pashow KML, McCabe DJ, Nash CA (2017) Tc removal from the waste treatment and immobilization plant low-activity waste vitrification off-gas recycle. J Sep Sci Technol:1–10
Trellu C, Mousset E, Pechaud Y, Huguenot D, Hullebusch ED, Esposito G, Oturan MA (2016) Removal of hydrophobic organic pollutants from soil washing/flushing solutions: A critical review. J Hazard Mater 306:149–174
Van L, Soler JM, Müller W, Bradbury MH (2004) Anisotropic diffusion in layered argillaceous rocks: a case study with Opalinus Clay. Environ Sci Technol 38:5721–5728
Virkutyte J, Sillanpää M, Latostenmaa P (2002) Electrokinetic soil remediation-Critical overview. Sci Total Environ 289:97–121
Yang JW, Lee YJ (2009) Electrokinetic removal of PAHs. In: Reddy KR, Cameselle C (eds), Electrochemical remediation technologies for polluted soils, sediments and groundwater. Wiley, Hoboken, pp 195–217
Yasunari TJ, Stohl A, Hayano RS, Burkhart JF, Eckhardt S, Yasunari T (2011) Cesium-137 deposition and contamination of Japanese soils due to the Fukushima nuclear accident. Proc Natl Acad Sci USA 108:19530–19519
Yu JW, Neretnieks I (1997) Theoretical evaluation of a technique for electrokinetic decontamination of soils. J Contam Hydrol 26:291–299
Yuan X, Qing Q, Zhang S, Lu X (2017) Vitrification of radioactive contaminated soil by means of microwave energy. AIP conference proceedings, vol 1820.
Yun SM, Kang CS, Kim J, Kim HS (2015) Evaluation of soil flushing of complex contaminated soil: An experimental and modeling simulation study. J Hazard Mater 287:429–437
Zhou J, Zheng X, Flury M, Lin G (2009) Permeability changes during remediation of an aquifer affected by sea-water intrusion: A laboratory column study. J Hydrol 376:557–566

Sorption Methods in Rehabilitation of Radioactively Contaminated Lands Using Surface-Modified Sorbents Based on Natural Aluminosilicates

Anna V. Voronina, Marina O. Blinova, Vladimir S. Semenishchev, and Dharmendra K. Gupta

1 Introduction

Sorption method is a prospective technique for rehabilitation of radioactively contaminated lands including decontamination of natural waters and soils from radionuclides and decrease of radionuclides transfer into agricultural vegetation. Available and effective sorption materials are necessary for decontamination of large volumes of radioactively contaminated waters and soils, creation of geochemical barriers and decrease of radionuclides migration in the lithosphere. Among these sorbents, natural aluminosilicates are the most interesting ones for radionuclides separation from natural waters.

Use of natural aluminosilicates for rehabilitation of radioactive contaminated landscapes especially developed after the Chernobyl disaster. Zeolites from Sokirnitskoye deposit (Ukraine) and some deposits from Siberia and Transcaucasia as well as bentonite and clays from Russia and Belarus showed a high effectiveness for ^{137}Cs, ^{134}Cs, ^{90}Sr and ^{239}Pu sorption from various natural waters including drinking water. Zeolites and clayey minerals (Ovchinnikov and Bezdenezhnykh 1996; Campbell and Davies 1997; Bondar 1998; Misaelides 2011) as well as composite sorbents (Budarkov et al. 1994) are used for the decrease of cesium and strontium radionuclides transfer from soil to agricultural vegetation. As a result of addition of natural sorbents, the decrease of ^{137}Cs and ^{90}Sr transfer from radioactive soils to plants was by the factor of 1.0–1.5 (Katsnelson et al. 1983). Furthermore,

A. V. Voronina (✉) · M. O. Blinova · V. S. Semenishchev
Radiochemistry and Applied Ecology Department, Physical Technology Institute, Ural Federal University, Ekaterinburg, Russia

D. K. Gupta
Institut für Radioökologie und Strahlenschutz (IRS), Gottfried Wilhelm Leibniz Universität Hannover, Hannover, Germany

D. K. Gupta, A. Voronina (eds.), *Remediation Measures for Radioactively Contaminated Areas*, https://doi.org/10.1007/978-3-319-73398-2_3

the sorbents improved agrochemical characteristics of the soils promoting the development of root systems and plants growth.

As compared with other aluminosilicates, clinoptilolite possess a high sorption affinity for cesium and strontium radionuclides, higher mechanical characteristic as well as stability against influence of acids and alkali. In the works (Tarasevich 1981, 1996), selectivity of clinoptilolite for ions of heavy and transition metals is studied. Glauconite, a widespread mineral, can be used as a natural sorbent for the tasks of rehabilitation of the environment. Tretyakov (2002) studied sorption of ^{90}Sr and ^{137}Cs by a glauconite concentrate from drinking (tap) water and water from the cooling pond of the Beloyarskaya Nuclear Power Plant (Russia). The degree of ^{137}Cs and ^{90}Sr sorption of glauconite from aqueous solutions depends of the ratio between mica and montmorillonite phases in the structure of the mineral. Strontium sorption increases with the increase of the percent of swelling interlayers, whereas cesium sorption decreases in this case.

Natural sorbents are cheap and available sorption materials. However, they possess low sorption capacity and mechanical strength. Furthermore, sorption of radionuclides by natural sorbents is reversible that may result in leaching of radionuclides back to soil or water. That's why, development and improvement of methods of synthesis of cheap, effective and biocompatible sorbents is very interesting for rehabilitation of radioactively contaminated lands.

This chapter systematizes data on characteristics of surface-modified ferrocyanide sorbents based on natural aluminosilicates (glauconite and clinoptilolite) that is important for using these sorbent for rehabilitation of radioactively contaminated lands. Mixed nickel-potassium ferrocyanides (NPF) and iron-potassium ferrocyanides (IPF) were obtained by the surface modification of natural glauconite (Nat-Gl) from Karinskoye deposit (Chelyabinsk region, Russia) and natural clinoptilolite (Nat-Cl) from Shivertooyskoye deposit (Chita region, Russia).

2 Requirements for Sorption Materials Being Used for Rehabilitation of Radioactively Contaminated Lands

Sorbents being used for rehabilitation should meet the following special requirements:

1. Sorbents should be eco-friendly and compatible with the environment.
2. Sorbents should possess high exchange capacity and specificity for radionuclides in presence of analogue ions because concentrations of analogue ions in natural waters and soil solutions are high and significantly exceed radionuclides concentration; therefore, competitive sorption of radionuclides and analogue ions will occur.
3. Sorbents should possess higher rates of radionuclides sorption as compared with the rate of radionuclides interception by soils and plants.

4. Sorbents should be a reliable matrix for radionuclides retention in order to minimize radionuclides leaching back to decontaminated media.
5. Technology of sorbents production should be easy and cheap; this will allow using these sorbents for decontamination of large volumes of radioactive soils and waters.

Sorbents possessing optimal combination of these characteristics will be effective for rehabilitation.

3 Specificity and Capacity of the Sorbents Based on Natural Aluminosilicates with Respect to Cesium and Strontium Radionuclides

Specificity and capacity of the sorption materials should be taken into account when sorbents are used for decontamination of natural waters and remediation of soils. High distribution coefficients for radionuclides provide high degrees of decontamination of aqueous media. Sorption capacity of the sorbents is also important for media containing elevated concentrations of macro constituents (Na^+, K^+ and Ca^{2+} cations) if sorption of radionuclides is not selective.

Table 1 presents basic sorption characteristics of natural aluminosilicates and ferrocyanide sorbents based on them. Distribution coefficients (K_d) of cesium and strontium as well as static exchange capacities (SEC) were obtained in a weakly

Table 1 Distribution coefficients of cesium and strontium as well as static exchange capacities for natural aluminosilicates and ferrocyanide sorbents based on them

Sorbent	Number of sorption sites	Henry region, mg L^{-1}	lg K_d, mL g^{-1}	SEC, mg g^{-1}
Sorption characteristics for cesium				
Nat-Gl	1	10^{-5}...1	3.1 ± 0.4	15 ± 4
NPF-Gl	2	10^{-7}...1	5.0 ± 0.3	220 ± 23
		10^2...10^3	2.4 ± 0.6	
Nat-Cl	1	10^{-5}...10	4.4 ± 0.5	21 ± 6
NPF-Cl	3	$1\cdot10^{-7}$...$1\cdot10^{-5}$	7.0 ± 1.0	500 ± 60
		$5\cdot10^{-5}$...1	5.7 ± 0.4	
		10^2...10^3	3.2 ± 0.7	
IPF-Cl	2	10^{-7}...1	4.9 ± 0.3	500 ± 25
		10^2...10^3	3.6 ± 0.6	
Sorption characteristics for strontium				
Nat-Gl	1	10^{-5}...1	1.9 ± 0.2	21 ± 2
NPF-Gl	1	10^{-5}...1	2.1 ± 0.3	37 ± 4
Nat-Cl	1	10^{-5}...1	3.4 ± 0.4	63 ± 6
NPF-Cl	1	10^{-5}...1	3.9 ± 0.4	560 ± 39

Adopted from Voronina et al. (2015a, b)

mineralized water (tap water) at pH 7.8 ± 0.2. Concentrations of the main ions in tap water were (mg L^{-1}): K^+ – 3.6; Na^+ – 25.5; Ca^{2+} – 40.0; Mg^{2+} – 8.0; Fe – 1.8; Cl^- – 16.5; SO_4^{2-} – 52.5.

Ferrocyanide sorbents showed significantly higher specificity and capacity with respect to cesium as compared with natural glauconite and clinoptilolite. Experiments on cesium sorption also showed that ferrocyanide sorbents based on natural aluminosilicates are polyfunctional (Orlov et al. 2017). The Table 1 also shows cesium distribution coefficients obtained at various initial concentrations of cesium in a solution. The NPF-Gl and IPF-Cl sorbents possess two types of sorption sites, whereas three types of sorption sites were found in the NPF-Cl sorbent. All sorbents showed the maximal values of cesium distribution coefficients at the initial concentration of cesium of up to 1 mg L^{-1} that corresponds to typical concentrations range of stable cesium in natural waters (Kulsky et al. 1980; Polyakov 2007; Polyakov et al. 2015).

Both natural aluminosilicates and modified sorbents showed approximately the same specificity for strontium within the error limits; however, modified sorbents possessed significantly higher static exchange capacity for this radionuclide.

4 Selectivity of Cesium and Strontium Sorption in Presence of Analogue Elements

Salt composition of natural waters with different mineralization may vary in a strong degree. Seawater highly mineralized underground waters and soil solutions containing radionuclides are especially different media for deactivation. Use of selective sorbents separating radionuclides and saving the same salt composition allow obtaining high decontamination factors in these media.

Selectivity of cesium sorption by ferrocyanide sorbents based on natural aluminosilicates (clinoptilolite and glauconite) was studied in KCl, NaCl, NH_4Cl and $CaCl_2$ solutions. Capacity of frayed edge site (FES) of natural and modified aluminosilicates with respect to cesium was determined under the conditions of blocking nonselective sorption sites by Ca^{2+} ions. Radiocesium interception potential (RIP) was determined in presence of 0.0005M KCl or NH_4Cl. Table 2 presents parameters of selective sorption of cesium by natural and modified aluminosilicates. Static

Table 2 Parameters of selective sorption of cesium by natural and modified aluminosilicates

			RIP(K), mmol kg^{-1}		RIP(NH_4), mmol kg^{-1}	
Sorbent	SEC, mg g^{-1}	[FES], mg g^{-1}	1 day	7 days	1 day	7 days
Nat-Gl	114 ± 11	14.8 ± 1.5	$1.6 \cdot 10^3$	$8.6 \cdot 10^3$	$4.3 \cdot 10^3$	$6.0 \cdot 10^3$
NPF-Gl	211 ± 15	138 ± 14	$3.8 \cdot 10^3$	$6.1 \cdot 10^5$	$1.3 \cdot 10^3$	$4.9 \cdot 10^3$
Nat-Cl	224 ± 28	135 ± 10	$8.8 \cdot 10^2$	$2.1 \cdot 10^3$	$1.6 \cdot 10^3$	$2.3 \cdot 10^3$
NPF-Cl	402 ± 28	136 ± 12	$3.3 \cdot 10^3$	$5.0 \cdot 10^6$	$2.9 \cdot 10^3$	$7.0 \cdot 10^3$

Adopted from Kulyaeva and Voronina (2016) and Kulyaeva and Voronina (2017)

exchange capacities and [FES] were determined for the same samples of sorbents; therefore, it is possible to assess changes of capacity as a result of surface modification of clinoptilolite and glauconite samples by ferrocyanides. Tables 1 and 2 present data for samples from different parties of aluminosilicates, therefore the values of static exchange capacities are different too. Usually, natural aluminosilicates (such as glauconite and clinoptilolite) are significantly inhomogeneous that may result in a difference of their sorption characteristics. In homogeneousness of the clinoptilolite from Shivertooyskoye deposit was mentioned by Slavkina and Babayev (2015). The value of capacity given in Table 1 is very similar for the values being obtained by the research team of Tomsk Polytechnical University (Russia) for the clinoptilolite from Shivertooyskoye deposit (0.19 mg-eq g^{-1}).

Surface modification of the glauconite by the mixed nickel-potassium ferrocyanide resulted in an increase of [FES], whereas in case of the clinoptilolite modification only increase of SEC was observed. Furthermore, modified aluminosilicates showed significantly higher values of RIP(K) as compares with natural ones. For the NPF-Cl sorbent, RIP(K) reached $5.0 \cdot 10^6$ mmol kg^{-1} that is 1000 times higher than for the Nat-Cl. In case of the NPF-Gl sorbent, RIP(K) was $6.1 \cdot 10^5$ mmol kg^{-1} that is 100 times higher than for the natural glauconite.

Selectivity factor α of ion exchange of trace amounts of cesium and monovalent ion K^+ was calculated in accordance with the Eq. (1) (Popov et al. 2011):

$$\alpha(\mathrm{Cs/M}) = \mathrm{RIP(M)}/[\mathrm{FES}], \quad (1)$$

where α(Cs/M) is selectivity factor of cesium and a monovalent ion M^+; [FES] is capacity of frayed edge sites, mmol kg^{-1}.

Simultaneous sorption of ^{137}Cs and ^{22}Na from NaCl solutions were studied in order to evaluate selectivity of cesium sorption in presence of macro amounts of sodium. Separation factors F_s were calculated in accordance with the Eq. (2):

$$F_s(\mathrm{Cs/Na}) = \frac{K_{d,Cs}}{K_{d,Na}}. \quad (2)$$

where $K_{d,\ Cs}$ is distribution coefficient of cesium, L kg^{-1}; $K_{d,\ Na}$ is distribution coefficient of sodium, L kg^{-1}.

Table 3 presents values of selectivity factors and separation factors obtained for natural and modified samples of glauconite and clinoptilolite.

Table 3 Selectivity of cesium sorption in presence of K^+ and Na^+

Sorbent	F_s (Cs/K)	F_s (Cs/Na)
Nat-Gl	78	–
NPF-Gl	58	–
Nat-Cl	2	2
NPF-Cl	4892	250

Adopted from Voronina et al. (2017)

The NPF-Cl sorbent showed the highest selectivity for cesium in presence of potassium and sodium. The dependences of cesium distribution coefficients by natural and modified aluminosilicates on concentrations of K^+, Na^+, NH_4^+ and Ca^{2+} in a wide concentration ranges are presented in a series of publications (Voronina et al. 2013, 2017).

5 Stability of the Sorbents Based on Natural Aluminosilicates Against Radionuclides Leaching

Reversible sorption of cesium and strontium radionuclides is possible for natural aluminosilicates and sorbents based on them. For example, Petrova et al. (2008) reported that strontium is almost totally desorbed from montmorillonite clay. Therefore, the study of chemical stability of spent sorbents against radionuclides leaching is very important for practical applications. Under the conditions of decontamination of natural waters or soils remediation, a sorbent may contact with natural waters of various mineralization, soil solutions and rainwaters that may result in radionuclides leaching. This process depends mainly on the mineral composition of solutions.

Cesium leaching from natural aluminosilicates and sorbents based on them was studied using a long-term leaching technique (Russian State Standard 52126-2003). The following types of water were used as a leaching agent: mineral water (mineralization = 1.5 g L^{-1}), fresh water and rainwater. Saturated samples of sorbents containing 600 Bq g^{-1} of ^{137}Cs were transferred into a container and filled up with 25 mL of a leaching solution. Portions of the leaching solution were changed in 1, 3, 7, 10, 14, 21 and 28 days after the start of the experiment. Activities of ^{137}Cs were analyzed in leaching and rinsing solutions. Using the data obtained, leaching degrees and leaching rates of cesium from sorbents were calculated.

The values of leaching rate R were calculated according to the Eq. (3):

$$R = \frac{a}{A \cdot S \cdot t} \tag{3}$$

where a is activity (Bq) or weight (g) of a radionuclide leached;

A is initial activity (Bq g^{-1}) or concentration (g g^{-1}) of a radionuclide in the sample before leaching;
S is the surface area of the sample, cm^2;
t is the duration of leaching since the last sampling of a leaching agent, days.

Table 4 presents total degrees and rates of cesium leaching from saturated samples of sorbents.

The natural glauconite showed the lowest chemical stability against cesium leaching by various water solutions. Surface modification of the glauconite and clinoptilolite resulted in a significant decrease of cesium leaching. In case of the NPF-Cl sorbent, degree of cesium leaching did not exceed 2% for all leaching

Table 4 Integral degrees of leaching and leaching rates of cesium from saturated samples of sorbents

	Degree of leaching, %			Leaching rate, g cm^{-2} day^{-1}	
Sorbent	Rain water	Fresh water	Mineral water	Fresh water	Mineral water
Nat-Cl	–	6.5	24.7	$2.4 \cdot 10^{-18}$	$9.0 \cdot 10^{-18}$
NPF-Cl	–	1.6	2.0	$2.0 \cdot 10^{-19}$	$2.5 \cdot 10^{-19}$
IPF-Cl	–	1.8	17.8	$2.6 \cdot 10^{-19}$	$2.1 \cdot 10^{-18}$
Nat-Gl	28.8	41.6	63.4	$6.7 \cdot 10^{-18}$	$9.9 \cdot 10^{-18}$
NPF-Gl	6.6	14.6	1.5	$3.7 \cdot 10^{-18}$	$3.9 \cdot 10^{-18}$

Adopted from Voronina et al. (2015b)

agents. Furthermore, the effect of salt content and composition in the leaching agent on degrees of cesium leaching was observed. The highest degrees and rates of leaching were observed in mineral water due to the competitive influence of analogue ions. The IPF-Cl sorbent showed lower stability against cesium leaching as compared with the NPF-Cl and NPF-Gl sorbents; probably, this in conditioned by a lower stability of the phase of mixed iron-potassium ferrocyanide.

6 Ecological Safety of Using the Sorbents Based on Natural Aluminosilicates for Rehabilitation Radioactively Contaminated Lands

The sorbents being used for decrease of radionuclides transfer from soil to agricultural plants will be in contact with soil solutions; therefore, these sorbents should not be a source of toxic compounds due to leaching. The assessment of nickel and ferrocyanide ions leaching is important when mixed nickel-potassium and iron-potassium ferrocyanides based on natural aluminosilicates are used for rehabilitation. According to Hygienic norms 2.1.7.2041-06 (2006), the maximal permissible concentration of nickel in a soil is 4.0 mg per kg of soil, whereas concentration of iron and ferrocyanides is not regulated. Therefore, Hygienic norms 2.1.5.1315-03 (2003) for drinking water was used for an approximate assessment of ecological safety of the modified sorbents from the point of view of ferrocyanide ions leaching. According to these Norms, concentration of $[Fe(CN)_6]^{4-}$ should not exceed 1.25 mg L^{-1}.

The quantity of nickel and ferrocyanide ions leached from the modified sorbents was evaluated using the long-term leaching technique described above under the conditions near to those of rehabilitation of radioactively contaminated lands. Weakly mineralized water was used as a leaching agent; time of leaching was 28 days. A sample of the sorbent (m = 200 mg) was in contact with 25 mL of the leaching agent. Change of leaching agent was performed after 1, 7, 14, 21 and 28 days from the beginning of the experiment. Concentrations of nickel and ferrocyanide ions were determined in each sample of the leachate. Determination of nickel concentration was performed by colorimetric method with 1% solution of dimethylglyoxime as a coloring agent and potassium persulfate as an oxidant. All solutions being analyzed were warmed in a drying oven at 80 °C for 15–20 min in order to express the reaction rate (Sandell 1959). Determination of ferrocyanide concentration was performed also by colorimetric method with iron (III) as colloidal solution of $Fe_4[Fe(CN)_6]_3$ (Lurye 1973). For a comparison, nickel and ferrocyanide ions leaching was studied also for another two sorbents, a thin-layer ferrocyanide sorbent based on clinoptilolite (TL-NPF-Cl) and iron-potassium ferrocyanide based on marl (IPF-Marl).

The results of nickel leaching from samples of ferrocyanide sorbents based on aluminosilicates by a weakly mineralized water is presented in Table 5.

The results have shown that total degree of Ni leaching from the modified sorbents NPF-Cl and NPF-Gl to a soil for the whole period of use will not exceed the maximal permissible concentration and will be 3.4 and 3.1 mg of Ni per kg of a soil respectively. By contrast, the total degree of Ni leaching from the thin layer sorbent TL-NPF-Cl will result in thrice maximal permissible concentration in a soil; therefore, this sorbent cannot be used for rehabilitation of radioactively contaminated lands.

The results on ferrocyanides leaching from sorbents to weakly mineralized water are presented in Table 6 and at Figs. 1 and 2.

The maximal values of leaching for one day were 0.06 mg g^{-1} for the NPF-Cl sorbent and 0.03 mg g^{-1} for the NPF-Gl sorbent. In the case of humidity of a soil after rain or irrigation as high as 1 L of water per kg of a soil, the maximal concentration of ferrocyanides in soil solution will be 0.6 and 0.3 mg L^{-1} respectively that does not exceed the maximal permissible concentration.

Table 5 Leaching of nickel from samples of ferrocyanide sorbents

Sorbents	Time of leaching, days					Weight of nickel leached from a sorbent, mg of Ni g^{-1} of sorbent
	1	7	14	21	28	
NPF-Cl	0.05	0.08	0.09	0.08	0.05	0.34
NPF-Gl	0.03	0.08	0.08	0.07	0.05	0.31
TL-NPF-Cl	0.44	0.42	0.15	0.11	0.06	1.20

Adopted from Blinova (2017)

Table 6 Leaching of ferrocyanide ions from samples of ferrocyanide sorbents, mg g^{-1} of sorbent

Sorbent	Time of leaching, days				
	1	7	14	21	28
NPF-Gl	0	0.241	0.131	0	0
NPF-cl	0.055	0.345	0.138	0	0
IPF-cl	0.442	6.540	2.359	0.014	0.007
TL-NPF-cl	0	0.159	0.097	0	0
IPF-marl	0.304	0.538	0.938	0.842	6912

Adopted from Blinova (2017)

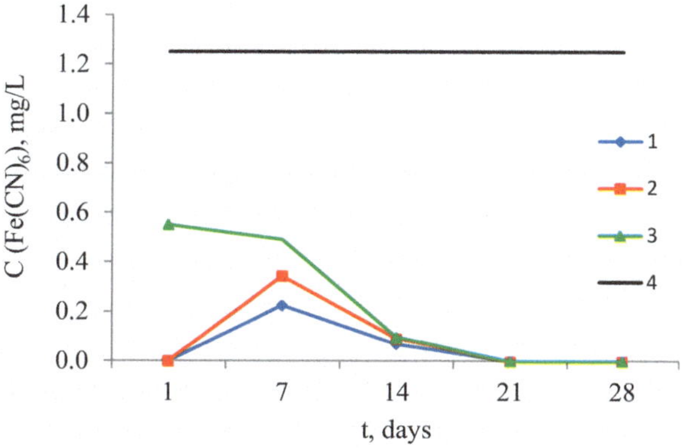

Fig. 1 Ferrocyanides leaching from the samples of modified aluminosilicates (*1* – TL NPF-Cl, *2* – NPF-Gl, *3* – NPF-Cl, *4* – maximum permissible concentration for ferrocyanide ions)

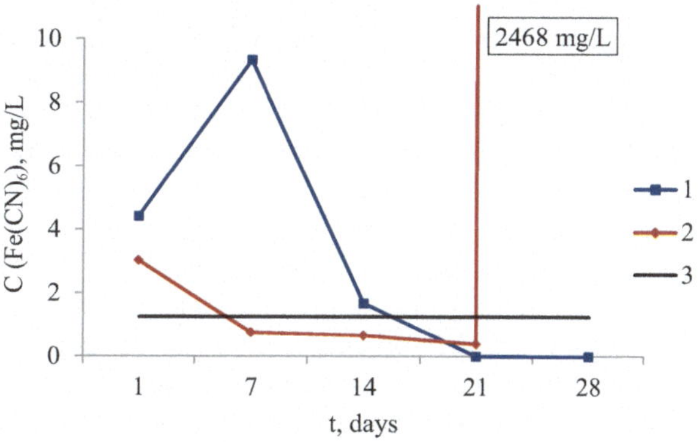

Fig. 2 Ferrocyanides leaching from the samples of modified aluminosilicates (*1* – IPF-Cl, *2* – IPF-marl, *3* – maximum permissible concentration for ferrocyanide ions)

In case of sorbents containing iron-potassium ferrocyanide, the concentration of ferrocyanide leached to water after a long contact significantly exceeded the maximal permissible concentration. The maximal concentrations of ferrocyanide leached for one day were 9.3 mg L^{-1} and 2468 mg L^{-1} for the IPF-Cl IPF-Marl sorbents respectively. Thus, the sorbents containing iron-potassium ferrocyanide showed higher leaching of both radiocesium and ferrocyanide to a solution that indicates low chemical stability of these sorbents. This fact is probably conditioned by the processes of oxidation of iron in the ferrocyanide ion from Fe(II) to Fe(III). Thus, the mixed iron-potassium ferrocyanides based on clinoptilolite and marl are not recommended for rehabilitation of radioactive soils.

7 Decrease of Cesium Radionuclides Transfer from Soil to Agricultural Vegetation

The assessment of effectiveness of rehabilitation is based on accumulation factors (K_a) of cesium in plants and cesium transfer decreasing factors ($K_{tr.\ dec}$) from soil to plants in presence of sorbents that were calculated according to Eqs. (4 and 5) respectively:

$$K_a = \frac{A_{veg}}{A_{soil}} \quad (4)$$

where

A_{veg} is specific activity of radiocesium in vegetation, Bq g^{-1},
A_{soil} is specific activity of radiocesium in soil, Bq g^{-1}.

$$K_{tr.dec} = \frac{A_{control}}{A_{veg}} \quad (5)$$

Where $A_{control}$ was specific activity of radiocesium in plants from control group of plants, grown in soil, contaminated by ^{137}Cs, without addition of sorbents, Bq g^{-1}.

Surface-modified sorbents based on natural aluminosilicates can be used for the decrease of cesium and strontium radionuclides transfer from soil to agricultural vegetation. Mineral composition of a soil, quantity and composition of organics in a soil, humidity and granulometric composition of a soil, physicochemical characteristics of a sorbent as well as the species of a plant will affect radionuclides distribution among soil, soil solution, plant and sorbent. A clayey soil with a high cation exchange capacity of 54.5 ± 1.3 mg-eq per 100 g of soil and organic matter concentration of 5.5% was used in the experiments on the effect of addition of surface-modified sorbents on the decrease of ^{137}Cs transfer from soil to agricultural vegetation. Activity of cesium in the soil was 300 ± 20 Bq g^{-1}. In the experiments, two methods of the sorbents addition to the soil were used. In the first method (Mixing), the sorbent was regularly mixed with the contaminated soil; after this seeds of oats (*Avena sativa*) were sowed in the soil. In the second method (Surface addition), the sorbent was added onto the surface of the soil after the seeds were sowed. For the comparison, a control group of the same plants was grown at the same conditions without addition of a sorbent. After this, the soil was irrigated in order to obtain the stable humidity. The oats have grown at the contaminated soil for 21 days. Oat was harvested after 21 days of growth. After harvest, samples were dried in a drying oven at 65 °C for 1 h and averaged; cesium specific activity in plants was measured by γ-spectrometry. The values of accumulation factors (K_a) of cesium in oats and cesium transfer decreasing factors ($K_{tr.\ dec}$) from soil to oats are presented at Figs. 3 and 4.

The results have shown that cesium transfer decreasing factors were significantly lower in presence of natural aluminosilicates as compared with modified aluminosilicates. Addition of natural aluminosilicates does not provide effective decontamination of soil solutions independently on the method of sorbent addition (Mixing or Surface addition). The low $K_{tr.\ dec}$ for natural aluminosilicates are conditioned by their low specificity for cesium, competitive sorption of macro elements from the soil solution as well as reversibility of cesium sorption. The obtained results agree well with the results of practical use of clayey minerals for rehabilitation of radioactively contaminated lands. Addition of these minerals to light soils resulted in $K_{tr.\ dec}$ values of 1.5–3.0; no decrease of cesium transfer was obtained on other types of soils (Sanzharova et al. 2005; Zubets et al. 2011).

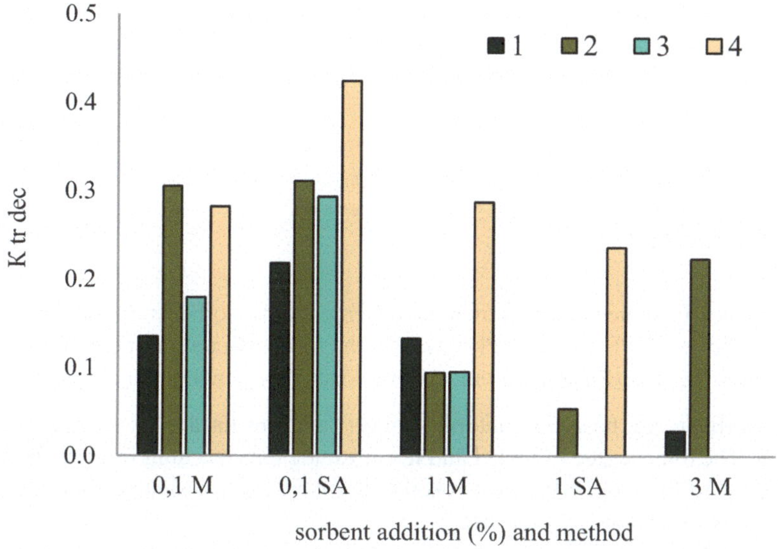

Fig. 3 The dependence of accumulation factor of ^{137}Cs in oats on content and method of addition of a sorbent to a soil (*1* – NPF-Gl, *2* – Nat-Gl, *3* – NPF-Cl, *4* – Nat-Cl). The methods of sorbents addition to soil are the following: M is the mixing method; SA is the Surface Addition method

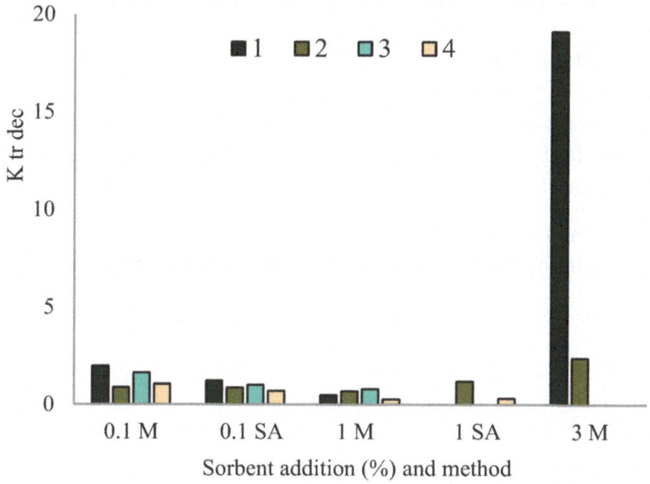

Fig. 4 The dependence of cesium transfer decrease factor to oats on a sorbent concentration in a soil (*1* – NPF-Gl, *2* – Nat-Gl, *3* – NPF-Cl, *4* – Nat-Cl). The methods of sorbents addition to soil are the following: M is the mixing method; SA is the Surface Addition method

The highest values of cesium transfer decreasing factors (19.1 ± 0.8) were obtained in case of addition of 3% wt. of NPF-Gl by the Mixing method. The effectiveness of this method exceeds the effectiveness of other methods described in literature.

8 Conclusions

The results of studies have shown that ferrocyanide sorbents based on natural aluminosilicates (NPF-Cl and NPF-Gl) meet the following requirements for sorption materials being used for rehabilitation of radioactively contaminated lands:

1. The sorbents are based on natural supports and not hazardous for the environment. The total degree of nickel and ferrocyanide ions leaching from the modified samples NPF-Cl and NPF-Gl to a soil will not exceed Russian permissible limits for the whole period of their use.
2. The sorbents possess high sorption capacity and selectivity over analogue ions. The values of RIP(K) were $6.1 \cdot 10^5$ mmol kg^{-1} for the NPF-Gl sorbent and $5.0 \cdot 10^6$ mmol kg^{-1} for the NPF-Cl sorbent. For the same sorbents, static exchange capacities for cesium were 220 ± 23 and 500 ± 60 mg g^{-1} respectively; for strontium these values were 37 ± 4 and 560 ± 39 mg g^{-1} respectively.
3. The sorbents possess low degrees and rates of radiocesium leaching. The degrees of cesium leaching from saturated sorbents into natural water with the mineralization of 1.5 g L^{-1} were 0.05% per day after 7 days of leaching; total degrees for 30 days of leaching were not exceed 2%.
4. The technology of producing these sorbents is simple and low cost.
5. Besides deactivation of natural waters, the sorbents can be used for the decrease of cesium transfer from soils to agricultural vegetation. As it was shown on clayey soils, the factors of decrease of ^{137}Cs transfer to oats were 19.1 ± 0.8.

Thus, the obtained results have shown that the NPF-Gl and NPF-Cl sorbents are interesting for realizing the sorption method of rehabilitation of radioactively contaminated lands.

References

Blinova MO (2017) Ferrocyanide sorbents based on natural aluminosilicates foe rehabilitation of radioactively contaminated territories. PhD thesis, Ural Federal University, Ekaterinburg

Bondar PF (1998) About the assessment of efficiency of sorbents as means of radionuclides retention in soils. Rad Biol Radioecol 2:267–272

Budarkov VA, Mayakov EA, Torubarova AA, Kalinin NF, Gelis VM, Milyutin VV, Penzin RA (1994) Method of caesium radionuclides transfer decreasing from soil to vegetation. Russian Federation Patent ⍰ 2013913 from 15.06.1994

Campbell LS, Davies BE (1997) Experimental investigation of plant uptake of caesium from soils amended with clinoptilolite and calcium carbonate. Plant Soil 189:65–74

Hygienic norms 2.1.5.1315-03 (2003) Maximal permissible concentrations of chemical agents in water of water bodied used for drinking and domestic purposes

Hygienic norms 2.1.7.2041-06 (2006) Soil, decontamination of populated areas, industrial and domestic waste, sanitary protection of soils. Maximal permissible concentrations of chemical agents in a soil

Katsnelson YY, Sklyarova ES, Likhachyov VA, Zelenshchikov GV, Kapustyan AS (1983) Using of natural sorbents from Rostov region deposits for ecological remediation of lands being under high anthropogenic impact. Radiochemistry 2:111–121

Kulsky LA, Goronovsky IT, Koganovsky AM, Shevchenko MA (1980) Handbook on properties, analytical methods and decontamination of water. Naukova Dumka, Kiev, USSR

Kulyaeva IO, Voronina AV (2016) Parameters of selective sorption of ^{137}Cs by natural and modified aluminosilicates. Abstracts of the Symposium «Chernobyl: 30 years after the accident» the International Conference of Students, Post-Graduates and Young Scientists «Lomonosov-2016»: 54

Kulyaeva IO, Voronina AV (2017) Selectivity of sorption of ^{137}Cs by natural and modified aluminosilicates in presence of Na^+ ions. Abstracts of the IV international youth scientific conference: physics technology innovation PTI-2017:121–122

Lurye YY (1973) Unified methods of water analysis. Chemistry, Moscow, USSR

Misaelides P (2011) Application of natural zeolites in environmental remediation: a short review. Micropor Mesopor Mat 144:15–18

Orlov PA, Voronina AV, Blinova MO, Yakovlev GA, Semenishchev VS (2017) Modeling statics and kinetics of ^{137}Cs sorption by modified aluminosilicates from aqueous solutions of various chemical compositions. AIP Conf Proc 1886:020058 (1–7)

Ovchinnikov NA, Bezdenezhnykh VS (1996) Method of soils rehabilitation. Russian Federation Patent ▯ 2064748 from 10.08.1996

Petrova MA, Flowers AG, Krip IM, Shimchuk TV, Petrushka IM (2008) Sorption of Sr on clay minerals modified with ferrocyanides and hydroxides of transition metals. Radiochemistry 50:502–507

Polyakov EV (2007) Behavior of ionic and colloid forms of microelements in colloidal chemical extraction from humic acid solutions. Radiochemistry 49:432–438

Polyakov EV, Volkov IV, Khelbnikov NA (2015) Competitive sorption of cesium and other microelements onto iron(III) hexacyanoferrate(II) in the presence of humic acids. Radiochemistry 57:161–171

Popov VE, Il'icheva NS, Stepina IA, Maslova KM (2011) Influence of the potassium and ammonium ion concentrations on the selective sorption of ^{137}Cs by illite and clinoptilolite. Radiochemistry 53:97–102

Russian State Standard R 52126-2003 (2003) Radioactive waste. Determination of chemical stability of solidified high-level radioactive waste using the continuous leaching

Sandell EB (1959) Colorimetric determination of traces of metals, 3rd edn. Revised and Enlarged. Interscience Publishers Inc., New York

Sanzharova NI, Sysoeva AA, Isamov NN, Alexakhin RM, Kuznetsov VK, Zhigareva TL (2005) Chemistry role in remediation farming land being under radioactive pollution. Russ Chem J 49:26–34

Slavkina OV, Babayev BB (2015) Obtaining and study of characteristics of selective inorganic sorbents. In: Proceedings of the Russian scientific conference of physicists students and young scientists (VNKSF-21). http://asf.ural.ru/VNKSF/tezis_v6/4/1.html

Tarasevich YI (1996) Use of natural sorbents as a deactivation agent during the elimination of the consequences of Chernobyl disaster. Water Chem Technol 2:127–131

Tarasevich YI (1981) Prirodnye sorbenty v protsessakh ochistki vody (Natural sorbents in processes of water treatment). Naukova Dumka, Kiev

Tretyakov SY (2002) Sorption of ^{90}Sr and ^{137}Cs on natural sorbents in model environmental systems. Radiochemistry 43:93–95

Voronina AV, Semenishchev VS, Savchenko MO, Bykov AA, Kutergin AS, Nedobuh TA (2013) Approaches to rehabilitation of radioactive contaminated territories. J Chem Technol Biotechnol 88:1606–1611

Voronina AV, Blinova MO, Kulyaeva IO, Sanin PY, Semenishchev VS, Afonin YD (2015a) Sorption of cesium radionuclides from aqueous solutions onto natural and modified aluminosilicates. Radiochemistry 57:446–452

Voronina AV, Blinova MO, Semenishchev VS, Gupta DK (2015b) Returning land contaminated as a result of radiation accidents to farming use. J Environ Radioact 144:103–112

Voronina AV, Gorbunova TV, Semenishchev VS (2017) The effect of the synthesis method on the parameters of pore structure and selectivity of ferrocyanide sorbents based on natural minerals. J Radioanal Nucl Chem 312:241–254

Zubets MV, Prister BS, Alexakhin RM, Bogdevich IM, Kashparov VA (2011) Urgent problems and tasks of scientific support of farming in radioactively contaminated zone of Chernobyl NPP. Agroecol J 1:5–20

Competitive Adsorption as a Physicochemical Ground for Self-Sufficient Decontamination Areas from Radioactive Pollutants

Evgeny V. Polyakov, Aleksey A. Ioshin, and Ilya V. Volkov

1 Introduction

The contamination of natural objects, soil and surface water with technogenic radionuclides is a result of global fallout induced by nuclear tests and the use of nuclear weapons in the twentieth century (González 1998), as well as by imperfect and becoming ever more sophisticated atomic power engineering (Gonzalez 1996; Yadigaroglu 2012). In response to the threat of worldwide radioactive contamination, there appeared approaches to the decontamination of natural objects, first of all, fresh water and soil. By now, vast experience in removal of radionuclides from aqueous media has been gained. However, the decontamination of solid materials is still a rather difficult economic and technological problem (Warner and Harrison 1993; Ding et al. 2016). A number of methods for the removal of radionuclides and heavy metals from soil have been proposed: reagent less (Nikulina 2016), chemical reagent treating methods (Hamby 1996; Streletskaya 2003) and their combination (Valcke et al. 1997; Maslova et al. 2013). However, no efficient and relatively cheap green chemistry friendly decontamination technology has been created hitherto (Ding et al. 2016). The most promising in this respect techniques are the reagent less methods, by using which it is possible to purify clayey soils with a decontamination index of ~10 by concentrating the most sorption-active clay fraction (its content is up to 15% of the total mass of soil) (Nikulina 2016). At a low level of radiation pollution, ploughing and digging-in of topsoil are used. In order to reduce the transfer of long-lived radionuclides into plants, soil is limed (fixation of Sr-90), and enhanced quantities of potassium fertilizers are introduced (suppression of ion-exchange sorption of Cs-137) (Guillen et al. 2008). The chemical reagent

E. V. Polyakov (✉) · A. A. Ioshin · I. V. Volkov
Laboratory for Physicochemical Analytical Methods, Institute of Solid State Chemistry, Ural Branch Russian Academy of Sciences, Ekaterinburg, Russia
e-mail: polyakov@ihim.uran.ru; mr.kj@uralweb.ru; Ilyavolkov@ihim.uran.ru

D. K. Gupta, A. Voronina (eds.), *Remediation Measures for Radioactively Contaminated Areas*, https://doi.org/10.1007/978-3-319-73398-2_4

treating methods include high-pressure water handling of soil (Kato 2015); flotation treatment of soils with surface-active substances and electro migration separation of radionuclides from soils (Veshev et al. 1996; Chirkst et al. 2001; Sanzharova et al. 2005). Since ^{90}Sr and ^{137}Cs radionuclides are fixed mainly by clay minerals, of interest is the deactivation technique based on ion-exchange desorption of these radionuclides with salt solutions containing cations with a high exchange capability (for example, Fe(III)) (Chirkst et al. 2001; Streletskaya 2003). There are examples of using microbe strains and phyto-sorbents in the processes of bioaccumulation of radionuclides in soils (Vandana et al. 2000; Sasaki and Takeno 2014). Because of the large scale of soil deactivation technologies, of increasingly greater importance for radionuclide behavior control in soils is the method of sorption fixation of radionuclides by introduction of an additional sorbent into soil (Wauters et al. 1996a, b; Valcke et al. 1997).

The role of sorbent here is to provide a sorption competition with plants for the absorption of radionuclides in the system "soil material – soil solution – sorbent – plant root age". To understand the quantitative processes in such multiphase and multicomponent system, physicochemical modeling of equilibrium (stationary) and dynamic interactions is used. The modeling is based on the application of heterogeneous ion exchange laws, selectivity constants, as well as the notions about sorption inhomogeneity of minerals – sorbents in the framework of the Langmuir theory (Langmuir 1916; Brunauer et al. 1938; Sips 1948; Elovich and Larionov 1962).

The development of countermeasures against radioactive contamination is based on the understanding of sorption phenomena with participation of soil components. The works (Cremers et al. 1988; Wauters et al. 1996a, b) gave an important insight into the sorption regularities in the distribution of radioactive pollutants in the natural environment. They are based on the generalization of sorption properties of different types of soils, which allowed an *a priori* estimation of the cesium distribution coefficient value in soils (Cremers et al. 1988). The selectivity discreteness of sorption centers of soils and the existence of highly selective sorption centers toward cesium ions in illite let them introduce the notion of ^{137}Cs radiocesium interception potential – RIP (Cremers et al. 1988). RIP is a product of the capacity of silicates relative to the high selectivity centers and the selectivity coefficient of Cs(I) with respect to K(I) during the exchange on common and high selectivity centers (Cremers et al. 1988; Wauters et al. 1996a; Sanzharova et al. 2005). Using RIP it is possible to estimate the distribution coefficient (*Kd*) of cesium ions between the soil solution and soil on the basis of their chemical composition. RIP characterizes the competition of cations (Cs^+-K^+, Cs^+-NH_4^+) for the sorption centers, generally highly selective centers (Wauters et al. 1996b). This allows one to consider the decontamination processes in terms of competitive sorption of microelement in a sorption system containing several competitive sorbents: soil (sorbent 1) – bentonite (sorbent 2) – soil solution of cesium radionuclides – plant (Vandenhove et al. 2005). In order to avoid the transfer of radionuclides from soil to plants, a competitive sorbent (bentonite) is introduced into contaminated soil, which is superior to the competitive sorbents (soil and plants) in the RIP value of (Cs/K) pair. The competitive sorbent suppresses the transfer of radionuclides ("intercepts" it) in the chain "soil – solution – plant".

The "price" for the positive effect of sorption suppression by plants is deeper radioactive contamination of soil. After such decontamination, for the removal of radionuclide from soil it is necessary to separate bentonite particles with adsorbed radionuclide from soil. Inevitable degradation of sorbent particles in soil (dispersion, variation in the phase and chemical composition under the action of humic and microbial species of soil, atmospheric moisture and fertilizers) can lead to secondary contamination of the soil. The separation of radionuclide from sorbent with a high sorption affinity is a more complicated sorption problem than decontamination at the initial stage of this scheme (Vandenhove et al. 2005).

In order to make a general idea of the potential of the competitive sorption method as an instrument of decontamination of materials, natural soils or artificial materials, it is expedient to consider the decontamination of solid in terms of heterogeneous equilibria and kinetics of microelement transfer in the simplest model of competitive system. The model includes material with sorbed microelement, aqueous solution of electrolyte providing the transport of microelement from the material to ambient environment, artificial sorbent capable of competing with contaminated material for the microelement. The use of the decontamination factor as a dependent variable will allow simulating competitive processes in this and more complicated systems and will provide a more comprehensive insight into the organization of natural and synthetic ion-exchange material decontamination technologies on self-sustained competitive sorption and green-chemistry principles (Polyakov 2012; Polyakov et al. 2015a).

2 Competitive Sorption: Statics

Consider as a model the statics and kinetics of competitive sorption in a system, for which the interphase distribution of microelement obeys the Henry's distribution law, and follows the Langmuir equation for medium and high sorbate concentration, Fig. 1 (Langmuir 1916; Egorov 1975; Benes and Majer 1980; Polyakov 2003). The contamination of a material with microelement ions we will consider as a reversible sorption process that can be arbitrarily close to thermodynamic equilibrium. According to the mass action law, the condition for effective desorption of microelement ions from contaminated material by an arbitrary competitive sorbent is a higher value of the distribution coefficient, Kd^*, of the sorbent in comparison with the material, Kd^m. We consider a competitive sorption in the system: "material (for example, illite) with sorbed ions – sorbate (desorbing solution containing electrolyte, simple and complex ions, true, and/or adsorption colloids of this microelement) –

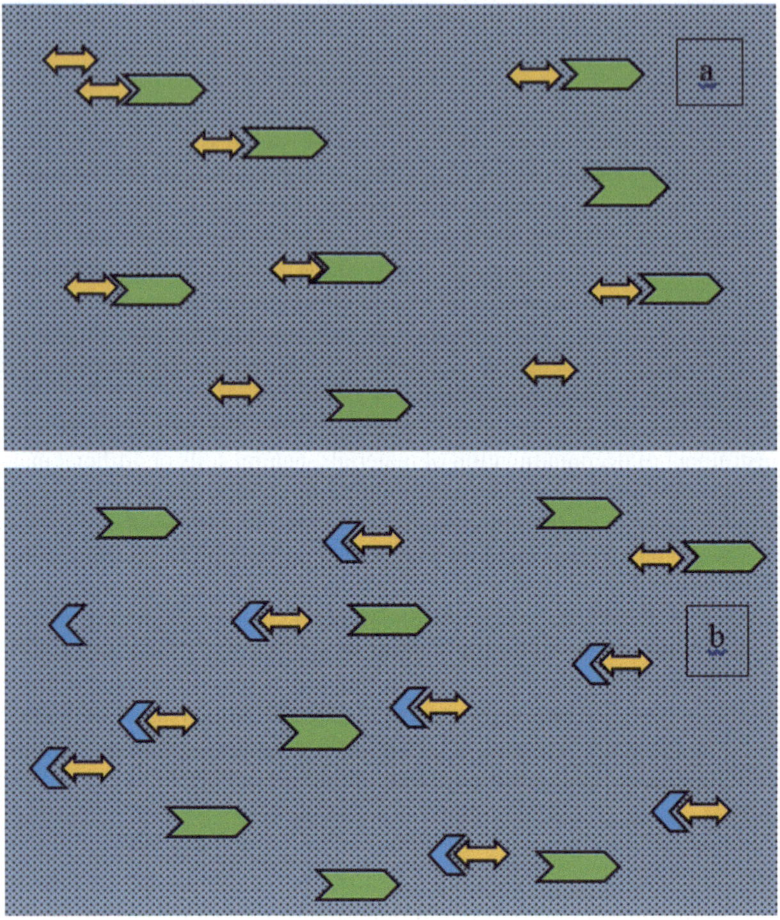

Fig. 1 The scheme of equilibrium competitive sorption of arbitrary microelement in the system "material – sorbate – sorbent": (**a**) the initial state, contaminated material; (**b**) the final state, the contaminated material is brought into contact with the sorbent. *Designations*: long yellow arrow –a particle of the material; short blue arrow –a particle of the sorbent; double-headed yellow arrow – microelement *M*. Combination of arrows, respectively: the state of contaminated material with sorbed microelement (green–yellow) and the state of microelement absorbed by the sorbent (blue–yellow); background–electrolyte solution. (Adopted from Polyakov et al. 2015b)

sorbent (for example, bentonite, Prussian blue, KU2 (Dowex-50)". The desorbing solution in the considered scheme acts as both complex solution and transport medium. It provides the interphase transfer of sorbate ions between the material and the sorbents. The introduction of a complexing agent (for example, humic acid, *HA*) or a competitive ion (for example, K(I) ions) into this solution displaces the equilibrium in the sorption system. This provides deeper desorption of both this microelement and other microelements sorbed by contaminated material owing to their transfer onto sorbate (Valcke and Cremers 1994; Polyakov et al. 2015b).

Represent an elementary model of desorption of a microelement from material onto foreign sorbent as an equilibrium competitive system "material – sorbate – sorbent" (1) (Polyakov et al. 2015b).

$$A \overset{K^m}{\rightleftarrows} B \overset{K^*}{\rightleftarrows} C. \tag{1}$$

Here, A is the state of adsorbed microelement M in material "A"; B is the state of microelement M in the sorbate, which is a solution of electrolyte "B" contacting with material "A" and artificially introduced sorbent "C"; and C is the state of adsorbed microelement M in the sorbent (Polyakov 2003). Let the coefficients K^m and K^* characterize the concentration constants of sorption equilibria of microelement M with participation of material (index m) and sorbent (index*), Fig. 1.

The driving force for the chemical distribution of desorbed microelement in (1) is the difference in its chemical potentials (constants K^m, K^*, distribution coefficients Kd^m, Kd^*) in each of the contacting equilibrium subsystems "material – sorbate", "sorbent – sorbate". The microelement distribution coefficient value can be additionally affected by introducing a competitive ion (Valcke and Cremers 1994), or a complexing agent like *HA* (Polyakov et al. 2015a; Volkov et al. 2013), into the suspension. The scheme in Fig. 1 differs from the schemes of desorption by a homogeneous solution of complexing agent in the fact that the result of competitive sorption here is the accumulation of microelement ions in the competitive sorbent phase, rather than in the solution phase. After completion of desorption, the sorbent can be left in the medium of the decontaminated material (Vandenhove et al. 2005) or can be removed from it and used as the final form for subsequent long-term storage (Polyakov et al. 2015a; Nikulina 2016). The latter possibility considerably improves the efficiency and ecological safety of the whole decontamination process. Soils, samples of building, roadway, fabric and other materials possessing surface ion-exchange groups can be considering as decontamination objects (Cremers et al. 1988; Sanzharova et al. 2005; Stepina et al. 2013). If a microelement in contaminated material is not in the ionic form but in the form of true or adsorption colloid, ultrafiltration and colloid-chemical extraction can be used for competitive sorption (Polyakov et al. 2000; Polyakov 2000).

From the methodological standpoint, there is a good reason to compare the results of thermodynamic estimation of the consequences of chemical desorption of microelement M from sorption-decontaminated material due to complexing and sorption of sorbent suspensions in the solution on the basis of the decontamination coefficient K_o (Maslova et al. 2013; Cremers et al. 1988; Stepina et al. 2013; Polyakov et al. 2015b). By definition, the decontamination coefficient K_o of a material is the ratio

$$K_o = \frac{M^{in}}{M^m} \quad \text{rel.units.,} \tag{2}$$

where M^{in} and M^m are the initial and equilibrium mass of a microelement in the material after desorption, mol. (Moskvin 2004).

Below we consider a heterogeneous equilibrium (1) for the case when sorption of a microelement by the material is accompanied by a complex formation in the solution. This is a known case of the sorption, which was first formulated in (Shubert 1948) and analyzed in detail in (Egorov 1975). In the framework of this approach, the evaluation of the decontamination efficiency of a material can be reduced to the description of competitive sorption of microelement *M* by this material in the presence of competitors for this microelement – monodentate complexing agent (*HA*) and monofunctional sorbent (index *) – in the Henry's law region, Fig. 1. Write the independent conditions of microelement distribution between material, solution and sorbent as the sorption ratio (ε) for the whole system (1) and for its subsystems: "material – sorbate" (ε^m), "sorbent – sorbate" (ε^*):

$$\varepsilon = \frac{M^m}{M + M^*} = \frac{\varepsilon^m}{1 + \varepsilon^*}, \quad \varepsilon^m = \frac{M^m}{M}, \quad \varepsilon^* = \frac{M^*}{M}, \quad M^{tot} = M + M^m + M^*, \tag{3}$$

$$\varepsilon = \frac{Kd^m[m^m]}{1 + Kd^*[m^*]},$$

$$[m^m] = \frac{m^m}{V}, [m^*] = \frac{m^*}{V}, \quad Kd^m = \frac{[M^m/m^m]}{[M/V]}, \quad Kd^* = \frac{[M^*/m^*]}{[M/V]}.$$

Here, *M* is the equilibrium mass of microelement in solution of volume *V*, ml. M^m and $M^*$are the equilibrium masses of microelement in material (index *m*) and sorbent (index *). M^{tot} is the total mass of microelement in system (1), mol; *m* and *m* * are the masses of material and sorbent, respectively, g (a sorbate solution may contain both a microelement and a complexing agent *HA*); Kd^m, Kd^* are the microelement distribution coefficients between material and solution (index *m*) and between sorbent and solution (index*), ml g^{-1}. The square brackets are the symbol of microelement concentration in phases.

Isotherm (3) describes the simplest situation of competitive sorption of a microelement in a system with two competitive sorbents and allows one to estimate at what ratio of the sorption affinity of the material and the sorbent to adsorbed microelement (Kd^m, Kd^*) the desired decontamination degree of this material should be achieved.

At the initial stage of analysis of competitive system (2, 3) it is interesting to evaluate the role of complexing agent *HA* as a desorbing agent in (1) in the absence of sorbent. Write the equation of the complexing reaction of microelement *M* with *HA* molecule as a homogeneous reaction

$$M + HA + MHA, \beta = \frac{[MHA]}{[M][HA]},$$

Assuming that the molecules of complexing agent *HA* and complex *MHA* are not adsorbed by the material, we write the microelement distribution coefficient dependence between the material and the solution (Kd^m) as a function of the equilibrium concentration of the complexing agent [*HA*]

$$Kd^m = \frac{Kd_0^m}{(1 + \beta[HA])}, \quad \varepsilon^m = Kd^m[m]. \tag{4}$$

Here, Kd_0^m is the distribution coefficient for a material in the absence of a complexing agent. When desorption is due to chemical binding of the microelement by a complexing agent *HA* according to (2), the decontamination coefficient of the material is:

$$K_o = \frac{M^{in}}{M^m} = \frac{M + M^m}{M^m} = \frac{1 + \varepsilon^m}{\varepsilon^m}, \quad Kd^* = 0. \tag{5}$$

It is obvious that the decontamination coefficient will be greater than unity only if desorption leads to reduction of Kd^m as compared with Kd_0^m when the complexing agent concentration grows according to (4). The expression for the decontamination coefficient during homogeneous sorption by a complexing agent solution (5) takes on the form:

$$K_o = \frac{1 + \varepsilon^m}{\varepsilon^m} = \frac{1 + Kd^m[m]}{Kd^m[m]}. \tag{6}$$

Figure 2 displays the characteristic dependences of the decontamination coefficient according to (6) for three different values of the microelement distribution

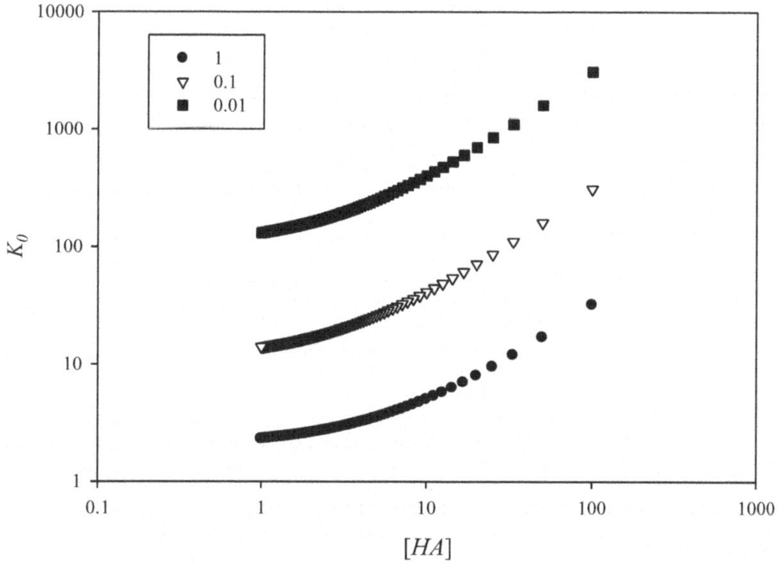

Fig. 2 The dependence between the decontamination coefficient of a material (K_o, rel. units), complexing agent concentration ([*HA*], rel. units) and sorption properties of decontaminated material (Kd_0^m, rel. units) according to Eqs. (2, 6, and 8). The values in the inset are used in the calculation of Kd_0^m at $\beta = 0.3$

coefficient Kd_0^m between a material and a sorbate solution. Desorption efficiency growth with increasing concentration of complexing agent in desorbing solution $[HA]$. This dependence characterizes the methods for chemical decontamination (Moskvin 2004).

The application of complexing agent solutions in the practice of decontamination gives rise to secondary radioactive solutions having a high concentration of desorbed microelement in a chemically more stable state. This leads to incremental costs for hardening of desorbing solution and for its transformation into the form suitable for long-term storage of isolated microelement (radionuclide). The modern methods of surface decontamination by polymer coatings also have some restrictions as to the decontamination of powder materials (Hamby 1996).

In terms of technology, a more appropriate procedure is desorption, when a contaminated material is brought into contact with a competitive sorbent suspension in solution according to Scheme (1) (Valcke et al. 1997; Polyakov et al. 2015b). Due to a high difference in chemical potentials, the sorbent concentrates the microelement absorbed by the material. After the required decontamination coefficient has been attained, the sorbent is left together with the material (Valcke et al. 1997) or is removed from it by any convenient phase separation method (Chirkst et al. 2001; Streletskaya 2003; Ioshin et al. 2016; Nikulina 2016).

Consider the microelement sorption equilibrium according to Eq. (1). Expression (2) for the decontamination coefficient K_o in this case takes the form:

$$K_o = \frac{M^{in}}{M^m} = \frac{M + M^m + M^*}{M^m}. \tag{7}$$

The difference between (5) and (7) is in the term M^* which expresses mass balance after the contaminated material is brought to sorption equilibrium with the sorbent suspension. For convenience of analysis, we assume that there is no complexing agent HA in the solution and the distribution coefficient Kd^* is not already equal to zero. The expression for the decontamination coefficient (7) is written as

$$K_o = \frac{1 + M^m/M + M^*/M}{M^m/M} = \frac{1 + Kd^m[m] + Kd^*[m^*]}{Kd^m[m]}. \tag{8}$$

$$K_o = \frac{1}{Kd^m[m]} + 1 + \frac{Kd^*[m^*]}{Kd^m[m]}. \tag{9}$$

This relation is the desired form of the decontamination coefficient dependence on the sorption parameters of contaminated material, sorbent and solution. The first term in (9) determines the sorption contribution of the material to the expression for K_o. The higher is the distribution coefficient for a material, the smaller is its decontamination coefficient as a result of contacting of the material with a certain volume of desorbing solution. When the distribution coefficient and the mass of the material ($Kd^m \cdot [m]$) increase, the first term of the decontamination coefficient asymptotically tends to zero. It is obvious that the effect of sorbent on sorption-contaminated material according to (9) is analogous to the action of complexing agent in model (4, 5 and 6). The decontamination coefficient increases with the concentration of complexing agent [*HA*] or the sorbent mass [*m**], Fig. 3. However, in the context of technological consequences, the effect of a liquid desorbing substance and of sorbent in the form of an aqueous suspension is different. In the former case, desorption of a microelement from the material is due to the formation of a more stable ion-molecular complex with *HA* molecules in the solution and a transfer of the microelement from the material to the solute. This solution is left for storage or is decontaminated, with the desorbed microelement being transferred into solid state. In the latter case, the microelement transfers, with the participation of the solution, from the material to a more thermodynamically stable sorption form in the sorbent and then it can be stored as a solid product.

Isotherm (9) shows that the third term – the ration of microelement distribution coefficients between the material and the sorbent, $Kd^*[m^*]/Kd^m[m]$ – plays the most important role in the considered decontamination model. It determines the technologically acceptable strategy of sorbent application as a decontaminating suspension, the attainment of a specified decontamination coefficient and the mass ratio of sorbent and decontaminated material at the same volume of desorbing solution. For example, according to (9), the decontamination coefficient 10^2 can be reached for a contaminated material with $Kd^m = 10$ rel. u. at the sorbent to material mass ratio 0.1 rel. u. if a sorbent with $Kd^* = 10^3$ rel. u. is used (Fig. 3).

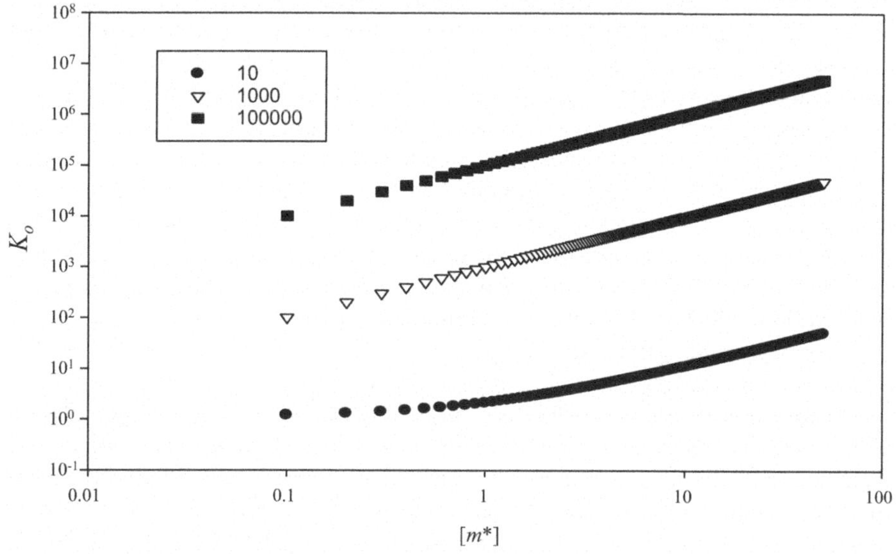

Fig. 3 The variation in the decontamination coefficient (K_o) with the sorbent distribution coefficient (Kd^*) growth in model (9) as a function of the sorbent mass per unit volume of solution [*m**] (rel. u.). The results of simulation for $Kd^m = 1$, $m = 1$ and $Kd^* = 10$, 10^3, 10^5 in (9), respectively, are given

Soils are one of the most difficult objects in terms of technology for Cs(I) and Sr (II) radionuclide decontamination (Sanzharova et al. 2005). According to the modern concepts about the sorption character of contamination of soils with microelements, in particular ^{137}Cs radionuclides, soils as natural sorbents have three types of exchange centers, whose distribution coefficients with respect to cesium ions differ approximately by an order of magnitude (Cremers et al. 1988). Type (I) is represented by low-selectivity exchange centers constituting the greatest fraction in the total exchange capacity of soil, ~95–98%. They are characterized by Kd^m (I) ~ 10^2 ml g^{-1}. The centers of the second type (II) have intermediate selectivity with respect to cesium ions. Their fraction in the total capacity is ~2–5% and the distribution coefficient is Kd^m(II) ~ 10^4 ml g^{-1}. Finally, there are super-selective centers (III), the fraction of which in the total capacity is about 0.02–0.06%, and Kd^m (III)~10^5 ml g^{-1}. The presence of exchange centers with Kd^m(II) and Kd^m(III) in soils is attributed to the presence of illite mineral from hydromica subgroup and to a special crystal structure of "frayed-edge" sites of hydromica microcrystals (Cremers et al. 1988).

In the considered model, the ratio between the distribution coefficients of the soil (material) and competitive sorbent mass, $Kd^*[m^*]/Kd^m[m]$, determines the decontamination coefficient value. Estimate the result of equilibrium sorption interaction in the competitive system "soil – Cs(I) – Prussian blue (PB)". Let a sample with exchange centers of chosen types (Kd^m(I) ~ 10^2 ml g^{-1}, Kd^m(II) ~ 10^4 ml g^{-1} and Kd^m(III) ~ 10^5 ml g^{-1}) and mass $[m]$ is brought to equilibrium with the sorbent, for example, PB powder, having a distribution coefficient of $Kd^* = 6.3 \cdot 10^6$ ml g^{-1} and sorbent mass $[m^*]$ (Polyakov et al. 2015b). The process of sorption with participation of soil, PB and soil solution includes competitive sorption equilibria of Cs(I) and ion analogues, in particular, K(I), NH_4^+, Mg(II), Ca(II) and their *HA*-complexes (Valcke and Cremers 1994; Vandenhove et al. 2005; Polyakov et al. 2015a). Further, for the sake of simplicity, we assume that the electrolyte composition remains constant, the sorbent capacity changes insignificantly as a result of sorption (Henry's region) and the ratio of the concentrations of Cs(I) and competitive cations in the solution and in the sorbents is considered in the Kd^*, Kd^m distribution coefficient values. From the soil decontamination coefficient (for example, $K_O = 3$ or 50) it is possible, using (9), to estimate the fraction of soil decontaminated by the sorbent to a preset decontamination degree in equilibrium conditions.

According to the estimated results, if soil is characterized only by nonselective centers of the type (I), the unit mass of the sorbent can provide the specified decontamination degree for 100 ($K_o = 50$) and 2600 ($K_o = 3$) mass units of soil, respectively (Polyakov et al. 2015b). The introduction of competitive cations, for example potassium or ammonium ions, into the decontaminated solution makes it possible to additionally decrease the cesium distribution coefficient by soil, thereby increasing the decontaminated mass (Sanzharova et al. 2005; Stepina et al. 2013). Considering that the estimation of Kd^* of ^{137}Cs for construction materials, such as concrete, granite, limestone, brick, asphalt, keeps within 10^3 ml g^{-1} (Stepina et al. 2013), the decontamination effectiveness of these materials will be similar to our estimates obtained for soil with nonselective exchange centers of type (I). So, the cesium ion decontamination coefficient $K_o = 3$, according to (9), for soil with nonselective exchange centers, Kd^m(I), leads to the decontaminated soil to sorbent mass ratio of ~$2 \cdot 10^3$ kg kg^{-1} if a sorbent with $Kd^* = 6.3 \cdot 10^6$ ml g^{-1} (PB) is used. Assuming the mass of a 10 cm layer of soil with an area of 1 m^2 and a volume density of ~1.1 g cm^{-3} (Yu et al. 2000) to be ~110 kg, we find that almost 20 m^2 soil can be decontaminated with the use of one kg of the chosen sorbent. Accordingly, the decontamination of 1 ha of soil at a depth of 10 cm requires ~ 500 kg of sorbent.

Sr(II) ions are characterized by reversible sorption on low-specificity (RES) functional groups of clay particles (Cremers et al. 1988). The yield of ^{90}Sr(II) into soil solution and the ^{90}Sr(II)/Ca(II) ratio in soil solutions varies from 0.49 to 0.78 of their ratio in soil and is due to more stable sorption of Sr(II) compared with Ca(II). The cation (^{90}Sr/Ca) exchange selectivity coefficient for most soils changes in the range 1–2 (Wauters et al. 1996a; Sanzharova et al. 2005; Comans et al. 1997). The estimation of the possibility of sorption decontamination of soil from ^{90}Sr radionuclides with the use of a highly selective sorbent gives the result similar to that for the system "soil – Cs(I) – PB". We can use Kd^m (~20 ml g^{-1}) for the system "illite – ^{85}Sr – 0.01M $CaCl_2$" at pH = 4–7 (Wissocqa et al. 2017) and Kd^* (~10^6 ml g^{-1}) for the sorbent – polyantimonous acid $Sb_2O_5 \cdot nH_2O$ (PAA) (Belinskaya and Militsina 1980; Zhang et al. 2016). It can be shown, that for the decontamination of 1 ha of soil from ^{90}Sr with $K_o = 3$ and $K_o = 50$ at a depth of 10 cm one needs 43 and ~1100 kg of PAA respectively, Fig. 4.

From the foregoing it is seen that the desorption of Cs(I) and Sr(II) radionuclides by selective sorbents (for example, (Belinskaya and Militsina 1980; Zhang et al. 2016; Remez 1994) is a technologically feasible method for the remediation of areas with a high level of radionuclide contamination (50 Cu km^{-2} and more (Kryshev et al. 2013). The high distribution coefficients are necessary, but not sufficient for radionuclide desorption. For successful application, the sorption technique shall have proper kinetics. It is necessary to have the information about the kinetic regularities of competitive sorption, as well as the time, during which the equilibrium decontamination coefficients are achieved in an elementary system (Eqs. 1, 7, and 9).

3 Competitive Sorption: Kinetics

In the previous section, we discussed the decontamination processes based on competitive sorption of radionuclides from contaminated material (material) on decontaminating sorbent (sorbent). The statics of desorption of a microelement from the adsorbed state in material is considered as a spontaneous process of competitive sorption in the system "material – aqueous solution – sorbent". The thermodynamics of equilibria in this system is indicative of formal equivalence of the achieved decontamination coefficients of material during equilibrium desorption by a complexing agent solution and sorbent aqueous suspension. The potential of competitive sorption as a processing method for the decontamination of soils was shown on the example of cesium ion sorption by iron-potassium cyanoferrate

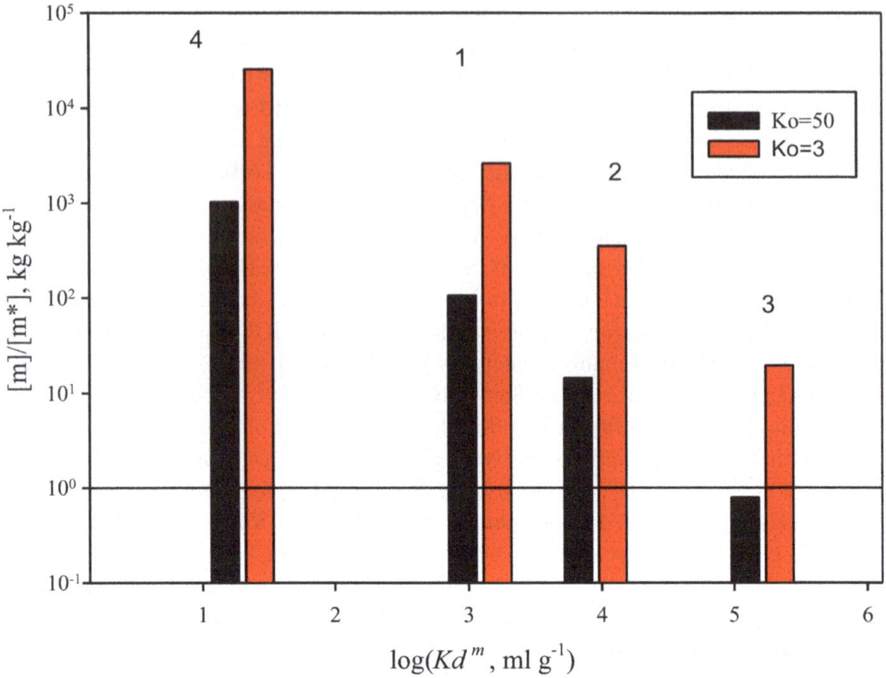

Fig. 4 The mass ratio $[m]/[m^*]$ of decontaminated soil to a mass unit of sorbent as a function of the Kd^m of soil and of decontamination coefficient K_o according to Eq. (9). Figures above the bars denote a set of sorption centers of the soil toward the radionuclide: 1–3: soil –^{137}CsCl – PB, $Kd^* = 6.3*10^6$ ml g^{-1}; 4: – "illite –^{85}Sr – 0.01 M $CaCl_2$ $Sb_2O_5nH_2O$ (PAA), pH = 4–7, $Kd^* = 10^6$, ml g^{-1}. (Adopted from Belinskaya and Militsina 1980; Zhang et al. 2016)

(PB) (Polyakov et al. 2015a). As one of possible options for the realization of competitive sorption, we proposed to use a new method for the decontamination of solutions and solid materials, which consists in the application of hermetic pockets with a sorbent having semi-permeable track membrane walls. Such sorption "mini-reactors" allow combining the spontaneous decontamination of material on the principles of competitive sorption and the possibility of facile removal of mini-reactors with the sorbent from decontaminated material powder both in natural and men-made conditions (Ioshin et al. 2016).

The driving force of the interphase transfer of contaminant microelement ions from material (sorbent "A", Fig. 1) via ionic state in aqueous solution "B" to the adsorbed state in sorbent (sorbent "C") is the difference in the chemical potentials of the microelement in sorbents "A" and "C". The time necessary for the achievement of equilibrium values of distribution coefficients of microelement "B" between "A" and "C" phases (relaxation time) should be technologically admissible for the decontamination in static or dynamic states in different-scale sorption systems.

To estimate the order of the relaxation time values of the reaction presented in Fig. 1, analyze the kinetic model of a competitive sorption system

$$A \underset{b1}{\overset{k1}{\rightleftarrows}} B \underset{b2}{\overset{k2}{\rightleftarrows}} C, \tag{10}$$

which will be considered as first-order consecutive chemical reactions between the ions of microelement M in an ultra-diluted solution and the sorption positions in material and sorbent in the framework of the Henry's law. When equilibrium in the reaction chain (10) is achieved, the system transfers to state (1) (Yablonskii et al. 1983; Polyakov and Betenekov 1988).

Kinetic model (10) is an elementary model (with the minimal number of components) of chemical kinetics of interphase transfer of microelement ions from material "A" via ionic state in aqueous solution "B" to the adsorbed state "C (Polyakov 2003). The chosen model has an advantage – it allows one to take into account the effect of the composition of solution "B" and the side-reactions of microelement complexing on the rate of elementary stages of competitive sorption. It can be used also for diffusion-controlled sorption in solution (Kokotov and Pasechnik 1970; Betenekov et al. 1999). As in previous section, we assume that the sorption behavior of system (1, 10) obeys the Henry's law; i.e. the microelement forms an ideal diluted solution in all phases.

The coefficients $k_{1,2}$ and $b_{1,2}$ of linear adsorption mechanism (10) characterize the specific microelement transformation rates in direct ($k_{1,2}$) and inverse direction ($b_{1,2}$) in competitive sorption kinetics. Analytical solution of the system of equations characterizing the kinetic mechanism (10) is given for the initial conditions, when the microelement concentration in material "A" at $t = 0$ is equal to C_0, and in solution "B" and sorbent "C" it is equal to zero (Rodigin and Rodigina 1960). As with the equilibrium system (1), consider the dependence of the kinetic analogue of the decontamination coefficient $K_o(t)$ on the independent sorption parameters of model (10): masses $[m]$, $[m^*]$, coefficients of direct and inverse reaction rates. For this purpose, we turn from equilibrium sorption characteristics Kd^m, Kd^* to their non-equilibrium analogues depending on the phase contact time, $Kd^m(t)$, $Kd^*(t)$. We use the known analytic solution of kinetics equations for mechanism (10) having a form of the time (t) dependence of the microelement concentration in "A, B, C" states (Rodigin and Rodigina 1960). Represent (9) in the form (11):

$$K_o(t) = \frac{1}{Kd^m(t)[m]} + 1 + \frac{Kd^*(t)[m^*]}{Kd^m(t)[m]}. \tag{11}$$

In (11), $K_o(t)$ is a non-equilibrium decontamination coefficient, whose dependence on the phase contact time (t) is determined by mechanism (10) and gives an equilibrium decontamination coefficient, $K_o = K_o(t \to \infty)$. Equation (11) solution with the initial conditions $A(t = 0) = A_0$, $B(t = 0) = C(t = 0) = 0$ can be finding in the

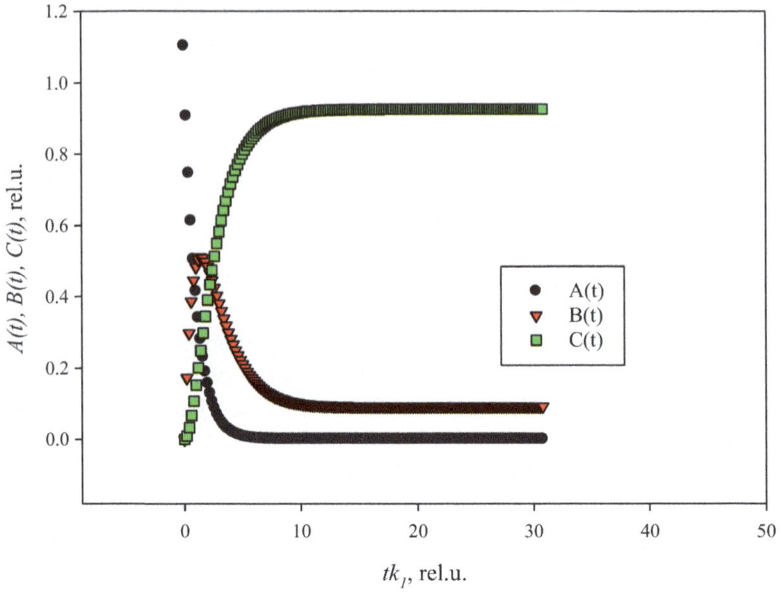

Fig. 5 An example of the dependence between the concentration of microelement *M* in "A", "B" and "C" states and the relative time of phase contact $t/\tau_1 = t \cdot k_1$ according to linear mechanism (10). In Eqs. (11 and 12), the initial microelement concentration in material, $A_o = 1$, the rate coefficients (rel. u.): $k_1 = 0.2$, $k_2 = 0.1$, $b_1 = 0.001$, $b_2 = 0.01$, $[m] = 1$, $[m^*] = 0.1$; equilibrium distribution coefficients (rel. u.) $Kd^m = 5 \cdot 10^{-4}$, $Kd^* = 10$

analytical form elsewhere (Rodigin and Rodigina 1960). We present the obtained results omitting the conclusion and analysis of sorption–desorption kinetics equations.

The dependences of the non-equilibrium decontamination coefficient $K_o(t)$ and non-equilibrium microelement distribution coefficients in "material – sorbate" and "sorbent – sorbate" subsystems on variables $[m]$, $[m^*]$, $k_{1,2}$, $b_{1,2}$ have the form

$$Kd^m(t) = \frac{A(t)}{B(t)}, \quad Kd^m(t \to \infty) = Kd^m, \quad Kd^*(t) = \frac{C(t)}{B(t)}, \quad Kd^*(t \to \infty) = Kd^*. \tag{12}$$

Their numerical analysis performed with the use of the Mathcad 14 program showed the following.

We consider a situation when a material is saturated with a microelement to the concentration A_0 (contamination stage) and then it is brought into contact with a suspension of sorbent "C" in solution "B". The kinetic process of decontamination of the material takes place spontaneously due to competitive interaction of the sorbate with the sorbent and the material. Typical isotherms of microelement concentration variation in the material, solution and sorbent versus the contact time according to (10, 11, and 12) are shown in Fig. 5. The reduction in the

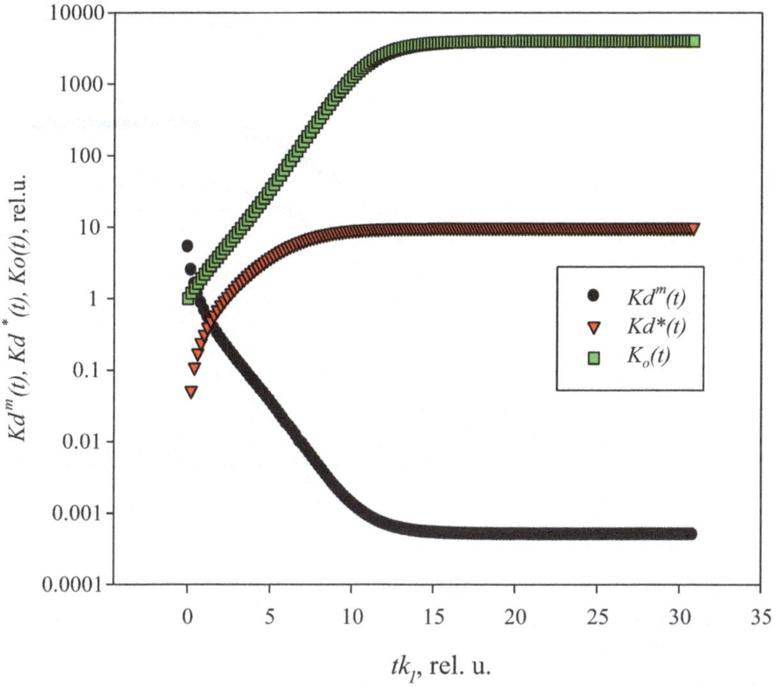

Fig. 6 An example of variation of the distribution coefficients and decontamination coefficients $Kd^m(t)$, $Kd^*(t)$, $K_0(t)$ with the relative phase contact time $(t.k_1)$ in model (11 and 12). The parameter values (rel. u.) are similar to those in Fig. 5, except $[m] = 0.1 = [m^*]$

M concentration in the material caused by competitive sorption on sorbent leads to the non-equilibrium concentration *B(t)* in the solution, which is sometimes much higher than the equilibrium one and passes through the maximum. The dependence of the non-equilibrium distribution coefficients on the contact time is generally monotonic reaching equilibrium Kd^m, Kd^* values in time $\sim 10 \cdot t \cdot k_1$. The non-equilibrium decontamination coefficient increases with time also monotonically, Fig. 6.

Equations (12) are complex functions of direct and inverse elementary reaction rate coefficients. The exponential factors in these equations contain two kinetic coefficients, $-x_{1,2}$, which can be determined from the experimental data on microelement sorption by the material and/or sorbent (Emanuel and Knorre 1984). The numerical analysis of the dependences of coefficients $x_{1,2}$ on the elementary stage rate constants k_1, k_2, b_1, b_2 according to (11 and 12) revealed that both the direct and inverse reaction constants affect $x_{1,2}$ if they differ in each elementary stage by more than one order of magnitude. The coefficients $x_{1,2}$ increase with the growth of the constants of any inverse reaction. If the direct reaction constants are larger than the inverse reaction constants, the coefficient x_1 no longer differs from k_1 and x_2 no longer depends on k_2.

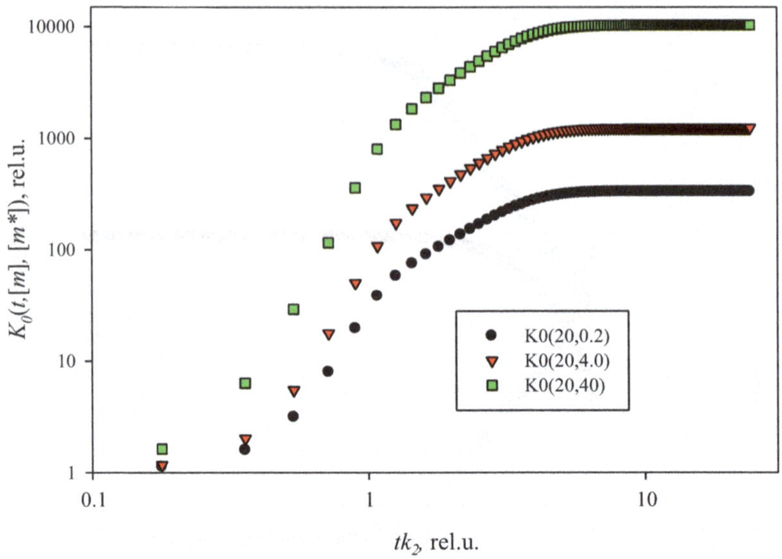

Fig. 7 An example of the kinetics of variation of the non-equilibrium decontamination coefficient $K_0(t, [m], [m^*])$ with time $(t{\cdot}k_2)$ calculated by Eqs. (11 and 12) for competitive system "material (SiO_2) – CsCl – sorbent (PB)". The figures in brackets at K_0 are the values of the "mass concentration" of material $[m]$ and sorbent $[m^*]$, ml g^{-1}, in Eq. (12); $Kd^m = 50$ ml g^{-1}, $Kd^* = 1{\cdot}10^5$ ml g^{-1}; $(t{\cdot}k_2)$ is the relative phase contact time

The $K_o(t)$-t isotherms feature the possibility of appearance of an inflection point, when in time $(3–5){\cdot}t{\cdot}k_1$ the non-equilibrium decontamination coefficient reaches the stationary value and further the equilibrium value, Fig. 7. The reasons are the intermediate stage "B" in mechanism (10), when microelement aqua-complexes are formed and accumulated in the solution, and the effect of reversibility of stages in kinetic mechanism (10). The degree of reversibility of elementary stages is an important factor of decontamination effectiveness in kinetic mode. It affects the decontamination coefficient value. When the mass of the material changes by one or two orders of magnitude with the sorbent mass, remaining constant, the kinetic isotherms retain their shape. The results of numerical analysis (10) reveal that increasing irreversibility leads to enhancement both of non-equilibrium and equilibrium decontamination coefficients.

To compare the model results with the behavior of a real sorption system, we used the data on the statics and kinetics of Cs(I) ion sorption on two sorption materials. First of them is silica SiO_2, 200–500 μm fraction (Silica Pearl) we considered as a model contaminated material. The second one is sorbent (PB, 40–80 μm fractions) (Polyakov et al. 2015a). A silica gel powder was preliminarily washed with distilled water and dried in air. High purity CsCl stock solution and 0.1 mmol l^{-1} HCl solutions were used for the preparation of solutions. The sorption isotherms were obtained in polyethylene test tubes with a phase contact time of 24 h and stirring of

Table 1 Initial data for modeling of competitive sorption kinetics in the system SiO_2-CsCl-PB in the region of Henry's law; rel. error < 0.4

	K_d, ml g^{-1}	k_1, min^{-1}	k_2, min^{-1}	b_1, min^{-1}	b_2, min^{-1}	F
PB	$1.0 \cdot 10^5$	–	$6.6 \cdot 10^{-2}$	–	$2.0 \cdot 10^{-3}$	180
SiO_2	50.0	$2.0 \cdot 10^{-2}$	–	$42.2 \cdot 10^{-2}$	–	2200

F is regression variation from the dependent variable mean to residual variation about the regression ratio. Regression model is first-order reversible reaction kinetics equation (*Table Curve 2D*), the number of measurements 10. $[m] = 0.02$ g ml^{-1}, $[m^*] = 0.0004$ g ml^{-1}

the suspensions at 50 min^{-1}. The sorbent mass in each test tube and the solution volume corresponded to $[m] = 20.0$ g l^{-1} (SiO_2) and $[m^*] = 4.0$ g l^{-1} (PB). Sorption and separation of the sorbent powder by filtration followed by solutions analysis by the ICP-MS method on an Elan 9000 (Perkin Elmer) device in quantitative mode to determine the cesium content. The sorption statics isotherms in the form of the logarithmic Langmuir equation were used to estimate K_d and the concentration region, in which the Henry's law was fulfilled, Table 1 (Volkov et al. 2017). The sorption kinetics isotherms of cesium ions in SiO_2 and PB are well approximated by the first-order reversible reaction with respect to Cs(I) cations, which allowed us to determine the constants of these reactions for the material and sorbent, Table 1 and Fig. 7 displays the decontamination kinetics isotherms of material (silicon oxide) on sorbent – PB in a closed system described by Eqs. (10, 11, and 12) with sorption parameters taken from the Table 1 for individual subsystems. The non-equilibrium decontamination coefficient $K_o(t)$ depends considerably on the material to sorbent mass ratio ($[m^*]/[m]$) and the phases contact time. In the range $[m^*]/[m] = 20/0.2$–$20/40$, K_o reaches 400–8000 in a time of $\sim 10 \cdot t \cdot k_2$, i.e. in 3–5 h under laboratory conditions. This is close to the modeling results on real soils (Valcke and Cremers 1994; Johnson and Dortch 2014). Such are the preliminary estimates of the behavior of competitive system (10) containing silica gel, electrolyte and Prussian blue solutions in the conditions close to the chemical or mixed-diffusion sorption kinetics mode. In order to establish the real kinetic peculiarities of the competitive sorption mechanism in static and dynamic systems, further studies are required.

4 Strontium (II) Mass Transfer in the System "HA-Water"

Humic acids (*HA*) play an important role in the mass transfer of microelements and radionuclides in the environment. They largely determine the mobility of radionuclides in geochemical systems, the ability to bind metal ions and to assimilate them by plants (Twardowska and Kyziol 2003). This provides the biological protective action of *HA*, reduces the toxicity of elements (Hg, Cd, Pb, As) due to complexing (Varshal and Buachidze 1983; Varshal et al. 1993) and promotes the transfer and fixation of ions of contaminated aqueous media by soils (Orlov 1990).

The adsorption regularities in the chemistry of humate complexes of metal ions have not been exhaustively examined because of objective complexity and dual nature of *HA* and their salts. Depending on the pH, ionic composition of chemical components of solution and the concentration of *HA*, they can transfer from mainly anionic state into colloidal state due to protonation, dimerization (Bergelin 2001) and interaction with cations (Davis et al. 2002). *HA* are able to coexist in the natural pH range in both molecular-ionic and colloidal states. The transfer from the molecular state of *HA* into the colloidal state begins, depending on the total concentration, at pH 3–4, when the anion protonation degree approaches 0.5 mol mol^{-1} (Fukushima et al. 1996).

The acid-base properties of *HA* play a major role in binding technogenic radionuclides, first of all Sr(II) ions, in natural aqueous solutions (Nash et al. 1981; Paulenova et al. 2000; Ozaki et al. 2003; Qiu et al. 2013). Strontium ions form stable complexes with *HA* in neutral and alkaline media. Their durability lowers with the growth of acidity (Paulenova et al. 2000). Complexing with natural organic ligands suppresses strontium sorption on minerals in neutral and alkaline media, but intensifies strontium transfer into solid phase together with solid *HA* (Paulenova et al. 2000; Qiu et al. 2013; Yu et al. 2015). The sorption isotherm of cesium (Cs-137) and strontium (Sr-85) ions on K and Ca forms of montmorillonite in humate solutions is a combination of Langmuir and Freundlich equations (Shaban and Macasek 1988; Sips 1948, 1950). Here, the stability constant of ^{85}Sr(II) – *HA* complex according to (Shaban and Macasek 1988) differs considerably (by three orders of magnitude) from the stability constants of humate complexes of other alkaline-earth element ions (Paulenova et al. 2000).

The information about the sorption behavior of humate complexes concerns mainly mineral or synthetic types of *HA* (Choppin 1999) produced by thermal or chemical treatment of *HA* (Kemdorff and Schnitzer 1980; Čežíková et al. 2001; Celebi et al. 2009). Although the processes of coprecipitation of Sr(II) with *HA* determine strontium transfer into bottom sediment, they are less well understood (Volkov et al. 2017).

The experimental data characterizing the sorption equilibria "^{90}Sr – HA" at variable sorbent (*HA*) and sorbate (Sr(II) + ^{90}Sr) concentrations showed that the sorption capacity (*E*, mg Sr g^{-1}) and the distribution coefficient (*Kd*, ml g^{-1}) during coprecipitation of humate complexes of Sr(II) with *HA* at pH = 2 depend on the concentration of *HA*, but do not depend on the initial pH of the sorbate. Strontium ion sorption during coprecipitation with HA (Fig. 8) obeys the behavior of the sorbent with two energetically non-equivalent sorption centers according to Langmuir isotherm (Kassandrova and Lebedev 1970; Adamson 1979; Čežíková et al. 2001; Volkov et al. 2017). This is confirmed by the data of IR and NMR spectroscopy of *HA* precipitates and by the acid-base properties of *HA* (Polyakov et al. 2015a). The "two-position" Langmuir isotherm determines the overall concentration of Sr(II) in the sorbent phase *C(s)* as a sum of element concentrations in each of the

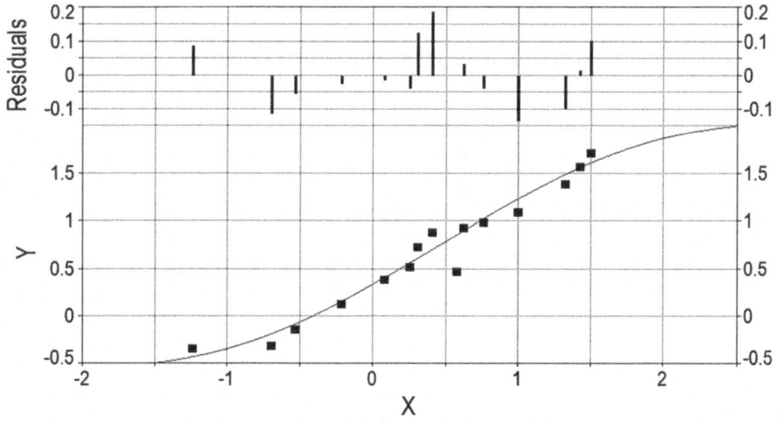

Fig. 8 The sorption isotherms of Sr(II) ions onto fresh HA precipitate (I) at pH = 2.0 ± 0.3, 22 C. *Y = log(C(Sr, HA))*, *X = log(C(Sr, aq.))*. *C(Sr, HA)* is the strontium concentration in *HA*, mg g^{-1}. *C(Sr, aq.)* is the strontium concentration in solution over *HA* precipitate, mg l^{-1}. The regression line is the Langmuir sorption equation on energetically homogeneous centers of two types, Eq. (13)

two *i*-th energetically and sorption non-equivalent and chemically unbound to each other positions $C(s)_i$, mg g^{-1}, in the form of (13) (Volkov et al. 2017):

$$\begin{aligned} C(s) &= \sum_i C(s)_i,\ \ C(s)_i = a_i/\left(1 + b_i C\left(Sr_{aq}\right)\right)_i, i = 1, 2. \\ Y &= X - \log\left(a1/\left(1 + b1 \cdot 10^X\right) + a2/\left(1 + b2 \cdot 10^X\right)\right). \end{aligned} \tag{13}$$

In (13), the coefficients $a_{1,2}$ (ml g^{-1}) = *Kd*(1,2) are the distribution coefficients of the centers with indices "1" and "2"; the ratio of coefficients $(a_{1,2}/b_{1,2})$ = *E*(1,2) is the sorption capacity of centers "1" and "2", mmol g^{-1}.

The validity of model (13) was proved by ANOVA techniques. The model parameters values calculated from the experimental data showed that at Sr(II) ion concentrations <0.1–1.0 mg l^{-1} the sorption of complex Sr(HA) by solid *HA* is characterized by the distribution coefficient log(*Kd*(1), ml g^{-1}) = 3.3 ± 0.6, the fraction of total capacity of *HA* is less than 0.02 wt.%, (Fig. 9). *Kd* (1) The remaining part of the *HA* precipitate capacity is characterized by log (*Kd*(2), ml g^{-1}) = 2.8 ± 0.2 (Volkov et al. 2017).

Considering the energy inhomogeneity of solid *HA* during co-precipitation of Sr(II), the distribution coefficient dependence on the *HA* concentration was studied at two initial concentrations of Sr(II) in solution – 20.0 μg l^{-1} and 1 mg l^{-1}. In the former concentration region, sorption is determined by "1" centers, while in the latter region – mainly by "2" centers, Fig. 9 (Volkov et al. 2017). The sorption interaction of strontium ions with *HA* during co-precipitation was simulated by the Shubert method considering that *Kd*(1) at low *HA* concentrations is determined by sorption of $Sr(HA)_{aq}$ complex (Paulenova et al. 2000), while at high concentrations – by

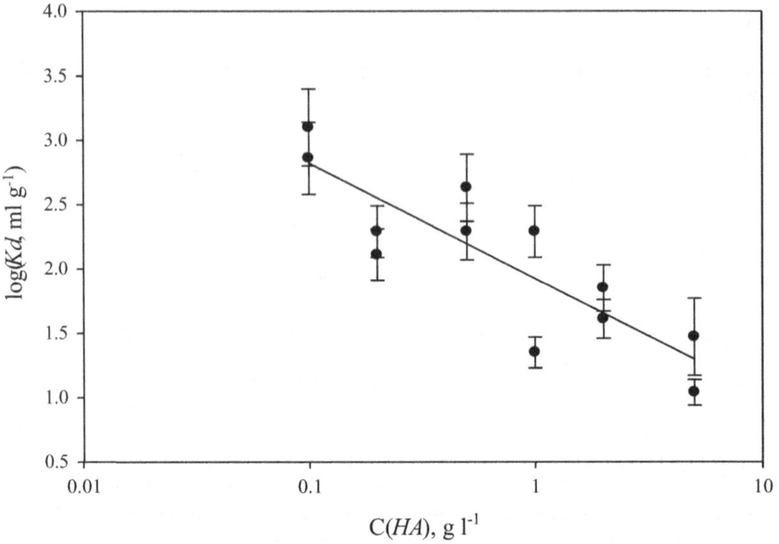

Fig. 9 The sorption isotherm of Sr(II) onto as-precipitated *HA* versus its equilibrium concentration; pH = 6.7–8.9, the initial concentration of Sr(II) – 20.0 μg l^{-1}, 22 °C. (Adopted from Volkov et al. 2017)

$Sr(HA)_{aq,2}$ complex (Paulenova et al. 2000; Kostić et al. 2013). The latter complex was treated as non-sorbable:

$$Sr(HA)_{aq} = \overline{Sr(HA)}, \qquad K1 = \left[\overline{Sr(HA)}\right] / \left[Sr(HA)_{aq}\right], \tag{14}$$

$$Sr(HA)_{aq} + HA = Sr(HA)_{2,aq}, \qquad \beta = \left[Sr(HA)_{2,aq}\right] / [Sr(HA)] \cdot [HA], \tag{15}$$

Here, $K1$ is the concentration constant of strontium ion sorption onto *HA*, and β is the concentration constant of formation of nonsorbable complex $Sr(HA)_{aq,2}$. The dependence of the overall distribution coefficient *Kd* of *HA* complexes $Sr(HA)_{aq}$ and $Sr(HA)_{aq,2}$ between the solid phase of *HA* (symbols with vinculum) and the solution according to model (14, 15, 16 and 17) in the concentration approximation has the form:

$$Kd = \left[\overline{Sr(HA)}\right] / \left(\left[Sr(HA)_{aq}\right] + \left[Sr(HA)_{2,aq}\right]\right) = K1/1(1 + \beta[(HA)]). \tag{16}$$

In the logarithmical form Eq. (16) is written as

$$\begin{aligned} \log(Kd) = \log(K1) - \log(1 + \beta[HA]^c) = a - \log(1 + b[HA]^c), \\ a = \log(K1), \qquad b = \beta, \end{aligned} \tag{17}$$

and the index c at the independent variable – the concentration of free humic acid $[HA]$ – was used as an additional parameter ($c = 1$ according to (16–17)) to verify

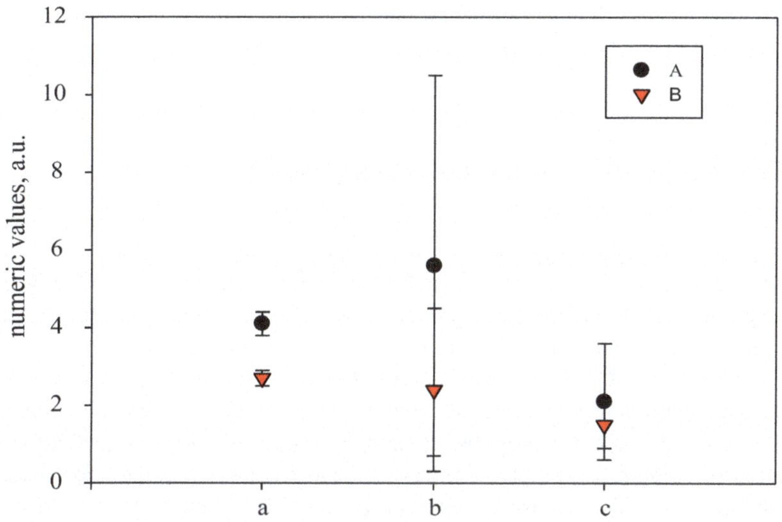

Fig. 10 The results of verification of Sr(II) sorption model onto humic acid in the co-precipitation mode in experiment with a variable concentration of HA, Eqs. (14 and 15). (HA) is the concentration of humic acid, g l^{-1}. The initial concentration of Sr(II) is 20.0 μg l^{-1} (A) and 1.0 mg l^{-1} (B). $a = \log(K1)$, $b = \beta$, $c = c$ in (16). (Adopted from Volkov et al. 2017)

eqs. (14–17). The comparison of the model coefficients performed by the least square method for sorption isotherms at two different initial concentrations of Sr (II) is given in Fig. 10. It is seen that for the isotherm at different Sr (II) concentrations, only the coefficients *a* exhibit a statistically significant difference, in agreement with the results of modeling of Sr(II) sorption by (16–17). The average value of the coefficient $b = (4.0 \pm 3.5)$ and the average value of $c = (1.8 \pm 1.3)$, which is a nonsignificant difference from 1. The value of the stability constant *b* of the strontium complex $Sr(HA)_{aq,2}$ determined by (17) is three orders of magnitude smaller than the formation constants of the $Sr(HA)_{aq}$ complex (Paulenova et al. 2000) suggesting a small stability of the $Sr(HA)_{aq,2}$ complex in concentrated *HA* solutions.

Thus, the sorption properties of freshly precipitated *HA* with respect to Sr(II) ions are affected by the strontium concentration in solution and the energy inhomogeneity of *HA* sorption centers. When strontium ions are co-precipitated onto *HA*, the distribution coefficient $\log(Kd(1), \text{ml g}^{-1}) = 3.3 \pm 0.6$ in ultradiluted solutions of Sr(II) and $\log(Kd(2), \text{ml g}^{-1}) = 2.8 \pm 0.2$ in solutions with a high ion content. In acid solutions, *HA* has a high affinity to the humate complex $Sr(HA)_{aq}$ and forms an unsorbed complex of the composition $Sr(HA)_{aq,2}$. This shows that it is possible to desorb Sr(II) radionuclides from soil and subsequently to isolate strontium from humate solution onto solid *HA* by co-precipitation. Since solutions of natural *HA* are relatively cheap and available in comparison with other, more selective, but toxic sorbents (polyantimonic acid and its salts), the sorption properties of *HA* with respect to Sr(II) can be used to develop the countermeasures against contamination of soils with ^{90}Sr as a result of technogenic accidents in the future (Davis et al. 2002).

5 New Competitive Sorption Techniques

The ways to realize the sorption extraction of microelements and their radionuclides from solutions are diverse: sorption in a limited-volume reactor, a flow-type reactor or in a sorption column. Of much importance for the decontamination practice are the efforts in creating new methods for sorption concentration of microelements as applied to decontamination of soil and other natural objects? The use of competitive sorption for the decontamination of materials requires a technique for its implementation, which would allow one to enjoy the advantage of selective sorption providing facile separation of sorbent from material after the desired decontamination coefficient has been attained. For this purpose we have proposed a new method for the decontamination of aqueous solutions from heavy metals and radionuclides, in which the sorbent and the reactor enclosing it are united in one isolated pocket – a thin-walled airtight mini-reactor, whose walls are made of a membrane material (for example, Nuclear Track Membranes) (Ioshin et al. 2016), Fig. 11. A sorbent/extractant or their solution is placed inside the mini-reactor. To prepare a pocketed membrane reactor, a track membrane is used to form a pocket (Fig. 11, stage I), a sorbent/extractant is brought into the pocket creating a mini-reactor with semipermeable walls (stage II), then the mini-reactor is made airtight (stage III) and is used for the sorption from liquid phase/soil (stage IV). Figure 11, at the right, shows a pocketed mini-reactor with PB powder placed into the aqueous electrolyte solution. If there were no membrane walls, PB would form a stable colloidal solution in the whole volume. In the offered design, the sorbent/liquid extractant and membrane pocket are integrated since the sorbent is isolated from the outside by the membrane material. Sorption by means of a mini-reactor creates some advantages. Among them are the simplification of the sorption

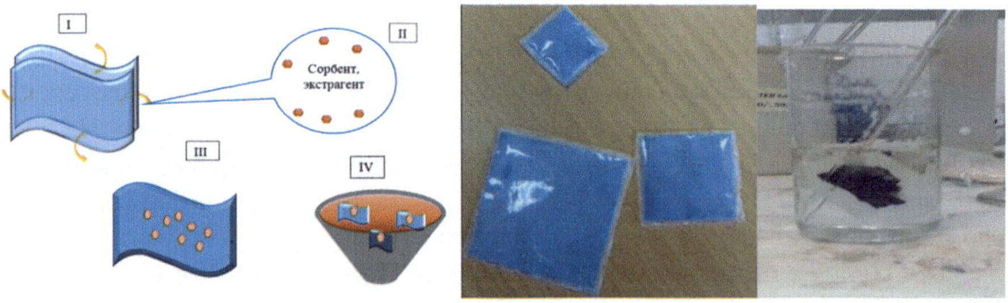

Fig. 11 A diagram of preparation (at the left); a view of a pocketed membrane mini-reactor with sorbent/extractant (at the center); a mini-reactor in a glass with a solution (at the right). Reactor walls – Nuclear Track Membranes with pore size 0.1 μm [Flerov Laboratory of Nuclear Reaction, Russian Federation (FLNR JINR, Dubna, RF)], sorbent – Prussian blue powder of mass 56.0 mg (PB) and size 0.2–0.9 μm

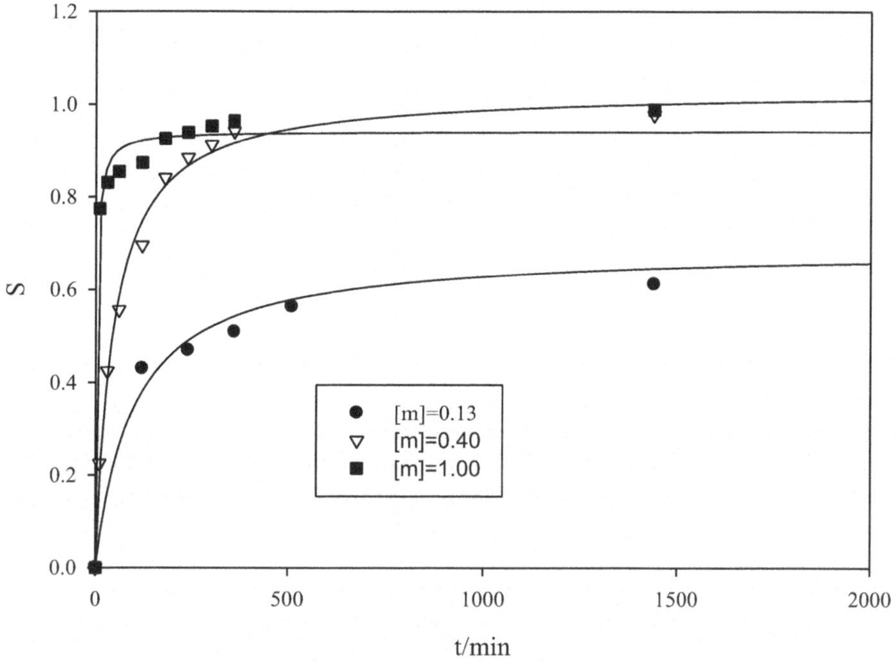

Fig. 12 The sorption kinetics of Cs((I) from tap water on a membrane mini-reactor filled with PB as a function of the ratio between the sorbent mass and the external solution volume [m] (shown in the inset, g ml^{-1}). The initial concentration Cs = 6.03 mg ml^{-1}. The sorption conditions are the same as in Fig. 8. Mixing was carried out by shaking up the vessel with the mini-reactor with a frequency of 0.1 min^{-1}. During the experiment, log(*Kd*, ml g^{-1}) = 5.0 ± 0.2 was reached, 22 °C

process and the possibility of sorption concentration in the diffusion-kinetic mode. The latter it is especially important for the decontamination of large volumes of soil, basin and river water under natural conditions, when time is not the decisive factor in comparison with the scale and degree of decontamination. There is no longer any need to use special chromatographic equipment, sorbate solution injection and ejection systems or sorbent loading/unloading procedures. In terms of technical results, of considerable importance are the performance characteristics of track membranes. The comparison of the sorption properties of mini-reactors filled with granular sorbents or extractants (silica gel, fraction 0.50–1.5 μm; KU2 cation exchanger, fraction 200–500 μm; Prussian blue (PB), fraction 0.2–0.9 μm, TBP, DEHPA, TOA) with individual sorbents revealed a similarity of their decontamination, distribution and separation coefficients for the majority of microelement ions. Figure 11 shows as an example a variation in cesium ion sorption from electrolyte solution (S) on a mini-reactor filled with PB sorbent versus solution contact time. The diffusion mass transfer mode determines the time, in the course of which the equilibrium characteristics are established if the membrane mini-reactor in the sorbate solution is not stirred. We have shown that the mass transfer rate in the conditions of Fig. 12 depends on the pocketed reactor area, the pore size of membrane wall and the stirring rate of the solution with the reactor. The time, during which the equilibrium sorption characteristics are reached, is units of months. These sorption relaxation time values are relatively high, but quite admissible in taking countermeasures to remedy the effects of man-made accidents on large areas. The described functional properties of mini-reactors with semipermeable walls, in our opinion, can be used in the competitive sorption technology for the decontamination of soil and ground from toxic/radioactive microelements (Ioshin et al. 2016).

6 Conclusion

The discussed model of microelement desorption from adsorbed state in material shows that the use of a competitive sorbent having a high selectivity with respect to cesium and strontium radionuclides can, under equilibrium conditions, be just as effective as desorption by complexing agents solutions. The kinetic properties of an elementary competitive system "material – sorbate – sorbent" considered on the example of cesium ion sorption reveal that during convective mass transport the region of equilibrium can be reached in hours or days depending on the material to sorbent mass ratio. This is comparable with the cesium ion spontaneous mass transfer rate in soil in the diffusion mode. The application of *HA* as a complexing agent for radionuclides with their subsequent co-precipitation concentrating together with a selective material, such as PB, allows one to localize as much as possible the level of interference of the sorption process in the physicochemical state of sorbate and to organize the decontamination as a self-sustaining process. The use of sorbents as part of pocketed mini-reactors with semipermeable walls offers additional advantages promoting self-sustaining sorbate mass transfer, as well as material and sorbate phase separation, which can considerably facilitate sorption decontamination of large soil, ground and water areas.

Acknowledgements The study has been performed in the framework of the state assignment (Project AAAA-A16-116122810210-1) with partial support of the RFBR (grant 18-03-00051), the Presidium of UB RAS (grant 18-3-3-5).

References

Adamson AA (1979) Physical chemistry of surface (trans: Abidor IG). Mir, Moscow.

Belinskaya FA, Militsina EA (1980) Inorganic ion-exchange materials based on sparingly soluble antimony (V) compounds. Uspekhi chimii 49:1904–1936

Benes P, Majer V (1980) Trace chemistry of aqueous solutions: general chemistry and radiochemistry/Published Amsterdam. Elsevier Scientific Publishing Company, New York

Bergelin A (2001) Acid-base properties and aggregation of humic materials. Dissertation Abstract International. Kungliga Tekniska Hogskolan, Royal Institute of Technology, Stokholm, Sweden

Betenekov ND, Kaftailov VV, Nedobukh TA, Egorov YV (1999) Influence of the true colloid state of sorbate on sorption kinetics and statics. Radiochemistry 41:253–257

Brunauer S, Emmett PH, Teller E (1938) Adsorption of gases in multimolecular layers. J Amer Chem Soc 60:309–319

Celebi O, Kilikli A, Erten HN (2009) Sorption of radioactive cesium and barium ions onto solid humic acid. J Hazard Mater 168:695–703

Čežíková J, Kozler j ML, Novák J, Janoš P (2001) Humic acids from coals of the North-Bohemian coal field. II. Metal-binding capacity under static conditions. React Funct Polym 47:111–118

Chirkst DE, Litvinova TE, Cheremisina OV, Streletskaya MI (2001) Experimental technology for deactivation of Sr-90 radionuclide contaminated ground. Radiochimiya 43:475–478. [in Russian]

Choppin GR (1999) Chemical separation technologies and related methods of nuclear waste management. Kluwer Publishers, Dordrecht

Comans RNJ, Hilton J, Cremers A, Geelhoed-Bonouvrie PA, Smith JT (1997) Interpreting and predicting in situ distribution coefficients of radiocaesium in aquatic systems. Edited by G. Desmet et al. Elsevier Science B.V.

Cremers A, Elsen A, De Preter P, Maes A (1988) Quantitative analysis of radiocaesium retention in soils. Nature 335:247–249.

Davis CJ, Eschenazi E, Papadopulos D (2002) Combined effect of Ca^{2+} and humic acid on colloid transport through porous media. Coll Polym Sci 280:52–58

Ding D, Zhang Z, Lei Z, Yang Y, Cai T (2016) Remediation of radiocesium-contaminated liquid waste, soil, and ash: a mini review since the Fukushima Daiichi Nuclear Power Plant accident. Environ Sci Pollut Res 23:2249–2263.

Egorov YV (1975) The statics of microcomponent sorption onto oxyhydrates. Atomizdat, Moscow

Elovich SY, Larionov OG (1962) Theory of adsorption from solutions of non electrolytes on solid (I) equation adsorption from solutions and the analysis of its simplest form, (II) verification of the equation of adsorption isotherm from solutions Izvestiya Akademii Nauk SSSR. Otdelenie Khimicheskikh Nauk 2:209–212

Emanuel NM, Knorre DG (1984) A course of chemical kinetics. Vysshaya shkola, Moscow

Fukushima M, Tanaka S, Nakamura H, Ito S (1996) Acid-base characterization of molecular weight fractionated humic acid. Talanta 43:383–390

Gonzalez AJ (1996) Chernobyl-Ten years after Global experts clarify the facts about the 1986 accident and its effects. IAEA Bull 4:2–13

González AJ (1998) Radioactive residues of the cold war period: a radiological legacy. IAEA Bulletin 3:2–11

Guillen J, Baeza A, Mocanu N (2008) Radioecology and environmental radioactivity. In: Browwn J, Jolle T, Strand P (eds) Analysis of the poltential disadvantages of chemical countermeasures against the transfer of radionuclides to plants. Proceedings of Norwegian Radiation Protection Authority, Bergen. Norway, pp 126–129

Hamby DM (1996) Site remediation techniques supporting environmental restoration activities. A review. Sci Total Environ 191:203–224

Ioshin AA, Polyakov EV, Volkov IV (2016) The possibilities of removing heavy metals and radionuclides from aqueous solutions in membrane reactor with pocketed sorbents. All-Russian Conference "Solid state chemistry and functional materials – 2016". Materials of the Conference. Ural Branch RAS, Ekaterinburg, pp 214–215

Johnson BE, Dortch MS (2014) Application of TREECS modeling system to Strontium-90 for Borschi Watershed near Chernobyl, Ukraine. J Environ Radioact 131:31–39

Kassandrova ON, Lebedev VV (1970) Analysis of observations. Nauka, Moscow

Kato K (2015) Radioactive contaminated soil cleaning apparatus and system for the same for efficient and safe handling of the radioactive contaminated soil. Japan Kokai Tokkyo Koho Patent JP 2015211956A, 20151126B

Kemdorff H, Schnitzer M (1980) Sorption of metals on humic acid. Geochim Cosmochim Acta 44:1701–1708

Kokotov YA, Pasechnik VA (1970) Ion exchange equilibrium and kinetics. Khimiy, Leningrad
Kostić IS, Anđelković TD, Nikolić RS, Cvetković TP, Pavlović DD, ALj B (2013) Binding strengths of heavy metals with humic acid. Hem Ind 67:773–779
Kryshev II, Vakulovskii SM, Katkova NM, Kim VM, Kryshev AI, Shershakov MV (2013) The radiation situation on the territory of Russia and bodering countries in 2012. In: Bulgakov VG (ed) FGBU Scientific Production Association, Taifun, Obninsk
Langmuir I (1916) The constitution and fundamental properties of solids and liquids. Part I. solids. J Amer Chem Soc 35:2221–2295
Maslova K, Stepina I, Konoplev A, Popov V, Gusarov A, Pankratov FR, Lee SD (2013) Fate and transport of radiocesium, radiostrontium and radiocobalt on urban building materials. J Environ Radioact 125:74–80
Moskvin AV (2004) New chemistry and technology handbook. Radioactive substances. Hazardous substances. Hygienic regulations. Part 11. Radioactive contamination and deactivation, vol 11. ANO NPO Professional, Moscow
Nash K, Frled S, Friedman AM, Sulllvan JC (1981) Redox Behavior, Complexing, and Adsorption of Hexavalent Actinides by Humic Acid and Selected Clays. Environ Sci Technol 15:834–837
Nikulina US (2016) Decontamination of technogenic soil and ground from radium series and mercury radionuclides by hydroclassification method. PhD thesis abstract. D. Mendeleev University of Chemical Technology of Russia, Moscow, pp 1–16
Orlov DS (1990) Humic acids of soil and the general theory of humification. MGU, Moscow
Ozaki T, Ambe S, Abe T, Arokiasamy JF (2003) Effect of humic acid on the bioavailability of radionuclides to rice plants. Anal Bioanal Chem 375:505–510
Paulenova A, Rajec P, Zemberyova M, Saskoiova G, Visacky V (2000) Strontium and calcium complexation Ьy humic acid. J Radioanalyt Nucl Chem 246:623–628
Polyakov EV (2000 Colloidal chemical extraction of microcomponents from aqueous phase Radiochimiya 42:423–426
Polyakov EV (2003) Reaction of ion-colloid forms of microelements and radionuclides in water solutions. Ural Branch RAS, Ekaterinburg
Polyakov EV (2012) Physicochemistry of humate complexes as a basis of green chemistry in radioecology. Book of Abstracts of the 7th Russian Conference on radiochemistry "Radiochemistry-2012". OOO VDV PAK, Dimitrovgrad, pp 43
Polyakov EV, Betenekov ND (1988) Linear sorption mechanism for a microcomponent sorbate by independent pathways. Kinet Catal 29:560–566
Polyakov EV, Ilves GN, Surikov VT (2000) Study of estepxtraction equilibria with participation of trace hydroxide colloids. Radiockhimiya 42:427–431
Polyakov EV, Volkov IV, Khelbnikov NA (2015a) Competitive sorption of cesium and other microelements onto iron(iii) hexacyanoferrate (ii) in the presence of humic acids. Radiochemistry 57:161–171
Polyakov EV, Volkov IV, Khlebnikov NA, Tsukanov RR, Ioshin AA (2015b) Competitive sorption as a method for decontamination of materials. Radiochemistry 57:172–177
Qiu Y, Yu S, Song Y, Wang Q, Zhong S, Tian W (2013) Investigation of solution chemistry effects on sorption behavior of Sr(II) on sepiolite fibers. J Mol Liq 180:244–251
Remez VP (1994) Composite sorbent and a method for the production of composite sorbents. RF Patent 2:021,009
Rodigin NM, Rodigina EN (1960) Consecutive chemical reactions. Mathematical analysis and calculation. Edited by Fyodorov GV. Moscow: AN SSSR.
Sanzharova NI, Sysoeva AA, Isamov NN, Aleksakhin RM, Kuznetsov VK, Zhigareva TL (2005) The role of chemistry in remediation of radioactive contaminated agricultural land. Ros Khim zh (J D Mendeleev Rus Chem Soc) 49:26–29
Sasaki K, Takeno K (2014) Removal of uranium, Cesium, strontium, toxic metals and radioactive cesium by a photosynthetic bacteria. Yosui to Haisui (Japan) 56:60–68
Shaban IS, Macasek F (1988) Influence of humic substances on sorption of cesium and strontium on montmorillonite. J Radioanalyt Nucl Chem 229:73–77

Shubert J (1948) Ion-exchanger for the detn. of phys.-chem. Properties of substances, esp. radiotracers, in soln. (I) theoretical. J Phys Coll Chem 52:340–344.

Sips R (1948) On the structure of a catalyst surface. J Chem Phys 16:490–495

Sips R (1950) On the structure of a catalyst surface II. J Chem Phys 18:1024–1026

Stepina IA, Maslova KM, Popov VE (2013) Potentials of radiocesium fixation in constructional materials. Radiochimiya 55:249–252

Streletskaya MI (2003) Physico-chemical substantiation of ion-exchange deactivation of grounds. PhD thesis, St. Petersburg State Mining University after GV Plekhanov

Twardowska I, Kyziol J (2003) Sorption of metals onto natural organic matter as a function of complexation and adsorbent – adsorbate contact mode. Environ Int 28:783–791

Valcke E, Cremers A (1994) Sorption-desorption dynamics of radiocaesium in organic matter soils. Sci Total Environ 157:275–283

Valcke E, Vidal M, Cremers A, Ivanov J, Perepelyatnikov G (1997) The use of zeolites as amendments in radiocaesium- and radiostrontium-contaminated soils: a soil-chemical approach. Part III: A soil-chemical test to predict the potential effectiveness of zeolite amendments. Zeolites 18:218–224

Vandana AP, More AK, Sachan SR, Gurg RP (2000) Bioaccumulation of 137Cs: deep soil decontamination. Rad Prot Environ 23:37–40

Vandenhove H, Cremers A, Smolders E, Van Hees M (2005) Effect of K and bentonite additions on Cs-transfer to ryegrass. J Environ Radioact 81:233–253.

Varshal GM, Buachidze NS (1983) Study of coexisting forms of mercury (II) in surface water. Zhurn analit Chimii 38:2156–2167

Varshal GM, Velyukhanova TK, Koshcheyeva IY (1993) Geochemical role of humic acids in element migration. Humic substances in biosphere. Nauka, Moscow

Veshev SA, Aleekseev SG, Dukhanin AS (1996) Study on radionuclide migation process in soil under the action of constant electric field. Geokhimiya 10:1005–1009

Volkov IV, Polyakov EV, Khlebnikov NA, Barysheva NM (2013) Sorption properties of silicate materials based on Ca_2SiO_4 in solutions of humic acids. Radiochimiya 55:425–430

Volkov IV, Polyakov EV, Denisov EI, Ioshin AA (2017) Sorption bahavior of strontium ions in humic acid solutions. Radiochimiya 59:63–70

Warner F, Harrison RM (1993) Radioecology after Chernobyl. Biogeochemical pathways of artificial radionuclides. Willey, New York

Wauters J, Elsen A, Cremers A, Konoplev AV, Bulgakov AA, Comans RNG (1996a) Part one: a simplified procedure for the solid phase characterization. Appl Geochem 11:589–594

Wauters J, Vidal M, Elsen A, Cremers A (1996b) Part two: a new procedure for solid phase speciation of radiocaesium. Appl Geochem 11:595–599

Wissocqa A, Beaucairea C, Latrillea C (2017) Ca and Sr sorption on Ca-illite: experimental study and modelling. Procedia Earth and Planetary Science. Proc Earth Planet Sci 17:662–665

Yablonskii GS, Bykov VI, Gorban AN (1983) Kinetic models of catalytic reactions. Nauka, Novosibirsk

Yadigaroglu G (2012) Safety and safety analysis: from CP1 to Fukushima. Internationale Zeitschrift fuer Kernenergie 57:86–92

Yu C, Loureiro C, Cheng JJ, Jones LG, Wang YY, Chia YP, Faillace E (2000) Data collection handbook to support modeling impacts of radioactive material in soil. Handbook. Environmental Assessment and Information Sciences Division Argonne National Laboratory, Argonne. Illinois, USA

Yu S, Mei H, Chen X, Tana X, Ahmad B, Alsaedi A, Hayat T, Wang X (2015) Impact of environmental conditions on the sorption behavior of radionuclide 90Sr(II) on Na-montmorillonite. J Mol Liq 203:39–46

Zhang L, Wei J, Zhao X, Li F, Jiang F, Zhang M, Cheng X (2016) Competitive adsorption of strontium and cobalt onto tin antimonite. Chem Engg J 285:679–689

Biological, Chemical and Nanosorption Approaches in Remediation of Metal Wastes

Santa Ram Joshi and Debajit Kalita

1 Introduction

1.1 Origin of Radioactive and Metal Waste

Metals and radioactive materials related waste are generated from activities where these materials are used as a part of the process or as ingredient of instruments and gadgets.

Due to huge differences of origin site of these waste, such as medical, research activities and industries, it is convenient to differentiate the activities of waste disposal (Radiation Safety General Regulations 1988; Méndez de Vigo 2000). Implementation of radioactive waste management necessitates protecting human health and the environment. There are several basic steps such as minimization, pretreatment, characterization, treatment, conditioning, transport, storage and disposal involved from waste generation to final disposal (IAEA 1970). To ensure the overall safety, security, quality, health, environmental and economic concerns, management system should be implemented in all steps of waste management, which is a primary importance of safety and environmental protection (IAEA 2006a, b, 2008).

Since the 1940s, the nuclear weapons materials production, operations and its associated disposal process are the possible contamination of water, air and soil (Chukanov et al. 1993). Sometimes liquid type of radioactive wastes, containing other chemicals disposed to water directly and/or sealed with metal container and buried under the ground resulting leakage into the surrounding contaminates the environment. Solid wastes like machinery, clothing etc. contaminated with radioactive compounds such as naturally occurring U, Th and man-made ^{60}Co, ^{137}Cs, ^{90}Sr,

S. R. Joshi (✉) · D. Kalita
Microbiology Laboratory, Department of Biotechnology and Bioinformatics, North-Eastern Hill University, Shillong, India

D. K. Gupta, A. Voronina (eds.), *Remediation Measures for Radioactively Contaminated Areas*, https://doi.org/10.1007/978-3-319-73398-2_5

^{239}Pu,^{241}Am, are generally disposed in trenches. Attributable to inhalation, external contacts of radiation to the body, ingestion of radioactive compounds through food and water consumption, the primary health risk arises constantly (IAEA 1999). Radioactive isotopes in the form of solid, liquid and gaseous wastes generated from industrial, research and medical applications in association with various routes like collection, processing, storage and waste disposal can lead to potential environmental contamination (IAEA 1994; Sobolev 1994). Rastogi et al. (2010) reported the impact of uranium mining soil on microbial diversity experimented by high-density 16S microarray and clone library.

Various processes are employed to treat aqueous radioactive wastes. Some widely used techniques are: ion exchange/sorption, precipitation, evaporation, ultrafiltration/reverse osmosis etc. According to the guideline of IAEA, before or after primary treatment, suspended particulate matter along with liquid must be removed before disposing the wastes (IAEA 2001; Abdel et al. 2011).

2 Types and Functional Properties of Sorbents

The terminology coined as sorption is utilized for absorption and adsorption both. Integration of a substance in one state into another of a different state is defined as absorption. The process that involves physical attachment or bonding of ions and molecules of a substance onto the surface of another molecule is defined as adsorption. The occurrence of adsorption resulting in the formation of a stable molecular phase at the interface leads to the formation of a surface complex (Stumm and Morgan 1996). The mass transfer process of a substance from liquid phase to the surface of a solid substance getting bound by physical and/or chemical interactions leads to adsorption (Kurniawan and Babel 2003). Sorption complex involved in molecular level determines the effect of sorption, which is divided into adsorption, absorption and surface precipitation mechanisms (Sposito 1982, 1986). The process of accumulation of adsorbate on the surface of a solid adsorbent where the ions, liquid, or gas molecules are held at the solid-water interface without the development of a three-dimensional molecular arrangement leads to occurrence of adsorption (Sposito 1986).

Adsorption is influenced by the distribution, the electrostatic properties and the reactivity of mineral surfaces of contaminants (Stumm 1992). Adsorption of the contaminant happens in various media, such as activated alumina used to treat radioactive elements and selective complexes. Activated carbon in granular size is medium for water treatment. In some cases, radon can be used in adsorption if gas is required to decontaminate the pollutant. Polyethylene coated activated carbon permeable for radon diffusion is utilized for the adsorption of radon gas. Similarly, natural substances can be useful for adsorption, for example, mixture of cretaceous sandstones and kaolinite matrix of the Sudety area (Poland) was shown to have efficient result as natural adsorbing system (IAEA 1999). In adsorption techniques, solid sorbents are extensively used to remediate chemical pollutants from waters, which are difficult to be removed by conventional biological wastewater treatments. Among all, activated carbon is most widely used to remove pollutants from wastewater (Ramakrishna and Viraraghavan 1997; Derbyshire et al. 2001; Babel and Kurniawan 2003).

2.1 Biosorbents

Biosorption is a process of accumulation of heavy metals and pollutants from wastewater by using biological materials which may be metabolically and/or physico-chemically mediated (Fourest and Roux 1992). In this physico-chemical biosorption process, metal ions adhere to the surface of the biomass (Drake and Rayson 1996). Oscik (1982) defines adsorption as "fundamental surface phenomena occurring at the boundary between two phases". Polymeric substances are extracellularly released by bacteria in biosorption process which have the properties of ion exchange reactions, complexation with negatively charged groups, and precipitation (Brown and Lester 1982; Rudd et al. 1984; Chang et al. 1995; Fukushi et al. 1996).

Biomass/adsorbents are attractive due to natural origin and can be used effectively and are a quick phenomenon of passive adsorption of metals through non-growing biomass acting as adsorbents.

A solid part like biosorbent, biological materials and a liquid phase containing adsorbate (metals) in dissolved condition are involved in the biosorption process. Because of higher affinity among adsorbent and adsorbate species, the adsorbate is attached and uses mechanisms to bound till the equilibrium is reached between adsorbate portion and its remaining part of the solution. Adsorbent and adsorbate affinity depends on the adsorbent distribution in the solid and liquid phases (Ramachandra et al. 2005). Various groups of microorganisms, bacteria, fungi, yeasts, cyanobacteria and algae are reported for their binding properties of variety of heavy metals, as well as biosorption, removal and recovery of heavy metal (s) (Greene and Darnall 1990; Volesky 1990; Wase and Foster 1997).

2.1.1 Factors Affecting Biosorption

Various factors that affect the biosorption are:

1. according to Aksu et al. (1992), temperature is not an important influencing factor in case of biosorption between of 20–35 °C.
2. pH is the most important parameter that affects the reaction between the metals, functional groups in the biomass and the solution (Friis and Myers-Keith 1998).

3. concentration of biomass is known to affect the specific uptake (Sao et al. 2014).
4. biosorption process is used to remove more than one metal ion from wastewater, where one metal ion influences another. According to the metal binding location these type of biosorption can be classified as: extra cellular accumulation/precipitation, cell surface sorption/precipitation, intracellular accumulation (Sao et al. 2014).

2.1.2 Plants

Some plant species have the capability to absorb contaminants by roots. This adsorption onto roots from the soil and ground water can accumulate or precipitate the contaminants within the root zone. Some plant species hyperaccumulate metal, metalloids, radionuclides, organic contaminants from soils, sediments and sludges in their roots and shoots (EPA 2000; Prasad and Oliveira Freitas 2003). 0.88 mg L^{-1} arsenic is reported to be accumulated by biomass of *Momordica charantia* (Pandey et al. 2009). Peanuts husks biosorbent exhibited Cr(III) sorption efficiency (Witek-Krowiak et al. 2010). *Potamogeton lucens, Salvinia herzogoi, Eichhornia crassipes, Myriophyllum brasillensis, Myriophyllum spicatum, Cabomba* sp., *Ceratophyllum demersum* are freshwater macrophytes and have been found to absorp of heavy metals extensively (Wang et al. 1996; Schneider and Rubio 1999; Schneider et al. 1999; Keskinkan et al. 2003).

2.1.3 Algae

At the optimum pH of 4.5, biosorption of Cd for *Sargassum* sp.1 is $q_{\max} = 0.90$ mmol g^{-1} and 0.66 mmol g^{-1} for *S. filipendula* I making almost 36% difference. With the difference of 16% among *S. vulgare, S. filipendula* I and *S. fluitans, the Cu uptake evaluation were* $q_{\max} = 0.93$, 0.89, and 0.80 mmol g^{-1} respectively (Davis et al. 2000). The biomass of *Palmaria tevera, P. palmate, Chondrus crispus* and *Ascophyllum nodosum* are reported to accumulate copper, zinc, cadmium, uranium (Kuyuack and Volesky 1988). Binding of Au^{3+}, Ag^{+}, and Hg^{2+} are totally pH dependent for *Chlorella Pyrenoidosa, Cyanidium caldarium* and *Spirulina platensis* (Darnall et al. 1988). Solutions containing Au^{3+}, Cu^{2+}, Fe^{2+}, Zn^{2+}, and Au^{3+} were selectively adsorbed by cyanobacterium, *Spirulina platensis* (Gee and Dudeney 1988). Lead uptakes up to 370 mg Pb g^{-1} were observed in *Fucus vesiculosus* and *Ascophyllum nodosum* in low pH medium (Holan et al. 1993). Waste biomass of *Scenedesmus obliquus* is reported to bioabsorb Cu (Mattuschka et al. 1993) (Table 1).

2.1.4 Fungi and Yeast

Many fungi and yeast, particularly the genera *Rhizopus, Aspergillus, Streptoverticullum* and *Xeromyces bisporus* have shown an excellent potential of

Table 1 Algae and associated functions used in metals biosorption approaches

Algae	Functions/activities reported of biosorption	Radioactive metal	References
Chlorella vulgaris (Algae)	Biosorption of Cm(III) on the surface	Curium	Ozaki et al. (2004)
Unicellular green algae	Removal of 95% of dissolved U from seawater	Uranium	Heide et al. (1973)
Chlorella regularis	Sorption capacities of uranium(VI)	Uranium	Horikoshi et al. (1979)
Sargassum sp.	Biosorption capacities of uranium(VI)	Uranium	Yang and Volesky (1999)
Peltigera membranancea	42 mg U gd/m biosorption capacity of U	Uranium	Haas et al. (1998)
Cladonia sp.	10 mg U gd/m biosorption capacity of U	Uranium	Haas (1998)
Umbilicaria sp.	29 mg U gd/m biosorption capacity of U	Uranium	Haas (1998)
Scenedesmus quadricauda	U accumulation occurred in two steps, a fast (temperature dependent) and a slow step	Uranium	Pribil and Marvan (1976)
Spirogyra	U(VI) crystals inside the algal cells	Uranium	Mann and Fyfe (1984, 1985)
Anacystis nidulans	U removal in the pH range between 3.0 and 5.0	Uranium	Liu and Wu (1993)
Cladonia rangiferina (lichen).	Remove U in the form of uranyl ions under acidic conditions	Uranium	Boileau et al. (1985)
Chara	U accumulated in the organic sediments via adsorption	Uranium	Koval'skiy and Vorotnitskaya (1965)

metal biosorption (Volesky and Tsezos 1982; Galun et al. 1984; Siegel et al. 1986; Polman and Breckenridge 1996). *Cladosporium resinae, Rhizopus arrhizus, Mucor miehei* and *Penicillium chrysogenum* are known for biosorption of Cu, Ni, Zn, Cd, Ag and Pb (Gadd et al. 1988; Fourest et al. 1992). Various reports suggests that *Pencillium chrysogenum* biomass was found to absorb Pb^{2+} (Niu et al. 1993); *Trichoderma viride* accumulates copper(Cu) (Townsley et al. 1986); *Absidia orchidis*, a fungus absorbs cobalts (Kuycak and Volesky 1988). The white-rot fungus *Polyporus versicolor* was used as biosorbent and removal of Ni(II) was investigated by Dilek et al. (2002). The uptake of uranium by immobilised microbial biomass of *Chaetomium distortum, Alternaria tenulis* has been reported in a study by Khalid et al. (1993). Among the yeast, *Candida tropicalis, Pichia guilliermondii* (Mattuschka et al. 1993); *Sacchromyces cerevisiae* (Volesky and Prasetyo 1994) have revealed excellent metal biosorbent capacity (Table 2).

Table 2 Fungi and associated functions used in metals biosorption approaches

Fungi	Functions/activities reported of biosorption	Radioactive metal	References
AM fungal strain	Uptake and translocate radiocesium *in vitro*	Cesium	Declerck et al. (2003) and Dupréde Boulois et al. (2005)
Saccharomyces cerevisiae	Biosorbing U (VI) ion on nonliving biomass of specie in batch system	Uranium	Dhankhar et al. (2011)
Kluyveromyces	180 mg U gd m^{-1} biosorption capacity of U	Uranium	Bustard et al. (1996)
Talaromyces emersonii	323 mg U gd m^{-1} biosorption capacity of U	Uranium	Bengtsson et al. (1995)
Aspergillus fumigatus	423 mg U gd m^{-1} biosorption capacity of U	Uranium	Bhainsa and D'Souza (1999)
Aspergillus flavus	40 mg U gd m^{-1} biosorption capacity of U	Uranium	Hafez et al. (1997)
Rhizopus sp.	More than 180 mg U gd m^{-1} biosorption capacity of U	Uranium	Bhainsa and D'Souza (1999)
Rhizopus arrhizus	42.3 mg U gd m^{-1} biosorption capacity of U	Uranium	McCready and Lakshmanan (1986)
Penicillium spp.	20.3 mg U gd m^{-1} biosorption capacity of U	Uranium	McCready and Lakshmanan (1986)
Rhizopus arrhizus	Th biosorption	Thorium	Tsezos and Volesky (1982)

2.1.5 Bacteria

The binding capacity of several metals in the order of Ag > Cr > Pb > Cu > Zn > Cd > Co > = Ni by waste biomass of *Streptomyces noursei* has been reported by Mattuschka et al. (1993). The cell wall of *Bacillus subtillis* is reported to uptake and retains metals (Beveridge 1986). A gram-negative bacterium *Pantoea* sp. TEM18 showed copper tolerance activity in the growth medium containing copper at concentrations of 180 mg L^{-1} (Guven Ozdemir et al. 2004). *Enterobacter cloaceae,* isolated from marine sediment of India, having exopolysaccharide can chelate copper (20%) and cobalt (8%) Iyer and Jha (2005). Liu et al. (2004) found that Zn(II) and Cu(II), and the combination of these two metals was effectively absorbed by *Thiobacillus thiooxidans* in the pH range of 2.0–6.0 and 4.0–5.0 respectively. Manganese oxidizing bacteria was utilized for the production of oxides using 1.6 g L^{-1} Mn^{2+} (Stuetz et al. 1993). The capability of biosorbtion (25–150 mg L^{-1}) of cadmium, chromium, copper, lead and nickel by *Bacillus thuringiensis* OSM29 has been isolated from cauliflower rhizospheric soil of industrial effluents (Oves et al. 2013). Other than these, there are various reports on the metal sorption by different strains of bacteria and actinomycetes such as *Pseudomonas* sp., *Zoogloea ramigera* and *Streptomyces* sp. (Norberg and Persson 1984;

Norberg and Rydin 1984; Mullen et al. 1989; Nakajima and Sakaguchi 1986; Addour et al. 1999) and extremotolerant *Acinetobacter radioresistens* 50v1 (McCoy et al. 2012), (Table 3).

Table 3 Bacteria and associated functions used in metals biosorption approaches

Bacteria	Functions/activities reported of biosorption	Radioactive metal	References
Bacillus sp. GTG – 83	Bio-absorption of Strontium from aqueous solution	Strontium	Ghazvini et al. (2007)
Shewanella halotolerans	Reduced U(VI) complexed with oxalate or citrate to U(IV) under anaerobic conditions with little precipitation of uranium	Uranium, Technetium	Ganesh et al. (1997)
Shewanella putrefaciens, S. oneidensis	Direct enzymatic reduction of Pu(VI) and Pu (V) to Pu(IV)	Plutonium	Neu et al. (2005)
Geobacter metallireducens	Direct enzymatic reduction of Pu(VI) and Pu (V) to Pu(IV)	Plutonium	Neu et al. (2005)
Shewenella putrefaciens	Reduced Np^{5+} to Np^{4+}	Neptunium	Lloyd et al. (2000)
Pseudomonas fluocesecns	Significant sorption Neptunyl (NpO^{2+})	Neptunium	Songakasiri et al. (2002)
Halomonas sp	Biosorption of Cm(III) on the surface	Curium	Ozaki et al. (2004)
Halobacterium salinarum	Biosorption of Cm(III) on the surface	Curium	Ozaki et al. (2004)
Microccous luteus	Sr-binding activity on cell envelope	Strontium	Faison et al. (1990)
Desulfovibrio desulfuricans	Reduce iodate to iodide	Iodine	Bors et al. (1991)
Citrobacter sp. N14	Removal of uranyl ion ($UO_2(2+)$) and its deposition in the form of polycrystalline hydrogen uranyl phosphate (HUO_2PO_4)	Uranium	Basnakova et al. (1998)
Citrobacter sp.	Accumulation of uranyl ion ($UO_2(2+)$) via precipitation with phosphate ligand liberated by phosphatase activity	Uranium	Macaskie et al. (2000)
Citrobacter sp.	Precipitates U(VI) on its cell surfaces as metal phosphates	Uranium	Roig et al. (1997)
Salmonella subterranea	Isolated from a low-pH, nitrate- and U(VI)-contaminated subsurface sediment from the US DOE FRC in Oak Ridge, Tennessee	Uranium	Shelobolina et al. (2004)
Citrobacter freudii	U(VI) biosorbtion	Uranium	Xie et al. (2008)
Firmicutes	U(VI) biosorbtion	Uranium	Haferburg et al. (2007)
Cystoseira indica	Adsorption of U(VI)	Uranium	Khani et al. (2005)
Serratia marcescens PKRS1	Uranium (U)-tolerant	Uranium	Kumar et al. (2011)

2.2 *Chemisorbents*

Using chemisorption methods for total remediation of rare metals from natural and industrial effluents is a rather effective method due to its chemical reactivity among the compounds and elements. Chemisorbents are used to remove individual ions from mixtures of pollutants under much specified conditions (Pozhidaev et al. 2003).

Used of graphite based chemical adsorbents connected with electrochemical regeneration has been developed for the removal of organic contaminants and destruction of radioactive oil around 95% in first steps and 99% of the emulsified oil (Brown and Roberts 2007). In this process, 80–90% of the radioactive species are converted to the liquid phase with minor release of tritium gaseous (Brown et al. 2013). Chemical bonding happening during the process of chemical adsorption is stronger than a physical bond between absorbents and adsorbate, which release heat by the formation of a mono-molecular layer of the adsorbate (Eckenfelder and Allen 1990).

Another type of promising adsorbent substance is heteropolyacids (HPAs) hooked into a support with the open pore morphology (Novacarb and SBA-15) and tungstophosphoric acid with sol–gel SiO_2 (Lapkin et al. 2003). As a sorbent, one of the new class is labile cations added metal sulfide ion exchangers (MSIEs) with extra-framework which is predominantly useful for the removal of heavy metal ions (e.g. Pb^{2+}, Hg^{2+}, Ni^{2+}, Cd^{2+} and Co^{2+}) and nuclear waste (UO_2^{2+}, Cs^+, and Sr^{2+}) (Manos et al. 2005). The cadmium ion sorption up to 150–160 mgg^{-1} was exhibited by chemisorption fibers having amine composite on grafted copolymers providing high extraction degrees (almost 99%) Druzhinina et al. (2011).

2.2.1 Chemical-Conjugate Absorbent

This type of absorbents and adsorbents depend on the developing of analytical signal on binding of metals, which is technologically more advanced as because it is reliable, easy to-function, cost effective, have high selectivity and sensitivity (Awual et al. 2013). Functional ligands present in N,N-disalicylidene-4, 5-dimethyl-phenylenedene (DDPD) was conjugated with mesoporous silica to detect ultra-trace Cu(II) ions and were successfully exploited in their removal (Awual et al. 2014). V(V), Mo(VI), W(VI), Th(IV), and U(VI) sorption properties are found to be very active in various class of organosilicon polymers, carbofunctional polyorganyl silsesquioxanes, which are very versatile and applicable (Voronkov et al. 1996, 2000). Shashkova et al. (2009) reported that a chemisorbent, made up of nickel phosphate and calcium carbonate has been used for extraction of iron(III) ions from mixed component of Ni^{2+}–Fe^{3+}–Pb^{2+}.

2.3 Nanomaterial

Various properties of nano-sorbents such as high surface area-to-volume ratio, porosity and size reduction, high surface reactivity to metal ions, enhancing the adsorption capacity and binding capacity make them suitable candidate for wide application in bioremediation approaches (Uheida et al. 2006; Zhou et al. 2009). Nanotechnology offers immense promise in the area of pollution sensing and prevention by exploiting novel properties of nanomaterials (Baruah and Dutta 2009).

Oxide based nanoparticles are one of emerging research areas for the removal of heavy metals from aqueous systems (Hua et al. 2012). Previously, Self-Assembled Monolayers on Mesoporous Supports (SAMMS) are effectively used as heavy metal sorbents from aquatic systems (Yantasee et al. 2010). Nano-composites of graphene oxide and magnetic graphene oxide have been chemically synthesized, characterized and also utilized for Am(III) and Pu(IV) isotopes sorption and Co(II), Ni(II), Cu (II) and Pb(II) sorption with variation in pH conditions (Lujaniene et al. 2016). An average size of 40 mm Cerium(III) molybdate nanoparticle have been found efficient for the removal of Tl(I), Pb(II), Th(IV), U(VI), and Cs(I) (Yousefi et al. 2012) (Table 4).

2.3.1 Mn and Zr

Manganese oxides (MnOs) nanoparticles showed high adsorptive activity than bulk manganese due to the polymorphic arrangement and specific surface area

Table 4 Nanomaterials and associated functions used in metals nanosorption approaches

Nanomaterial	Functions/activities reported of remediation	Metal/ Metalloids	References
Iron based nano-particles	High affinity towards different pollutants as absorbent	Cr^{3+}, Co^{2+}, Ni^{2+}, Cu^{2+}, Cd^{2+}, Pb^{2+}	Gupta et al. (2015)
Manganese oxides (MnO) nano-particles	Due to the development mechanism of inner-sphere by ion-exchange process, metals are absorbed onto surface	Pb(II), Cd (II) and Zn (II)	Gupta et al. (2015)
Zinc oxide (ZnO) nano-particles	Because of electropositive nature of nanomaterials, microporous nano-assemblies show high affinity towards adsorption of metals	Pb^{2+}, Hg^{2+} and As^{3+}	Singh et al. (2013)
Hybrid gold nano composite membrane	Efficient desalination and removal of radio-active iodine	Iodine (I)	Mushtaq et al. (2017)
Prussian blue-coated magnetic nanoparticles	Removed cesium from radioactive contaminated waste	Cesium (Cs)	Jang et al. (2014)
Titanate bio-nanocomposites	Efficient adsorbents with sorption capacity of 120 mg g^{-1} (1.75 meq. g^{-1})	Barium (Ba^{2+})	Xu et al. (2014)

(Luo et al. 2010). Nandi et al. (2013) synthesized Magnetic Mn-doped Fe(III) oxide nanoparticles, which have capability to sorb Cd(II) and Cu(II) from aqueous solution. It has been observed that hydrous ferric oxide (HFO) absorbs Cd^{2+} and hydrous manganese oxide (HMO) adsorbs Pb^{2+}, Zn^{2+}, and Cd^{2+} specifically, and by a complicated redox process T^{l+} adsorbs on HMO in an order of $Pb^{2+} > Zn^{2+} > Cd^{2+} > T^{l+}$ on the basis of the softness of these metals. In presence of H^+ and other adsorbing ions, the adsorption are reversible, i.e. the adsorption capability of 0.2 mole heavy metal ion/mole HMO has higher adsorption (0.56 mole/mole HMO), except lead (Gadde and Laitinen 1974; Fan et al. 2005). Inner-sphere complex formation occurs due to the sorption of Pb (II), Cd (II) and Zn (II) onto the Hydrous Manganese Oxide (HMO), which is an ion-exchange process (Parida et al. 1981).

The magnetic nanoparticles conjugated with actinide specific chelators MNP-Che separation process is very effective process which separates actinides (Np, Am, Cm) from aqueous nuclear waste to decrease radiation threats. Here, the MNPs are coated to absorb the pollutants very selectively onto the surface and can be collected through magnetic field (Kaur et al. 2013).

2.3.2 Fe and Fe^0

Sulfate-reducing bacteria can synthesize magnetic iron sulfide nanoparticles, which adsorb radioactive ions of Tc-99 from solution range of 400–500 m(2) g^{-1} (Watson et al. 2001). Radioactively aqueous waste containing Mo has been removed by Pb-Fe based adsorbents (Dodbiba et al. 2008). Nano-scaled carbon black increased the affinity of Cu(II),Cd(II) and Fe^0 nanoparticles towards As(III) binding in groundwater (Kanel et al. 2005). In the process of Magnetically Assisted Chemical Separation (MACS), 93.8% adsorption of U(VI) was achieved (Mansouri et al. 2014). Nanoscaled Zero valent Iron (NZVI) nanoparticles effectively function for absorption of As(V) concentrations from ~85–100% in ranges from 1–10 mg L^{-1} within 2 h (Bezbaruah et al. 2014).

2.3.3 Maghemite

Absorption studies were done on carbon-encapsulated magnetic nanoparticles (CEMNPs) to observe the removal capacity of Cu^{2+}, Co^{2+}, and Cd^{2+} ions (Bystrzejewski et al. 2009). Studies to evaluated and observed the low cost and efficient way of absorption and decontamination of Cs-137, Sr-90, Cd-115, Cr-51, Hg-203 etc. was by the use of nanocrystalline γ-Fe_2O_3 and also Co-60 radionuclides by Superparamagnetic iron oxide nanoparticles (IO-MNPs) (Bagla and Thakur 2017; El-Deen et al. 2017).

2.3.4 Carbon Nanomaterials (CNM)

Single-walled carbon nanotubes (SWCNTs), multi-walled carbon nanotubes (MWCNTs), nanoporous carbon, carbon beads and carbon fibers are various forms of carbon based nanomaterials. The efficiency of Carbon nanotubes (CNTs) has been exhibited by the adsorption of lead (49–97 mg g^{-1}), nickel (9–47 mg g^{-1}), zinc (11–43 mg g^{-1}) and cadmium (10–11 mg g^{-1}) which was is high in case of divalent metal ions in the order of $Pb^{2+} > Ni^{2+} > Zn^{2+} > Cd^{2+}$ (Dresselhaus et al. 2001; Li et al. 2002, 2003a, b, 2005, 2006; Chen and Wang 2006; Lu and Chiu 2006; Lu et al. 2006). Pb^{2+}, Cu^{2+}, Cd^{2+} is absorbed by CNTs nanoparticles mostly grown in Al_2O_3 and 67.11 mg g^{-1}, 33.78 mg g^{-1}, 11.23 mg g^{-1}, and 8.92 mg g^{-1} of Pb^{2+} was adsorbed by CNTs/Al2O3, PAC, commercial CNTs and Al_2O_3 respectively (Hsieh and Horng 2007). Stafiej and Pyrzunska (2007) have reported MWCNTs displaying the absorption of Cu^{2+}, Co^{2+}, Cd^{2+}, Zn^{2+}, Mn^{2+}, Pb^{2+}. SWCNTs, MWCNTs, and granular activated carbon sorbs Ni^{2+} in 47.85, 38.46, and 26.39 mg g^{-1} respectively (Lu et al. 2008). Chemical vapour deposition process (CVD) is a widely used process for synthesizing carbon nanoparticles, having properties of heavy metal sorption (Ruparelia et al. 2008). The ability of both oxidized CNTs and untreated CNTs showed difference in the adsorption of Cd^{2+}, Cu^{2+}, Ni^{2+}, Pb^{2+}, and, Zn^{2+} (Rao et al. 2007).

2.3.5 Ti, MgO and ZnO Nanoparticles

Nickel oxide and titanium dioxide based novel nanoparticles was synthesized and applied for ^{90}Sr, ^{137}Cs and $^{152\text{-}154}Eu$ removal from aqueous solution (Brynych et al. 2014). For As(III) removal, nanosized titanium dioxide is another promising material which is widely used and very effective (Visa and Duta 2013). The average size of titanium dioxide nanoparticles of 10–50 nm, has been reported for adsorption of Zn and Cd at 15.3 mg g^{-1} and 7.9 mg g^{-1} respectively (Liang et al. 2004). TiO_2-entrapped mesoporous MCM-41 removes Cr (VI) upto 91% from of a solution of 100 mg LG1 Cr (VI) metal at pH ~5.5 (Parida et al. 1981).

A flowerlike mesoporous fabricated MgO microspheres by a facile methods shows adsorption capabilities to Cd (II) and Pb (II) (Li et al. 2003a; Gao et al. 2009). Zinc oxide nanoparticles is excellent candidate to remove pollutants and for its application in nanoremediation. It has been researched that porous ZnO nanoplates and ZnO hollow microspheres have high efficiency in the removal of Cu (II), when it is exposed to porous nano-sheets (Wang et al. 2010). The porous ZnO nano-assemblies have capacity to remove toxic Co^{2+}, Ni^{2+}, Cu^{2+}, Cd^{2+}, Pb^{2+}, Hg^{2+} and As^{3+} from wastewater (Singh et al. 2013).

3 Conclusion

Remediation of metal wastes dumped into the environment has always been a challenge and research attempts have generated findings that have opened vistas for the exploitation of the newer approaches to environmental remediation. Though phytoremediation has been a success story in remediating metal wastes, but the newer approaches of using other biological components like algae, bacteria, fungi and actinomycetes have added potential candidates to the exiting technology. Chemical remediation though used extensively still remains concern for its wider applicability due to its after effects and derivative damages that are offshoots of its application in remediation. Modern research is now focusing on nano-remediation as an alternative where nanoparticles along with life forms are concurrently used for remediation of metal wastes. Promising results have already been generated for nanomaterials and their coentrapped synergistic effect with microorganisms in remediation of metal contaminated sites and waste waters and this could be the technology to look forward to in betterment of the remediation strategies available for metal contaminated sites.

References

Abdel RO, Rahman HA, Hung YT (2011) Liquid radioactive wastes treatment: a review. Water 3:551–565

Addour L, Belhocine D, Boudries N, Comeau Y, Pauss A, Mamer N (1999) Zinc uptake by *Streptomyces rimosus* biomass using a packed-bed column. J Chem Technol Biotechnol 74:1089–1095

Aksu Z, Sag Y, Kutsal T (1992) The biosorption of copper (II) by *C. vulgaris* and *Z. ramigera*. Environ Technol 13:579–586

Awual RM, Yaita T, El-Safty SA, Shiwaku H, Okamoto Y, Suzuki S (2013) Investigation of palladium(II) detection and recovery using ligand modified conjugate adsorbent. Chem Eng J 222:172–179

Awual MR, Rahman IMM, Yaita T, Khaleque MA, Ferdows M (2014) PH dependent cu(II) and Pd (II) ions detection and removal from aqueous media by an efficient mesoporous adsorbent. Chem Eng J 236:100–109

Babel S, Kurniawan TA (2003) Low-cost adsorbents for heavy metals uptake from contaminated water: a review. J Hazard Mater B 97:219–243

Bagla H, Thakur J (2017) Decontamination of radionuclides using γ-Fe2O3 as a nanosorbent. Geophys Res Abs 19:EGU2017–EG14952

Baruah S, Dutta J (2009) Nanotechnology applications in pollution sensing and degradation in agriculture: a review. Environ Chem Lett 7:191–204

Basnakova G, Stephens ER, Thaller MC, Rossolini GM, Macaskie LE (1998) The use of *Escherichia coli* bearing a phoN gene for the removal of uranium and nickel from aqueous flows. Appl Microbiol Biotechnol 50:266–272

Bengtsson L, Johansson B, Hackett TJ, McHale L, McHale AP (1995) Studies on the biosorption of uranium by *Talaromyces emersonii* CBS 814.70 biomass. Appl Environ Microbiol 42:807–811

Beveridge TJ (1986) The immobilization of soluble metals by bacterial walls. In: Ehrlich HL, Holmes DS (eds) Biotechnology and bioengineering symposium no. 16: biotechnology for the mining, metal-refining and fossil-fuel processing industries. Wiley, New York, pp 127–140

Bezbaruah AN, Kalita H, Almeelbi T, Capecchi CL, Jacob DL, Ugrinov AG, Payne AS (2014) Ca–alginate-entrapped nanoscale iron: arsenic treatability and mechanism studies. J Nanopart Res 16:2175

Bhainsa KC, D'Souza FS (1999) Biosorption of uranium(VI) by *Aspergillus fumigatus*. Biotechnol Tech 13:695–699

Boileau L, Nieboer JRE, Richardson DHS (1985) Uranium accumulation in the lichen *Cladonia rangiferina*. Part I. Uptake of cationic, neutral, and anionic forms of the uranyl ion. Can J Bot 63:384–389

Bors J, Erten H, Martens R (1991) Sorption studies of radioiodine on soils with special references to soil microbial biomass. Radiochim Acta 52–53:317–325

Brown MJ, Lester JN (1982) Role of bacterial extracellular polymers in metal uptake in pure bacterial culture and activated sludge – II effects of mean cell retention time. Water Res 6:1549–1560

Brown NW, Roberts EPL (2007) Electrochemical pre-treatment of effluents containing chlorinated compounds using an adsorbent. J Appl Electrochem 37:1329–1335

Brown NW, Campen AK, Wickenden DA, Roberts EPL (2013) On-site destruction of radioactive oily wastes using adsorption coupled with electrochemical regeneration. Chem Eng Res Des 91:713–721

Brynych V, Pospechova J, Prochazkova L, Cuba V, Szatmary (2014) Sorption of Cs(I), Sr(II) and Eu(III) on modified nickel oxide. Nanocon, November 5th–7th, Brno, Czech Republic, EU

Bustard M, Donnellan N, Rollan A, McHale L, McHale AP (1996) The effect of pulse field strength on electric field stimulated biosorption of uranium by *Kluyveromyces marxianus* IMB3. Biotechnol Lett 18:479–482

Bystrzejewski M, Pyrzynska K, Huczko A, Lange H (2009) Carbon-encapsulated magnetic nanoparticles as separable and mobile sorbents of heavy metal ions from aqueous solutions. Carbon 47:1201–2009

Chang D, Fukushi K, Ghosh S (1995) Simulation of activated sludge cultures for enhanced heavy metals removal. Water Environ Res 67:822–827

Chen C, Wang X (2006) Adsorption of Ni(II) from aqueous solution using oxidized multiwall carbon nanotubes. Ind Eng Chem Res 45:9144–9149

Chukanov VN, Volobuev PV, Poddubnyj VA, Trapeznikov AV (1993) Ecological problems of Ural. Russ J Nondestruct 29:516–521

Darnall DW, Greene B, Gradea-Torresday J (1988) Gold binding to algae in BioHydrometallurgy: In: Norris PR, Kelly DP (eds) Proceedings International Symposium, Science and Technology Letters, Kew, Surrey, UK, pp 487–498

Davis TA, Volesky B, Vieira RHSF (2000) *Sargassum* seaweed as biosorbent for heavy metals. Water Res 34:4270–4278

Declerck S, Duprède Boulois H, Bivort C, Delvaux B (2003) Extra radical mycelium of the arbuscular mycorrhizal fungus *Glomus lamellosum* can take up, accumulate and translocate radiocaesium under root-organ culture conditions. Environ Microbiol 5:510–516

Derbyshire F, Jagtoyen M, Andrews R, Rao A, Martin-Gullon I, Grulke E (2001) Carbon materials in environmental applications. In: Radovic LR (ed) Chemistry and physics of carbon, vol 27, Marcel Dekker, New York pp 1–66

Dhankhar R, Hooda A, Solanki R, Sainger PA (2011) *Saccharomyces cerevisiae:* a potential biosorbent for biosorption of uranium. Int J Engg Sci Technol 3:5397–5407

Dilek FB, Erbay A, Yetis U (2002) Ni(II) biosorption by polyporous versicolor. Process Biochem 37:723–726

Dodbiba G, Wu IC, Kikuchi T, Fujita T (2008) Adsorption of molybdenum in nitric acid solution by using Pb-Fe based adsorbents. In: Proceedings of the 2008 Global Symposium on Recycling, Waste Treatment and Clean Technology (REWAS 2008), Cancun, Mexico, 12–15 October, pp 63–68

Drake LR, Rayson GD (1996) Plant derived materials for metal ion-selective binding and pre-concentration. Anal Chem 68:22A–27A

Dresselhaus G, Dresselhaus MS, Avouris P (2001) Carbon nanotubes synthesis, structure, properties and applications. Springer, Berlin

Druzhinina TV, Kilyushik YA, Plotnikov DP (2011) Sorption of cadmium ions with chemisorbing polymer fibers. Theor Found Chem 45:482–486

Dupréde Boulois H, Delvaux B, Declerck S (2005) Effects of arbuscular mycorrhizal fungi on the root uptake and translocation of radiocaesium. Environ Pollut 134:515–524

Eckenfelder Jr WW, Allen PW (1990) Toxicity reduction of industrial effluents, Van Norstrand Reinhold, New York, pp 203–228

El-Deen GES, Imam NG, Ayoub RR (2017) Preparation, characterization and application of superparamagnetic iron oxide nanoparticles modified with natural polymers for removal of ^{60} Co-radionuclides from aqueous solution. Radiochim Acta 105:141–159

Faison BD, Cancel CA, Lewis SN, Adler HI (1990) Binding of dissolved strontium by *Micrococcus luteus*. Appl Environ Microbiol 56:3649–3656

Fan M, Boonfueng T, Xu Y, Axe L, Tyson TA (2005) Modeling Pb sorption to microporous amorphous oxides as discrete particles and coatings. J Colloid Interface Sci 281:39–48

Fourest E, Roux JC (1992) Heavy metal biosorption by fungal mycelial by-products: mechanisms and influence of pH. Appl Microbiol Biotechnol 37:399–403

Fourest E, Canal C, Roux JC (1992) Improvement of heavy metal biosorption by mycelial dead biomasses (Rhizopus arrhizus, Muchor miehei, and Pencillium chrysogenum): pH control and cationic activation. FEMS Microbiol Rev 14:325–332

Friis N, Myers-Keith P (1998) Biosorption of uranium and lead by *Streptomyces longwoodensis*. Biotechnol Bioeng 28:21–28

Fukushi K, Chang D, Ghosh S (1996) Enhanced heavy metal uptake by activated sludge cultures growth in the presence of bioplymer stimularors. Water Sci Technol 34:267–272

Gadd GM, White C, de Rome L (1988) Heavy metal and radionucleide uptake by fungi and yeasts, in bio hydrometallurgy. In: Norris PR, Kelly DP (eds), Proceedings International Symposium, Science and Technology letters, Kew, Surrey, pp 421–436

Gadde RR, Laitinen HA (1974) Studies of heavy metal adsorption by hydrous iron and manganese oxides. Anal Chem 46:2022–2026

Galun M, Keller P, Malki D, Feidstein H, Galun E, Siegel S, Siegel B (1984) Removal of uranium (VI) from solution by fungal biomass: inhibition by iron. Water Air Soil Pollut 21:411–414

Ganesh R, Robinson KG, Reed GR, Saylor GS (1997) Reduction of hexavlaent uranium from organic complexes by sulfate- and iron-reducing bacteria. Appl Environ Microbiol 63:4385–4391

Gao Z, Bandosz TJ, Zhao Z, Han M, Qiu J (2009) Investigation of factors affecting adsorption of transition metals on oxidized carbon nanotubes. J Hazard Mater 167:357–365

Gee AR, Dudeney AWL (1988) Adsorption and crystallisation of gold at biological surfaces, in bio hydrometallurgy. In: Norris PR, Kelly DP (eds) Proceedings International Symposium, Science and Technology letters, Kew. Surrey, pp 437–451

Ghazvini PTM, Ghorbanzadeh Mashkani S, Ghafourian H (2007) Bioabsorption of strontium from aqueous solution by new strain *Bacillus* sp. GTG-83. WM'07 conference, Tuscon, AZ

Greene B, Darnall DW (1990) Microbial oxygenic photoautotrophs for metal-ion binding. In: Ehrlich HL, Brierley CL (eds) Microbial mineral recovery. McGraw-Hill, New York, pp 277–302

Gupta VK, Tyagi I, Sadegh H, Shahryari-Ghoshekand R, Makhlouf ASH, Maazinejad B (2015) Nanoparticles as adsorbent a positive approach for removal of noxious metal ions: a review. Sci Technol Dev 34:195

Haas JR (1998) A comparison of U and lanthanide bioaccumulation by some common lichens. Geological Society of America, Annual meeting, Toronto 30:204–205

Haas JR, Bailey EH, Purvis OW (1998) Bioaccumulation of metals by lichens; uptake of aqueous uranium by *Peltigera membranancea* as a function of time and pH. Am Mineral 83:1494–1502

Haferburg G, Merten D, Buchel G, Kothe E (2007) Bio-sorption of metal and salt tolerant microbial isolates from a former uranium mining area. Their impact on changes in rare earth element patterns in acid mine drainage. J Basic Microbiol 47:474–484

Hafez N, Abdel-Razek AS, Hafez MB (1997) Accumulation of some heavy metals on *Aspergillus flavus*. J Chem Technol Biotechnol 68:19–22
Heide EA, Wagener K, Paschke M, Wald M (1973) Extraction of uranium from sea water by cultured algae. Naturwissenschaften 60:431
Holan ZR, Volesky B, Prasetyo I (1993) Biosorption of cadmium by biomass of marine algae. Biotechnol Bioeng 41:819–825
Horikoshi T, Nakajima A, Sakaguchi T (1979) Uptake of uranium by *Chlorella regularis*. Agric Biol Chem 43:617–623
Hsieh SH, Horng JJ (2007) Adsorption behavior of heavy metal ions by carbon nanotubes grown on microsized Al^2O^3 particles. J Univ Sci Technol Bejing Min Matel Mater 14:77–84
Hua M, Zhang S, Pan B, Zhang W, Lv L, Zhang Q (2012) Heavy metal removal from water/wastewater by nanosized metal oxides: a review. J Hazard Mater 211–212:317–331
International Atomic Energy Agency (1970) Standardization of radioactive waste categories. Technical reports series no. 101. International Atomic Energy Agency (IAEA), Vienna
International Atomic Energy Agency (1994) Handling, treatment, conditioning and storage of biological radioactive wastes, IAEA-TECDOC-775, Vienna
International Atomic Energy Agency (1999) Technologies for remediation of radioactively contaminated sites, Austria ISSN 1011-4289
International Atomic Energy Agency (2001) Handling and processing of radioactive waste from nuclear applications. Technical reports series no. 402. International Atomic Energy Agency (IAEA), Vienna
International Atomic Energy Agency (2006a) Application of thermal technologies for processing of radioactive waste. IAEA-TECDOC-1527. International Atomic Energy Agency (IAEA), Vienna
International Atomic Energy Agency (2006b) Fundamental safety principals. IAEA safety standards series. SF-1. IAEA, Vienna
International Atomic Energy Agency (2008) The management system for the processing, handling and storage of radioactive waste. Safety guide no. GS-G-3.3. International Atomic Energy Agency (IAEA), Vienna
Iyer A, Jha B (2005) Biosorption of heavy metals by a marine bacterium. Mar Pollut Bull 50:340–343
Jang SC, Hong SB, Yang HM, Lee KW, Moon JK, Seo BK, Huh YS, Roh C (2014) Removal of radioactive cesium using Prussian blue magnetic nanoparticles. Nano 4:894–901
Kanel SR, Manning B, Charlet L, Choi H (2005) Removal of arsenic(III) from groundwater by nanoscale zero-valent iron. Environ Sci Technol 39:1291–1298
Kaur M, Zhang H, Martin L, Todd T, Qiang Y (2013) Conjugates of magnetic nanoparticle actinide specific chelator for radioactive waste separation. Environ Sci Technol 47:11942–11959
Keskinkan O, Goksu MZL, Yuceer A, Basibuyuk M, Forster CF (2003) Heavy metal adsorption characteristics of a submerged aquatic plant (*Myriophyllum spicatum*). Process Biochem 39:179–183
Khalid AM, Ashfaq SR, Bhatti TM, Anwar MA, Shemsi AM, Akhtar K (1993) The uptake of microbially leached uranium by immobilized microbial biomass. In: Torma AE, Apel ML, Brierley CL (eds) Biohydrometallurgical technologies. In: Proceedings of the International Biohydrometallurgical Symposium, The Minerals, Metals and Materials Society, Warrendale, PA
Khani M, Keshtkar A, Meysami B, Zarea M, Jalali R (2005) Biosorption of uranium from aqueous solutions by nonliving biomass of marine algae *Cystoseira ical* heterogeneity in an in situ uranium bioremediation field site. Appl Environ Microbiol 71:6308–6318
Koval'skiy VV, Vorotnitskaya IY (1965) Biogenic migration of uranium in Lake Issyk-Kul. Geokhimiya 6:724–732
Kumar R, Acharya C, Joshi SR (2011) Isolation and analyses of Uranium tolerant *Serretia marcescens* strains and their utilization for aerobic Uranium U(VI) bioadsorption. J Microbiol 49:568–574

Kurniawan TA, Babel S (2003) A research study on Cr(VI) removal from contaminated wastewater using low-cost adsorbents and commercial activated carbon. In: Second International Conference on Energy Technology towards a Clean Environment (RCETE), vol 2. Phuket, Thailand, 12–14 February, pp 1110–1117

Kuycak N, Volesky B (1988) Biosorbents for recovery of metals from industrial solutions. Biotechnol Lett 10:137

Lapkin A, Bozkaya B, Mays T, Borello L, Edler K, Crittenden B (2003) Preparation and characterization of chemisorbents based on heteropolyacids supported on synthetic mesoporous carbons and silica. Catal Today 81:611–621

Li YH, Wang S, Wei J, Zhang X, Xu C, Luan Z, Wu D, Wei B (2002) Lead adsorption on carbon nanotubes. Chem Phys Lett 357:263–266

Li YH, Ding J, Luan Z, Di Z, Zhu Y, Xu C, Wu D, Wei B (2003a) Competitive adsorption of Pb^{2+}, Cu^{2+} and Cd^{2+} ions from aqueous solutions by multiwalled carbon nanotubes. Carbon 41:2787–2792

Li YH, Wang Z, Luan J, Ding J, Xu C, Wu D (2003b) Adsorption of cadmium(II) from aqueous solution by surface oxidized carbon nanotubes. Carbon 41:1057–1062

Li Y, Di Z, Ding J, Wu D, Luan Z, Zhu Y (2005) Adsorption thermodynamic, kinetic and desorption studies of Pb^{2+} on carbon nanotubes. Wat Res 39:605–609

Li YH, Zhu Y, Zhao Y, Wu D, Luan Z (2006) Different morphologies of carbon nanotubes effect on the lead removal from aqueous solution. Diam Relat Mater 15:90–94

Liang P, Shi T, Li J (2004) Nanometer-size titanium dioxide separation/pre concentration and FAAS determination of trace Zn and cd in water sample. Int J Environ Anal Chem 84:315–321

Liu HH, Wu JT (1993) Uptake and recovery of americium and uranium by *Anacystis* biomass. J Environ Sci Health A Tox Hazard Subst Environ Eng 28:491–504

Liu HL, Chen BY, Lan YW, Cheng YC (2004) Biosorption of Zn(II) and Cu(II) by the indigenous Thiobacillus thiooxidans. Chem Eng J 97:195–20

Lloyd JR, Yong P, Macaskie LE (2000) Biological reduction and removal of np(V) by two microorganisms. Environ Sci Technol 34:1297–1301

Lu C, Chiu H (2006) Adsorption of zinc(II) from water with purified carbon nanotubes. Chem Eng Sci 61:1138–1145

Lu C, Liu C (2006) Removal of nickel(II) from aqueous solution by carbon nanotubes. J Chem Technol Biotechnol 81:1932–1940

Lu C, Chiu H, Liu C (2006) Removal of zinc(II) from aqueous solution by purified carbon nanotubes: kinetics and equilibrium studies. Ind Eng Chem Res 45:2850–2855

Lu C, Liu C, Rao GP (2008) Comparisons of sorbent cost for the removal of Ni^{2+} from aqueous solution by carbon nanotubes and granular activated carbon. J Hazard Mater 151:239–246

Lujaniene G, Semcuk S, Kulakauskaite I, Mazeika K, Valiulis D, Juskenas R, Tautkus S (2016) Sorption of radionuclides and metals to graphene oxide and magnetic graphene oxide. J Radioanal Nucl Chem 307:2267–2275

Luo T, Cui J, Hu S, Huang Y, Jing C (2010) Arsenic removal and recovery from copper smelting wastewater using TiO2. Environ Sci Technol 44:9094–9098

Macaskie LE, Bonthrone KM, Yong P, Goddard DT (2000) Enzymically mediated bioprecipitation of uranium by a *Citrobacter* sp.: a concerted role for exocellular lipopolysaccharide and associated phosphatase in biomineral formation. Microbiology 146:1855–1867

Mann H, Fyfe WS (1984) An experimental study of algal uptake of U, Ba, V, Co and Ni from dilute solutions. Chem Geol 44:385–398

Mann H, Fyfe WS (1985) Uranium uptake by algae: experimental and natural environments. Can J Earth Sci 22:1899–1903

Manos MJ, Iyer RG, Quarez E, Liao JH, Kanatzidis MG (2005) {Sn[Zn4Sn4S17]}6−: a robust open framework based on metal-linked penta-supertetrahedral [Zn4Sn4S17]10− clusters with ion-exchange properties. Angew Chem 44:3552–3555

Mansouri N, Saberyan K, Noaparast M (2014) Adsorption of U(VI) from aqueous solution by Triocthylamine (TOA) functionalized magnetite nanoparticles as a novel adsorbent. J Adv Chem 10:2403–2414

Mattuschka B, Junghaus K, Straube G (1993) Biosorption of metals by waste biomass. In: Torma AE, Apel ML, Brierley CL (eds) Biohydrometallurgical technologies. Proceedings of the International Biohydrometallurgical Symposium, The Minerals, Metals and Materials Society, Warrendale, PA

McCoy KB, Derecho I, Wong T, Tran HM, Huynh TD, La Duc MT, Venkateswaran K, Mogul R (2012) Insights into the extremotolerance of *Acinetobacter radioresistens* 50v1, a gram-negative bacterium isolated from the Mars Odyssey spacecraft. Astrobiology 12:854–862

McCready RGL, Lakshmanan VI (1986) Review of bioabsorption research to recover uranium from leach solutions in Canada. In: Eclles S (ed) Immobilisation of Ions by bio-sorption. Canada Centre for Mineral and Energy Technology, Energy Mines and Resources Canada, Ottawa, pp 219–225

Méndez de Vigo GE (2000) Origin and radioactive waste management, In: Official College of Physicians Illustrious. 3rd edition. 105:188

Mullen MD, Wolfe DC, Ferris FG, Beveridge TJ, Flemming CA, Bailey GW (1989) Bacterial sorption of heavy metal. Appl Environ Microbiol 55:3143–3149

Mushtaq S, Yun SE, Yang JE, Jeong SW, Shim HE, Choi MH, Park SH, Choi YJ, Jeon J (2017) Efficient and selective removal of radioactive iodine anions using engineered nanocomposite membranes. Environ Sci Nano 4:2157–2163

Nakajima A, Sakaguchi T (1986) Selective accumulation of heavy metals by microorganisms. Appl Microbiol Biotechnol 24:59–64

Nandi D, Basu T, Debnath S, Ghosh AK, De A, Ghosh UC (2013) Mechanistic insight for the sorption of Cd(II) and Cu(II) from aqueous solution on magnetic Mn-doped Fe(III) oxide nanoparticle implanted graphene. J Chem Eng Data 58:2809–2818

Neu MP, Icopini GA, Boukhalfa H (2005) Plutonium speciation affected by environmental bacteria. Radiochim Acta 93:705–714

Niu H, Xu XS, Wang JH (1993) Removal of lead from aqueous solutions by *Penicillium* biomass. Biotechnol Bioeng 42:785–787

Norberg A, Persson H (1984) Accumulation of heavy metal ions by *Zoogloea ramigera*. Biotechnol Bioeng 26:239–246

Norberg A, Rydin S (1984) Development of a continuous process for metal accumulation by *Zoogloea ramigera*. Biotecnol Bioeng 26:265–268

Oscik J (1982) Adsorption. Chichester, Ellis Horwood Ltd. Publishers

Oves M, Khan MS, Zaidi A (2013) Biosorption of heavy metals by *Bacillus thuringiensis* strain OSM29 originating from industrial effluent contaminated north Indian soil. Saudi J Biol Sci 20:121–129

Ozaki T, Gillow JB, Kimura T, Ohnuki T, Yoshida Z, Francis AJ (2004) Sorption behavior of europium(III) and curium(III) on the cell surfaces of microorganisms. Radiochim Acta 92:741–748

Ozdemir G, Ceyhan N, Ozturk T, Akirmak F, Cosar T (2004) Biosorption of chromium(VI), cadmium(II) and copper(II) by *Pantoea* sp. TEM18. Chem Eng J 102:249–253

Pandey PK, Choubey S, Verrma Y, Pandey M, Chandrashekhar P (2009) Biosorptive removal of arsenic from drinking water. Bioresour Technol 100:634–637

Parida KM, Kanungo SB, Sant BR (1981) Studies on MnO_2-I, chemical composition, microstructure and other characteristics of some synthetic MnO_2 of various crystalline modifications. Electrochim Acta 26:435–443

Polman JK, Breckenridge CR (1996) Biomass-mediated binding and recovery of textile dyes from waste effluents. Text Chem Colour 28:31–35

Pozhidaev YN, Panezhda EV, Grigor'eva OY, Kirillov AI, Belousova LI, Vlasova NN, Voronkov MG (2003) Carbofunctional Polyorganylsilsesquioxanes as sorbents for some rare metals. Dokl Chem 389:97–100

Prasad MNV, Oliveira Freitas HMD (2003) Metal hyperaccumulation in plants–biodiversity prospecting for phytoremediation technology. Electron J Biotechnol 6:110–146

Pribil S, Marvan P (1976) Accumulation of uranium by the chlorococcal alga *Scenedesmus quadricauda*. Arch Hydrobiol Suppl 49:214–225
Radiation Safety General Regulations (1988) Official Journal of the Federation, November 22
Ramachandra TV, Ahalya N, Kanamadi RD (2005) Biosorption: techniques and mechanisms. CES Tchnical Rep 110:33
Ramakrishna KR, Viraraghavan T (1997) Dye removal using low cost adsorbents. Water Sci Technol 36:189–196
Rao GP, Lu C, Su F (2007) Sorption of divalent metal ions from aqueous solution by carbon nanotubes: a review. Sep Purif Technol 58:224
Rastogi G, Osman S, Vaishampayan PA, Andersen GL, Stetler LD, Sani RK (2010) Microbial diversity in uranium mining-impacted soils as revealed by high-density 16S microarray and clone library. Microb Ecol 59:94–108
Roig MG, Manzano T, Diaz M (1997) Biochemical process for the removal of uranium from acid mine drainages. Water Res 31:2073–2083
Rudd T, Sterritt RM, Lester JN (1984) Formation and conditional stability constants of complexes formed between heavy metals and bacterial extracellular polymers. Water Res 18:379–384
Ruparelia JP, Duttagupta SP, Chatterjee AK, Mukherji S (2008) Potential of carbon nanomaterials for removal of heavy metals from water. Desalination 232:145
Sao K, Khan F, Pandey PK, Pandey M (2014) A review on heavy metals uptake by plants through biosorption. Intl Proc Econ Develop Res 75:78
Schneider IAH, Rubio J (1999) Sorption of heavy metal ions by the non-living biomass of freshwater macrophytes. Environ Sci Tech 33:2213–2217
Schneider IAH, Rubio J, Smith RW (1999) Effect of some mining chemicals on biosorption of Cu (II) by the non-living biomass of the freshwater macrophyte *Potamogeton lucens*. Miner Eng 12:255–260
Shashkova IL, Rat'ko AI, Mil'vit NV, Muravitskaya EV (2009) The use of chemisorption technologies in separation of multicomponent mixtures of metal ions. Russ J Appl Chem 82:940–946
Shelobolina ES, Sullivan SA, O'Neill KR, Nevin KP, Lovley DR (2004) Isolation, characterization, and U(VI)-reducing potential of a facultatively anaerobic, acid-resistant bacterium from low-pH, nitrate- and U(VI)-contaminated subsurface sediment and description of *Salmonella subterranea* sp. nov. Appl Environ Microbiol 70:2959–2965
Siegel S, Keller P, Galun M, Lehr H, Siegel B, Galun B (1986) Biosorption of lead and chromium by *Penicillium* preparations. Water Air Soil Pollut 27:69–75
Singh S, Barick KC, Bahadur D (2013) Fe_3O_4 embedded ZnO nanocomposites for the removal of toxic metal ions, organic dyes and bacterial pathogens. J Mater Chem A 1:3325–3333
Sobolev LA (1994) Radiation anomalies in Moscow region. Ecomaz, Moscow 52
Songakasiri K, Reed DT, Rittmann BE (2002) Biosorption of neptunium(V) by *Pseudomonas fluorescens*. Radiochim Acta 90:785–789
Sposito G (1982) On the surface complexation model of the oxide aqueous solution interface. J Coll Interf Sci 91:329–340
Sposito G (1986) Distinguishing adsorption from surface precipitation. Geochem Process Miner Surf 323:217–228
Stafiej A, Pyrzynska K (2007) Adsorption of heavy metal ions with carbon nanotubes. Sep Purif Technol 58:49
Stuetz RM, Madgwick JC, Gee AR (1993) Immobilisation of biosorbed metal ions. In: Torma AE, Apel ML, Brierley CL (eds) Biohydrometallurgical technologies. Proceedings of the International Biohydrometallurgical Symposium, The Minerals, Metals and Materials Society, Warrendale, PA
Stumm W (1992) Chemistry of the solid-water Interface: processes at the mineral-water and particle-water Interface in natural systems. Wiley, Canada
Stumm W, Morgan JJ (1996) Aquatic chemistry: chemical equilibria and rates in natural waters. Wiley, New York

Townsley CC, Ross IS, Atkins AS (1986) Biorecovery of metallic residues from various industrial effluents using filamentous fungi. In: Lawrence RW, Branion RMR, Ebner HG (eds) Fundamental and applied biohydrometallurgy. Elsevier, Amsterdam, pp 279–289

Tsezos M, Volesky B (1982) The mechanism of thorium biosorption by *Rhizopus arrhizus*. Biotechnol Bioeng 24:955–969

U.S. Environmental Protection Agency (2000) Introduction to phytoremediation. National Risk Management Research Laboratory, EPA/600/R-99/107

Uheida A, Iglesias M, Fontas C, Hidalgo M, Salvado V, Zhang Y, Muhammed M (2006) Sorption of Palladium(II), Rhodium(III), and Platinum(IV) on Fe_3O_4 nanoparticles. J Coll Interf Sci 301:402–408

Visa M, Duta A (2013) TiO_2/fly ash novel substrate for simultaneous removal of heavy metals and surfactants. Chem Eng J 223:860–868

Volesky B (1990) Biosorption of heavy metals. CRC Press, Boston, November. ISBN 0849349176, 408

Volesky B, Prasetyo I (1994) Cadmium removal in a biosorption column. Biotechnol Bioeng 43:1010–1015

Volesky B, Sezos M (1982) Separation of uranium by biosorption. Biotechnol Bioeng 23:583–604

Voronkov MG, Vlasova NN, Pozhidaev YN (1996) Organosilicon ion-exchange and complexing sorbents. Zh Prikl Khim 69:705–718

Voronkov MG, Vlasova NN, Pozhidaev YN (2000) Organosilicon ion exchange and complexing adsorbents. Appl Organomet Chem 14:287–303

Wang TC, Weissman JC, Ramesh G, Varadarajan R, Benemann JR (1996) Parameters for removal of toxic heavy metals by water milfoil (*Myriophyllum spicatum*). Bull Environ Cont Toxicol 57:779–786

Wang X, Cai W, Lin Y, Wang G, Liang C (2010) Mass production of micro/nanostructured porous ZnO plates and their strong structurally enhanced and selective adsorption performance for environmental remediation. J Mater Chem 20:8582–8590

Wase J, Forster CF (1997) Biosorbents for metal ions. Taylor & Francis, London pp 238

Watson JHP, Croudace IW, Warwick PE, James PAB, Charnock JM, Ellwood DC (2001) Adsorption of radioactive metals by strongly magnetic iron sulfide nanoparticles produced by sulfate-reducing bacteria. Sep Sci Technol 36:2571–2607

Witek-Krowiak A, Szafran RG, Modelski S (2010) Biosorption of heavy metals from aqueous solutions onto peanut shell as a lowcost biosorbent. Desalination 265(1–3):126–134

Xie S, Yang J, Chen C, Zhang X, Wang Q, Zhang C (2008) Study on biosorption kinetics and thermodynamics of uranium by *Citrobacter freudii*. J Environ Radioact 99:126–133

Xu M, Wei G, Liu N, Zhou L, Fu C, Chubik MP, Chubik MP, Gromov A, Han W (2014) Novel fungus-titanate bio-nanocomposites as high performance adsorbents for the efficient removal of radioactive ions from wastewater. Nanoscale 6:722–725

Yang J, Volesky B (1999) Biosorption of uranium on *Sargassum* biomass. Water Res 33:3357–3363

Yantasee W, Rutledge RD, Chouyyok W, Sukwarotwat V, Orr G, Warner CL, Warner MG, Fryxell GE, Wiacek RJ, Timchalk C, Addleman RS (2010) Functionalized nanoporous silica for the removal of heavy metals from biological systems: adsorption and application. ACS Appl Mater Interf 2:2749–2758

Yousefi T, Khanchi AR, Ahmadi SJ, Rofouei MK, Yavari R, Davarkhah R, Myanji B (2012) Cerium(III) molybdate nanoparticles: synthesis, characterization and radionuclides adsorption studies. J Hazard Mater 215-216:266–271

Zhou YT, Nie HL, White CB, He ZY, Zhu LM (2009) Removal of Cu^{2+} from aqueous solution by chitosan-coated magnetic nanoparticles modified with aketoglutaric acid. J Colloid Interface Sci 330:29–37

Potential of Biochar as a Measure for Decreasing Bioavailability of ^{137}Cs in Soil

Aleksander Nikolaevich Nikitin, Olga Aleksandrovna Shurankova, Olga Il'inichna Popova, Ihar Anatol'evich Cheshyk, and Ruslan Kovsarovich Spirov

1 Introduction

The current situation in the world is characterized by an increase in quantity and sharpness of global problems: climate change, reduction of the area of agricultural lands, food security, environmental pollution and many others. It causes the necessity of search of means and methods of solution to these problems now or in the near future. It is clear that the uniform decision for all global problems doesn't exist. But the solution to one of them shouldn't be followed by an aggravation of another one, and ideally–reduce tension on the whole complex of problems at once.

One of the solutions to a complex of global challenges is biochar. This material is not only a valuable soil ameliorant but also the convenient tool for prevention of accumulation of carbon dioxide in the atmosphere. While adjust conditions for production of biochar, it is possible to achieve the production of significant amounts of biogas and biodiesel. These products are fuel and energy resources made from local raw materials and allow diversifying the power branch. Besides, these energy resources can be considered as "green" as during their burning carbon dioxide, which would also get to the atmosphere at the natural decomposition of fossils, is released there.

Application of the adsorbents into the soil, which allows reducing the availability of radionuclides to root absorption, is one of the methods of receiving standardly net crop production on contaminated lands. After the Chernobyl accident a wide range of ameliorants, which have the occluding properties in relation to ^{137}Cs, have been tested. Among them it is possible to find sapropels (silt deposits of freshwater reservoirs consisting of organic substances) (Perepel'atnikova et al. 1993;

A. N. Nikitin (✉) · O. A. Shurankova · O. Il'inichna Popova · I. A. Cheshyk · R. K. Spirov
State Scientific Institution, Institute of Radiobiology of the National Academy of Sciences of Belarus, Gomel, Republic of Belarus

D. K. Gupta, A. Voronina (eds.), *Remediation Measures for Radioactively Contaminated Areas*, https://doi.org/10.1007/978-3-319-73398-2_6

Burdakov et al. 1994; Kunovskij et al. 1996); ferrocyanide compounds (Ageets 2001), in particular winemaking waste such as ferrocyanide-bentonite sorbent (Azarenko 2012); extracts of brown coal (Ageets 2001); vermiculite and bentonite (Ratnikov et al. 1996); specially synthesizable polymer of poly-N, N-dimethyl-3,4-dimethylaminopyridine chloride (Bakunov and Archipov 1994; Krugliakov et al. 2008); calcium-containing ameliorants (bergmeal and clay mergel) (Ageets 2001); aluminosilicates (Krugliakov et al. 2008).

After 30 years of Chernobyl accident, it is relevant to search for economically and ecologically efficient protective measures aimed at transfer reduction of ^{137}Cs and ^{90}Sr from soil to crop production and increase in fertility of the soil (Ageets et al. 2003; Israel and Bogdevich 2009). This problem is especially relevant for peat and marsh soils which are widely presented in the region with the high content of organic substances. These soils reveal significantly higher transfer factors of ^{137}Cs in crops in comparison with soils of mineral origin. The research conducted on the territory of Belarus has proved the effectivness of traditional counter-measures (application of manure, phosphorus-potassium fertilizers in raised doses). However the systematic use of increased doses of mineral fertilizers hurts the ecological condition of soils and increases production cost. At the same time, the use of cesium-selective sorbents (bentonite, clinoptilolite, and zeolite) was ineffective. Long-time retention of cesium in the soil has led to a reduction of mobile forms of ^{137}Cs and application of the clay minerals had no significant effect on the decrease in intake of radioisotopes of cesium from the soil to a plant.

The characteristics of biochar include significant area of free surface per unit of mass and volume as well as high sorption ability. This fact has allowed us to make a hypothesis that application of biochar into the soil will allow not only increasing its fertility but also reducing accumulation of such pollutants like heavy metals and technogenic radionuclides in products of agriculture.

The purpose of the research was an assessment of a possibility of use of biochar (the biomass which has undergone pyrolysis processing) and soil-improving additives on its basis for decreasing the transfer of ^{137}Cs into crop production.

2 Biochar and Its Use in Agriculture

2.1 General Characteristics of Biochar

Biochar is the material with high content of carbon (up to 90%) obtained from the organic mass (wood, manure, grassy vegetation, foliage, etc.) which was subjected to heat treatment (temperature around 700 °C) without access (or at insignificant access) of oxygen. It should be noted that at the same time the whole complex of gaseous, liquid, and solid products is formed. Biochar is only one of them, but this product has been used by mankind since ancient times (Harris 1999).

Chemical properties of biochar strongly vary depending on raw materials used for its production and conditions of the technological process. The primary element here is carbon which can partially keep linear or aromatic links with separate atoms. Certain parts acquire the atomic structure characteristic of graphite. However, on the whole, the material has amorphous structure. The parts with regular crystal structure aren't numerous. Their representation grows at the increase in temperature at pyrolysis (Emmerich and Luengo 1996).

Brennan (Brennan et al. 2001) points to the existence in biochar of organic heterocycles including oxygen, nitrogen, phosphorus, and sulfur. Among the main functional groups of organic compounds, there are hydroxyl, carbonyl, carboxyl, amino-, and keto-groups. Microparts with the base or acid properties are present in one sample at a distance of several micrometers from each other.

Biochar includes oxygen and nitrogen (0.2–5.6%). Mineral macro-elements also presented in its composition (phosphorus – 0.3–4.8%, potassium – 0.1–5.8%) Schmidt and Noack (2000). Potassium and chlorine gain high volatility when heated to the temperature of several hundred degrees above Celsius. Ca and Si gain volatility at a much higher temperature. Magnesium presents in organic substance both in a mineral form and as a part of organic compounds (covalent bond). Biomass can lose it only at very high temperatures.

Phosphorus, sulfur, and nitrogen are bound with complex organic substances. But if the former two elements remain quite stable at low-temperature pyrolysis, nitrogen in these conditions becomes very mobile (Schnitzer et al. 2007). Other elements, such as iron, manganese, and others, being a part of various organic and inorganic compounds, are practically not lost in the pyrolysis.

Concerning microelements, as a part of biochar, the insufficient amount of research has been conducted. At the same time, it is known that this material can sorb heavy metals (Swiatkowski et al. 2004).

Most of the researchers incline that mineral elements aren't bound with biochar carbon. However, at least in some types of biochar, potassium, and calcium join in the carbon matrix and form so-called phenoxides and also can be found between graphene layers (Wornat et al. 1995). As a part of biochar such inorganic substances as sylvite, quartz, amorphous silicon, calcite, hydroxyapatite, phosphate, sulfate, and nitrate of calcium, oxides and hydroxides of calcium, magnesium, aluminum, titan, manganese, zinc, and iron are found (Parr 2006).

Schnitzer (Schnitzer et al. 2007) points to the existence of the remains of organic compounds in biochar; he has divided them into six classes: N-heterocyclic compounds, furan-like compounds, phenols and phenol-like compounds, gasoline and gasoline-like compounds, carbocyclic compounds, aliphatic compounds. For some of these compounds, positive influence on the speed of germination of seeds and growth of plants has been noted, others favorably affect the development of microorganisms (sesquiterpenes), but many of such substances are biocides. Trompowsky (Trompowsky et al. 2005) points at the opportunity of extraction from biochar of humic acids with H/C ratio which is similar to biochar in general. However, O/C relation in the extracted humic acids is much higher, in comparison with biochar.

The microstructure of biochar has features of that organic material from which it has been received. During the microscopic investigations the remains of the conducting system of plants, as well as smaller pores with a diameter of about

10 microns are found here (Fukuyama et al. 2001; Martínez et al. 2006). The microscopic analysis of biochar also points at the existence of many microcracks.

2.2 Influence of Biochar on Agrochemical Properties of Soil

Application of biochar into the soil can significantly change its characteristics such as structure, porosity, water-retaining ability, air permeability, the area of a free surface, distribution by sizes of particles, density and others. As a rule, it is connected with the improvement of conditions of growth for plants. Biochar has a high water-retaining ability. It can serve in the soil as a water buffer: to sorb excess moisture, providing an intake of oxygen to roots and gradually give water in a drought period. Besides, biochar has a high absorbing ability about basic elements of food of plants. At this expense, it allows extending the validity of mineral fertilizers and removing the toxic effect of their use in excessive doses (Woods et al. 2009).

Biochar differs in a quite high degree of chemical stability. It can remain in the soil for many years, practically without losing its properties (Shindo 1991; Cheng et al. 2008).

The absorbing capacity of biochar concerning metals is determined by the existence of a charge and pi-electronic clouds on its surfaces. At the same time, it can undergo essential changes if the pH of environment changes. We should also note the fact that the absorbing ability is determined by chemical properties of metals. Polyvalent metals and metals with amphoteric properties can modify the affinity to superficial structures of biochar during measurement of acidity and oxidation-reduction capacity of the soil, too.

2.3 Use of Biochar in Agriculture

Signs of use of biochar in agriculture can be traced in slash-and-burn and creation of artificial soils of Terra Preta in the Amazon basin. Its first records in scientific publications date back to the nineteenth century. Trimble (1851) wrote that he had noticed the use of dust from charcoal in all farms and in all countries he had been.

Impact of biochar on characteristics of soil and growth of cultural plants was for the first time studied by Retan (1915) and Tryon (1948). In the first half of the twentieth century, a lot of manuals on practical agriculture recommended biochar as a means for improvement of the water-retaining ability of soils, increase in productivity as a component of mixes for greenhouses, soil for lawns (Santiago and Santiago 1989; Morley 1927). In the 1980s biochar opportunities as an ameliorant were actively studied in Japan (Kishimoto and Sugiura 1980, 1985).

The next splash in interest in biochar is connected with the search for practical tools for prevention of accumulation of carbon dioxide in the atmosphere and opening of the fact that fertile soils of the Amazon basin were created by man with the use of this technology (Glaser et al. 2001). Petersen (Petersen et al. 2001) believes that the high level of development of civilization of pre-Columbian America was in many respects due to the use of biochar in agriculture.

The ability of biochar to increase the fertility of the soil and improve growth of plants is connected with a number of its properties. One of them is the presence of macro- and micro-elements in this material. The elements are necessary for the growth of plants. It must be kept in mind, however, that the general content of these elements doesn't depict the situation entirely. Their considerable part can be almost inaccessible for root systems of plants. Therefore it is necessary to consider only those forms of potassium, phosphorus, nitrogen, calcium, magnesium, iron and other elements which can freely pass into solution or exchange for other cations or anions from root allocations or those which come to the environment due to the activity of microorganisms.

In particular, despite rather a high percentage of nitrogen in biochar, its content in ammonium and nitrate forms is very insignificant (Bridle and Pritchard 2004). Contrary to this, potassium here is mainly in a mobile form.

The organic substances increase the area of a free surface on the sandy and sandy-loam soils. In the case of clay soils, organic substance promotes structuring and formation of air cavities. It is shown that application of biochar into soil performs the function similar to organic substance—this technique allows increasing significantly area of a free surface on sandstones and improves air permeability of clay soils (Chan et al. 2007). It is also necessary to point at one more analogy between the organic substance of soil and biochar—both of these factors contribute to the development of soil microflora.

Moreover, biochar can sorb the elements which are contained in the soil solution. It can reduce their availability a little. It is shown that this phenomenon sometimes takes place in respect of available nitrogen in soil (Rondon et al. 2007). But, on the other hand, this ability of biochar prevents washing away and loss of nutrition elements from the root horizon. Therefore, the availability of biochar in soil increases the period of validity of mineral fertilizers. One more factor of indirect impact of biochar on fertility is its ability to regulate the acidity of the environment and fix toxic substances which slow down the growth of plants on the surface.

Some authors suggest using biochar as a means for decreasing of biological availability of heavy metals and organic pollutants in soils (Zhang et al. 2013). Its high sorbing ability and big surface area have to reduce the mobility of pollutants in soil. However, it can also reduce the availability of mineral nutrition elements; therefore the application of biochar in a complex with composts and other ameliorants can become an optimal variant (Beesley et al. 2011). Laboratory research has shown that biochar possesses rather a high affinity in relation to Cs ions ($Kd > 100$) (Hamilton et al. 2016). This property, together with the high contents of soluble potassium in this material, has to promote a decrease in the root intake of Cs in plants when introducing biochar into the soil. For podsolic and peat soils of Polesie, it is also important that application of biochar leads to decrease in acidity of the soil.

3 Objective and Methods of the Experiment

3.1 Design of the Experiments

Vegetative experiment was conducted out for assessment of the influence of biochar and soil-improving additives on its basis on biological availability of ^{137}Cs and its transfer to agricultural plants. Model plant in this experiment was mangold (*Beta vulgaris var. cycla*) kind "Mangold red PC 1".

We used in the experiment the soil sampled in the exclusion zone of the Chernobyl NPP, mixed with peat in the ratio of 1:4 in weight. Dose rate on the surface of the soil mix was 0.8–1.0 $\mu Sv \cdot h^{-1}$. The specific activity of ^{137}Cs in the soil mix was 34.0–49.6 $kBq \cdot kg^{-1}$.

Not only biochar was tested in the research. Mineral sorbent bergmeal and bokashi – a product of microbiological processing of production wastes of wheat, rice, and buckwheat – was tested as soil ameliorants too. Biochar and bokashi were introduced in quantity 1 and 3% of the total weight of soil mix; bergmeal – in the amount of 0.1 and 0.4%. In options of experiment 2–5, 8–9 and 14–15 we used bokashi from a waste of processing of wheat; in option 10 – bokashi from a waste of processing of rice; in option 11 – bokashi from a waste of processing of buckwheat.

Additives were introduced separately and in mixes, in a percentage ratio from the total weight of soil mix. The whole scheme of vegetative experiment included 17 options:

1. Control (C);
2. Biochar 3% + Bokashi 1% (BC3% + B1%);
3. Biochar 3% + Bokashi 3% (BC3% + B3%);
4. Biochar 1% + Bokashi 1% (BC1% + B1%);
5. Biochar 1% + Bokashi 3% (BC1% + B3%);
6. Biochar 1% (BC1%);
7. Biochar 3% (BC3%);
8. Bokashi 1% (B1%);
9. Bokashi 3% (B3%);
10. Bokashi from waste of processing of Rice 3% (Br3%);
11. Bokashi from waste of processing of Buckwheat 3% (Bg3%);
12. Bokashi 3% + Bergmeal 0.1% (B3% + T0.1%);
13. Bokashi 3% + Bergmeal 0.4% (B3% + T0.4%);
14. Biochar 1% + Bokashi 3% + Bergmeal 0.1% (BC1% + B3% + T0.1%);
15. Biochar 1% + Bokashi 3% + Bergmeal 0.4% (BC1% + B3% + T0.4%);
16. Bergmeal 0.1% (T0.1%);
17. Bergmeal 0.4% (T0.4%).

Experimental soil mixture was humidified with water within seven days before the establishment of constant humidity of 70–80% of full moisture capacity. Further, the level of moistening of the soil was supported at this level and controlled by the weight method.

The experiment was done with five-time repetitions. Plastic pots were filled with soil mixture (860 g per pot). In each pot, 21 seeds of mangold were planted at a depth of 2 cm. On the 65th day after planting the plants were cut off. The number of plants, a weight of the elevated part and roots, length of green parts of plants was identified in all repetition of each option. Activity concentration of ^{137}Cs in the aboveground parts was measured after drying. The soil after the experiment was dried up and sent to the agrochemical analysis and measurement of activity concentration of ^{137}Cs.

In the experiments the following soil-improving additives were tested:

- Biochar – the product of pyrolysis processing of wood of hard-wooded broadleaved breeds. Before application into the soil, biochar was thinly crushed to the size of particles of 50–500 microns.
- Bergmeal from the "Stalnoye" deposit is a complex polymineral formation consisting of five disperse, thinly mixed phases: opal-cristobalite, X-ray amorphous opal, zeolites, calcite and clay minerals. The porosity of bergmeals fluctuates from 40 up to 60%. Due to the high content of calcium, the bergmeals from the "Stalnoye" field belong to limy bergmeals which are also characterized by the high content of zeolites and other natural minerals.
- Bokashi is a product of fermentation of waste of processing of grain crops: wheat, rice, and buckwheat. Per 10 kg of waste, we use 0.4 l of the microbiological preparation EM Konkur (the culture of bacteria *Lactobacillus plantarum*, *Lactobacillus casei*, *Streptococus lactis*, *Saccharomices cerevisiae* in special environments) and 0.4 l of molasses. Bokashi from a waste of processing of buckwheat was enriched with a fish meal for an increase in the content of phosphorus and other macro elements. The components mixed with water added, in the quantity which allowed the mixture to be molded in a ball quickly, but also easily break if it was in hands. The mixture is slightly condensed, filled up in densely tied plastic bags which are placed in the dark, warm place for 15–30 days. The mixture has to become covered by white bloom, which confirms the correct course of the microbiological process. Bokashi which are ready to use, have uniform consistency and a sweet-sour smell of kvass. For long storage of a bokashi it is dried up to the air and dry state, then, it is crushed to the size of no more than 1 cm.

3.2 Measurement Activity Concentration of ^{137}Cs in Soil and Plants and Assessing Its Distribution by Physical-Chemical Forms

Samples of the soil were dried up, sifted on a sieve with a cell of 1 mm. The studied plants were cut off and weighed. The phytomass was dried up to the air-dry state and grounded in the blender to a powdery state. The soil samples were placed in Marinelli's vessels with volume 1 L, and the plant samples were placed in vessels 'Denta' with volume 0.1 L. The filtrates, received during separation of physical-chemical forms of ^{137}Cs, were brought in Marinelli's vessels with volume 1 L.

Measurement of activity concentration of ^{137}Cs in samples was carried out according to the accepted manual with the use of a gamma-spectrometer Canberra-Packard (USA) with the coaxial semiconductor detector Ge(Li) with the expanded energetic range. The relative efficiency of registration for the energy of 1.33 MEV was 22.4%. The relative error of measurement of ^{137}Cs activity concentration in samples was in the range from 5 to 10% depending on the activity of a sample. The minimum detected specific activity of ^{137}Cs was 3 Bq in case of the duration of measurement equal to 1 hour in vessel 'denta' with volume 0.1 L.

Separation of physical-chemical forms of ^{137}Cs was carried out by the modified technique (Ure and Davidson 2002). The soil samples for separation were taken so that activity concentration of ^{137}Cs in it was equal to 300–800 Bq. We estimated a distribution of the radionuclide in the following forms:

A – Exchange and soluble.
B – Reduced, which is mainly connected with oxides of iron and manganese.
C – Oxidized, which is mostly connected with organic substance of the soil.
D – Not extracted rest.

Microsoft Excel and StatSoft Statistica 6 packages were used for processing obtained data. The main statistical characteristics were: mean, standard deviation, a confidence interval for mean, t-test and F-test. We consider difference statistically significant if a value of p was not less than 0.05.

3.3 Morpho-biological Features of Mangold and Ecological Conditions of Its Cultivation

Mangold is a biennial vegetable plant of the Orach family (Atriplexceae). It appeared as a result of long-term selection from the plant growing wild on the coast of the Mediterranean Sea under the name of sea beetroot. Mangold is a close relative to ordinary beetroot and is usually cultivated as a one-year crop. Mangold belongs to plants with an average consumption of nutrients; it prefers the soils, rich with organic substances with a deep cultural layer. It likes additives, friable, well fertilized with organic chemistry soils on open sunny sites.

Mangold prefers soils with the reaction close to neutral pH 6.5–7.0. The average yield from 1 running meter is around 1 kg. The optimum temperature for seeds germination is +25 °C, and for growth and development of plants– +15 to +23 °C. The plant is rather drought-resistant. Mangold needs plentiful watering when seeds sprout and when crowns of leaves are formed. In the rest of the time, it is a drought-resistant culture which loves light and heat. The appearance of plants in various options of the experiment on the 65th day after planting is presented in Fig. 1.

Fig. 1 Mangold on the 65th day of the experiment. The experiment options: control (*1*), biochar 3% + bokashi 1% (*2*), biochar 3% + bokashi 3% (*3*), biochar 1% + bokashi 1% (*4*), biochar 1% + bokashi 3% (*5*), 1% biochar (*6*); 3% biochar (*7*); bokashi 1% (*8*); bokashi 3% (*9*); bokashi (rice) 3% (*10*); bokashi (buckwheat) 3% (*11*); bokashi 3% + bergmeal 0.1% (*12*); bokashi 3% + bergmeal 0.4% (*13*); biochar 1% + bokashi 3% + bergmeal 0.1% (*14*); biochar 1% + bokashi 3% + bergmeal 0.4% (*15*); bergmeal 0.1% (*16*); bergmeal 0.4% (*17*)

4 Influence of Biochar and Its Composition with Soil Ameliorants on Growth of Plants and Accumulation of ^{137}Cs in Their Biomass

4.1 *Influence of Biochar on Agrochemical Indicators of Soil*

Agrochemical characteristics of soil mixtures are follows: pH_{KCL} 6.1–6.5 (in control-6.2); P_2O_5–1100–3000 mg·kg^{-1} of soil (in control-1320 mg·kg^{-1}); Ca^{++}-37.5–50.3 mmol·(100 g)$^{-1}$ of soil (in control-7600 mmol·(100 g)$^{-1}$; Mg^{++} 1000–1400 mmol·(100 g)$^{-1}$ of soil (in control- 1000 mmol·(100 g)$^{-1}$); ash-content 40–55% (in control 40%); hydrolytic acidity (Ha)- 8–15 mmol·(100 g)$^{-1}$ of soil (in control- 11 mmol·(100 g)$^{-1}$); the amount of absorbed bases (S)- 84–92 mmol·(100 g)$^{-1}$ of soil (in control- 90 mmol·(100 g)$^{-1}$); absorption capacity (T)-95–105 mmol·(100 g)$^{-1}$ of soil (in control- 100 mmol·(100 g)$^{-1}$); the degree of base saturation (V)- 81–92% (in control- 90%). Data of the agrochemical analysis indicate that application of the studied additives into soil mixtures led to the increase in exchange potassium and absorption capacity of soil mixtures (Table 1), which has to result in the decrease in a transfer of ^{137}Cs to the soluble form.

The durability of fixing of radionuclides in soil depends on the reaction of the environment, a capacity of absorption of soil, a structure of exchange cations, and the content of organic substances. Absorption of cesium in plants from soils with acid reaction comes easier than from neutral ones. Therefore, the agrochemical techniques, influencing these indicators, change the state of the radionuclide in the soil. The acidity of the soil mixture varies from 6.1 to 6.5 in the experiment options. The decrease in acidity of the soil to 6.5 was found in the following options which are written down in the decreasing order: 3% biochar– 1% biochar– 3% a bokashi–0.1% bergmeal– 0.4% bergmeal. Application of the corresponding additives into the soil neutralizing its acidity and creating agrochemical barriers has to reduce intake the radionuclide in plants. Taking into account the received indicators of the agrochemical analysis of peat soil mixture it should be noted that application of 3% of biochar separately or in combination with bokashi from waste of processing wheat leads to decrease in acidity of the soil from 6.2 to 6.5, it, in turn, promotes the increase in green material of mangold by 7–18% in respect to control. The acidity degree shift towards the neutral, with the growth of concentration of biochar in the peat soil in the experiment with mangold.

Low content of water-soluble potassium in soil increases root uptake of Cs. Therefore, increase of its concentration decrease accumulation of radioisotopes of cesium sharply in the situation of its concentration below 250 μM. In the condition of higher concentration of water-soluble K in soil its impact on the accumulation of Cs is much less. This phenomenon associated with two primary mechanisms of potassium uptake by roots: carriers and channels. Channels operate in the condition of high concentration of K in soil and transport it according to the electrochemical potential. K-channel has a pronounced selectivity K > Rb > N > Cs. They can even be blocked by Cs. K-transporters (for example, HKT1) transfer the ion in the root

Table 1 Agrochemical indexes of the soil mixtures

Option of the experiment	pH_{KCl}	Ca, mmol·(100 g)$^{-1}$	Mg, mmol·(100 g)$^{-1}$	P_2O_5, mg·g^{-1}	K_2O water-soluble, mg·kg^{-1}	S, mmol·(100 g)$^{-1}$	Ha, mmol·(100 g)$^{-1}$	T, mmol·(100 g)$^{-1}$	V, %
C	6.20	42.20	8.75	1324	66.1	90.60	12.83	103.40	87.63
BC3% + B1%	6.41	38.15	8.54	1431	90.4	91.80	10.30	102.10	89.95
BC3% + B3%	6.24	37.55	9.13	2004	110.8	90.35	12.80	103.15	87.60
BC1% + B1%	6.33	40.95	8.20	1558	74.3	91.40	11.80	103.20	88.58
BC1% + B3%	6.30	38.15	8.16	1974	86.5	90.35	11.60	101.95	88.68
BC1%	6.36	41.65	6.27	1338	70.2	91.45	10.70	101.90	89.53
BC3%	6.48	41.28	7.79	1160	93.5	92.55	10.30	102.85	90.03
B1%	6.29	43.30	8.10	1555	68.2	91.70	12.50	104.20	88.08
B3%	6.19	37.68	8.86	2124	79.4	88.00	13.10	101.10	87.03
Br3%	6.08	37.53	12.14	3014	84.1	85.20	14.30	99.50	85.68
Bg3%	6.18	41.20	8.13	1609	88.6	88.15	13.80	101.95	86.45
B3% + T0.1%	6.12	37.83	8.15	2202	114.3	86.80	15.05	101.85	85.33
B3% + T0.4%	6.22	42.35	10.08	2337	121.6	88.70	13.10	101.80	87.15
BC1% + B3% + T0.1%	6.27	40.88	8.80	2110	134.8	87.80	12.90	100.70	87.23
BC1% + B3% + T0.4%	6.32	44.10	8.91	2254	141.3	89.40	13.60	103.00	86.80
T0.1%	6.30	46.10	7.13	1060	92.4	92.20	12.90	105.10	87.73
T0.4%	6.32	50.33	8.39	1200	123.7	91.90	12.70	104.60	87.90

against the electrochemical gradient and activated at low concentration of water-soluble potassium in the soil. The HKT1 has high affinity to the Cs ion and transports Cs^+ in the root much efficiently then K-channels (Zhu 2000).

Any tested soil additives in the experiment increase a concentration of water-soluble potassium in the soil. But the concentration of water-soluble potassium in control is quite high, so the effect of increasing K level in soil can't be high. In some cases, it can be masked by other factors. Biochar in dose 3% increase concentration of water-soluble potassium around 30%, but the highest effect on this index has complex additive which consists of biochar, bokashi, and bergmeal in concentration 0.4%.

Soils with a bigger absorption capacity are characterized by smaller intake of radionuclides in plants. The maximum capacity of absorption was found when introducing bergmeal into soil mixture in the concentration of 0.1 and 0.4% of the mass of the soil mixture. The smallest value of the capacity of absorption of the soil was observed when introducing 3% of bokashi into the soil mixture.

From soils with a high degree of saturation with exchange cations, cesium comes less to plants. The maximum value in exchange calcium is observed in the decreasing order: 0.4% bergmeal–0.1% bergmeal–biochar 1% + bokashi 3% + bergmeal 0.4%.

4.2 *Influence of Biochar on Biological Productivity of Mangold*

Application of biochar to the peat soil in the concentration of 1 and 3% of the weight of the soil mixture leads to a various impact on the biological productivity of mangold. Data on the elevated biomass of plants are presented in Table 2.

Table 2 Biological productivity of the elevated mass of mangold during application of biochar into the peat soil

Option	Number of surviving plants	Height of plants, cm	Weight of aboveground organs (dry), g
Control	25.00 ± 6.9	14.25 ± 2.24	1.67 ± 0.29
Biochar 1%	23.00 ± 4.1	11.47 ± 0.71	1.47 ± 0.05
Biochar 1% + Bokashi 1%	19.50 ± 1.8	12.22 ± 0.26	1.50 ± 0.14
Biochar 1% + Bokashi 3%	18.75 ± 3.7	10.93 ± 1.26	0.83 ± 0.27
Biochar 1% + Bokashi 3% + Bergmeal 0.1%	20.00 ± 3.1	12.50 ± 1.22	1.47 ± 0.30
Biochar 1% + Bokashi 3% + Bergmeal 0.4%	17.75 ± 5.8	12.82 ± 1.67	1.51 ± 0.63
Biochar 3%	25.05 ± 2.3	11.80 ± 0.94	1.91 ± 0.29
Biochar 3% + Bokashi 1%	30.75 ± 3.3	11.39 ± 1.32	1.89 ± 0.39
Biochar 3% + Bokashi 3%	25.25 ± 3.7	10.98 ± 1.49	1.67 ± 0.34

The analysis of the data has shown that application of 1% biochar into the peat soil causes decrease in growth and development of plants by 12% regarding control. Increase in concentration of biochar in the peat soil up to 3% (from the mass) leads to increase in biological productivity of mangold by 15% concerning control values.

Application of 1% of biochar in combination with 1% of bokashi from waste of processing of wheat into the peat soil; 1% of biochar +3% of bokashi +0.1% of bergmeal; 1% of biochar +3% of bokashi +0.4% of bergmeal slightly reduces indicators of biomass of plants with reference to control. However, when introducing 1% of biochar in combination with 3% of bokashi from a waste of processing of wheat into the peat soil, there is oppression of growth and development of mangold. Biomass of plants in this option was twice lower than on control.

The use of biochar in the concentration of 3% of the mass of the soil mixture in combination with bokashi from processing waste of wheat of 1 and 3% have shown a different impact on the biological productivity of mangold. While adding to the peat soil of 3% of biochar in combination with 1% of bokashi from waste of processing of wheat, there is an insignificant increase in biomass of plants. The additive which consists of 3% of biochar and 3% of bokashi from a waste of processing of wheat didn't significantly change the biological productivity in comparison with control.

Application of bokashi from a waste of processing of wheat into the peat soil was carried out in the concentration of 1% and 3%. The analysis of the data is presented in Table 3 and shows that application of 1% of bokashi from a waste of processing of wheat into the peat soil lowers biomass of plants by 40% in comparison with control values. The same effect is observed when introducing bokashi into the peat soil in the concentration of 3%.

Application of mixture which consists of 3% of bokashi in combination with 0.1% of bergmeal or 0.4% of bergmeal inhibits growth of plants. Therefore biomass of plants decreases significantly; it up to 60% lowers in comparison with control.

After application of 3% of bokashi from a waste of processing of buckwheat into the soil the elevated mass of plants decreases twice concerning control. The biomass gain for 10% of control values is noted in the option with introducing of 3% of bokashi from a waste of processing of rice into the peat soil.

Table 3 Biological productivity of the aboveground biomass of mangold when introducing bokashi into the peat soil

Option	Number of surviving plants	Height of plants, in cm	Weight of aboveground organs (dry), in g
Control	25.00 ± 6.9	14.25 ± 2.24	1.67 ± 0.29
Bokashi 1%	14.25 ± 5.3	11.12 ± 1.19	1.02 ± 0.49
Bokashi 3%	17.25 ± 3.3	10.95 ± 0.29	1.11 ± 0.19
Bokashi (rice) 3%	28.25 ± 5.5	11.12 ± 1.39	1.82 ± 0.27
Bokashi (buckwheat) 3%	11.50 ± 3.4	12.91 ± 1.98	0.88 ± 0.43
Bokashi 3% + Bergmeal 0.1%	12.25 ± 4.4	11.23 ± 0.45	0.73 ± 0.16
Bokashi 3% + Bergmeal 0.4%	19.00 ± 3.5	13.37 ± 0.79	1.48 ± 0.20

Table 4 Biological productivity of mangold after application of bergmeal into the organic soil

Option	Number of surviving plants	Height of plants, in cm	Weight of aboveground organs (dry), in g
Control	25.00 ± 6.90	14.25 ± 2.24	1.67 ± 0.29
Bergmeal 0.1%	33.25 ± 2.30	13.15 ± 1.16	2.13 ± 0.26
Bergmeal 0.4%	27.25 ± 4.20	12.65 ± 1.20	1.87 ± 0.25

Thus, application of 3% of bokashi from a waste of processing of wheat into the peat soil separately and together with bergmeal leads to a reliable decrease in biological productivity of mangold by 11–56% of control values. The most reliable oppression of growth and development of plants in the organic soil are observed when using 3% of bokashi together with 0.1% of bergmeal.

Application of bokashi from a waste of processing of wheat and buckwheat into the soil in a concentration of 3% of the mass of the soil mixture leads to oppression of growth and development of mangold for 33 and 47% respectively concerning control. The maximum decrease in biomass of plants is observed after application of 3% of bokashi from waste of processing of buckwheat in the peat soil mixture. Application of 3% of bokashi from a waste of processing of rice into the organic soil promotes a reliable increase in biomass of mangold by 10% of control values.

Bergmeal was applied in the soil in the concentration of 0.1 and 0.4%. In all options with the application of bergmeal into the organic soil, the increase in biomass of plants (Table 4) is noted. The maximum increase in biomass of plants is noted when introducing the minimum concentration of bergmeal of 0.1% into the soil. Addition of bergmeal to the peat soil in the concentration of 0.1% has raised this indicator for 30%, and 0.4% of bergmeal– for 20%.

From the obtained data it is possible to draw a conclusion that application of bergmeal into the peat soil in the concentration of 0.1–0.4% favorably influences the growth of plants and, as a result, there is a biomass gain.

The analysis of the data has shown that the maximum effect is observed in the options with application into the peat soil bergmeal in concentration of 0.1%. In this case biomass of aboveground parts of the plants was 2.13 ± 0.26 g, which is 1.5 times higher, than in control. The minimum values of a biomass of aboveground organs of the plants were observed in the option with the application into the soil the mixture of bokashi (3%) + bergmeal (0.1%) – 0.73 ± 0.16 g.

Insignificant increase in aboveground biomass of plants in comparison to control was observed when there was application of the following additives into the peat soil mixture: 3% of biochar; mixture of 3% of biochar +1% of bokashi; 3% of bokashi from waste of processing of rice; 0.1% of bergmeal and 0.4% of bergmeal.

4.3 *Influence of Biochar on Accumulation of ^{137}Cs in Plants on Peat Soil*

The analysis of data has shown that application of 1% of biochar into the peat soil significantly reduces accumulation of ^{137}Cs in plants (Table 5). The effect of addition biochar in the concentration of 3% to the soil was much weaker (Fig. 2). ANOVA analysis has shown the reliability of these dependencies at significance value of $p < 0.05$.

Addition to the peat soil of 1% of biochar together with bokashi from a waste of processing of wheat in the concentration of 1% and 3% leads to increase in accumulation of ^{137}Cs in plants (Fig. 3). Moreover, the maximum values are observed during application of 1% of biochar into the soil together with 3% of bokashi from a waste of processing of wheat. After application of 1% of biochar into the soil in combination with bokashi and bergmeal an insignificant increase in accumulation of ^{137}Cs in plants is observed. As it follows from the obtained data application of the three-component additive on the basis of 1% of biochar in combination with bokashi and bergmeal into the organic soil causes an insignificant increase in accumulation of ^{137}Cs in plants.

After application of two-component additives on the basis of 3% of biochar in combination with bokashi from a waste of processing of wheat in a concentration of 1 and 3% into the peat soil, there is a doubtful increase in accumulation of ^{137}Cs in plants. Therefore, we conclude that in the maximum degree the complex additive consisting of a mixture of 1% of biochar and 3% of bokashi increased the accumulation coefficient of ^{137}Cs in plants. The minimum values of coefficients of accumulation of ^{137}Cs in plants are noted for plants in the experiment option with the application of 1% of biochar into the peat soil.

Application of 3% of bokashi without biochar into the soil mixture slightly increases the transfer of ^{137}Cs in plants (Table 6), and 1% of bokashi increases

Table 5 Influence of biochar on accumulation of ^{137}Cs in aboveground organs of plants

Option	Activity concentration of ^{137}Cs in soil, 10^3·Bq·kg^{-1}	Activity concentration of ^{137}Cs in plants, 10^3·Bq·kg^{-1}
Control	37.85 ± 0.99	8.98 ± 0.36
Biochar 1%	33.47 ± 0.58	6.02 ± 0.43
Biochar 1% + Bokashi 1%	34.75 ± 0.74	7.89 ± 0.46
Biochar 1% + Bokashi 3%	32.62 ± 0.82	9.75 ± 0.46
Biochar 1% + Bokashi 3% + Bergmeal 0.1%	39.19 ± 0.63	7.55 ± 0.49
Biochar 1% + Bokashi 3% + Bergmeal 0.4%	39.76 ± 0.67	7.98 ± 0.47
Biochar 3%	33.58 ± 0.80	7.28 ± 0.31
Biochar 3% + Bokashi 1%	29.45 ± 0.78	7.38 ± 0.32
Biochar 3% + Bokashi 3%	29.66 ± 0.74	7.26 ± 0.28

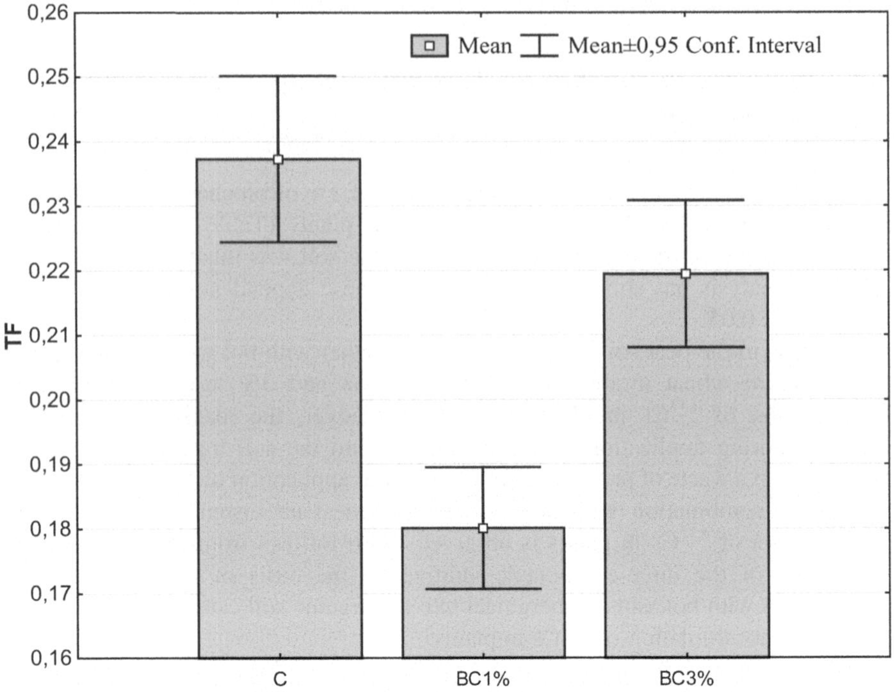

Fig. 2 Influence of biochar on transfer factor of ^{137}Cs into plants

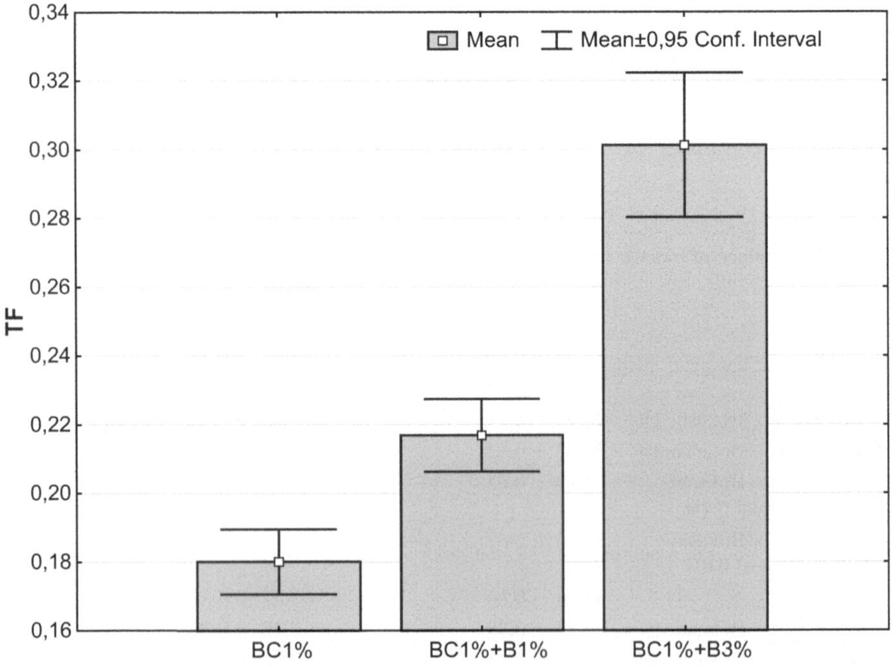

Fig. 3 Influence of biochar and its mixtures with bokashi on transfer factor of ^{137}Cs into plants

Table 6 Influence of bokashi on accumulation of ^{137}Cs in the aboveground parts of plants

Option	Activity concentration of ^{137}Cs in soil, $10^3 \cdot Bq \cdot kg^{-1}$	Activity concentration of ^{137}Cs in plants, $10^3 \cdot Bq \cdot kg^{-1}$
Control	37.85 ± 0.99	8.98 ± 0.36
Bokashi 1%	36.37 ± 0.72	11.18 ± 0.61
Bokashi 3%	39.84 ± 0.70	9.88 ± 0.60
Bokashi (rice) 3%	41.83 ± 0.58	8.49 ± 0.43
Bokashi (buckwheat) 3%	42.59 ± 0.69	16.23 ± 0.50
Bokashi 3% + Bergmeal 0.1%	42.36 ± 0.67	9.21 ± 0.37
Bokashi 3% + Bergmeal 0.4%	34.37 ± 0.67	5.34 ± 0.37

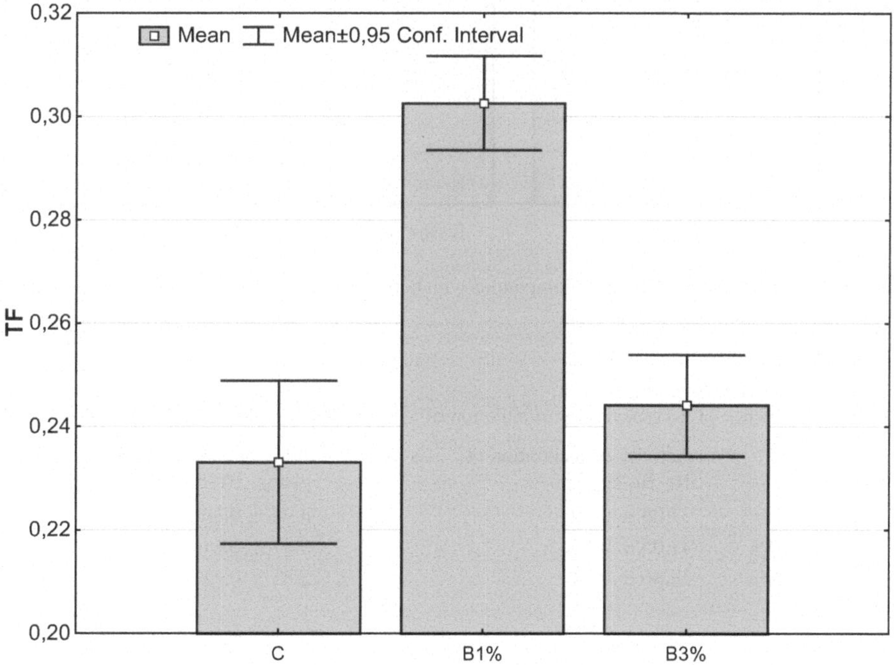

Fig. 4 Influence of the isolated application of bokashi on transfer factor of ^{137}Cs into plants

this index by 30% (Fig. 4). The carried-out statistical analysis has revealed the reliable dependence of these index at significance value of $p < 0.05$.

The analysis of the influence of application of 3% of bokashi from a waste of processing of wheat, rice, and buckwheat into the soil on the transfer of ^{137}Cs in plants has shown the highest coefficients of accumulation in plants in option with bokashi from buckwheat; the transfer factor exceeded control almost twice. This index slightly exceeded control in option with application bokashi produced from rice. However, these regularities are statistically doubtful.

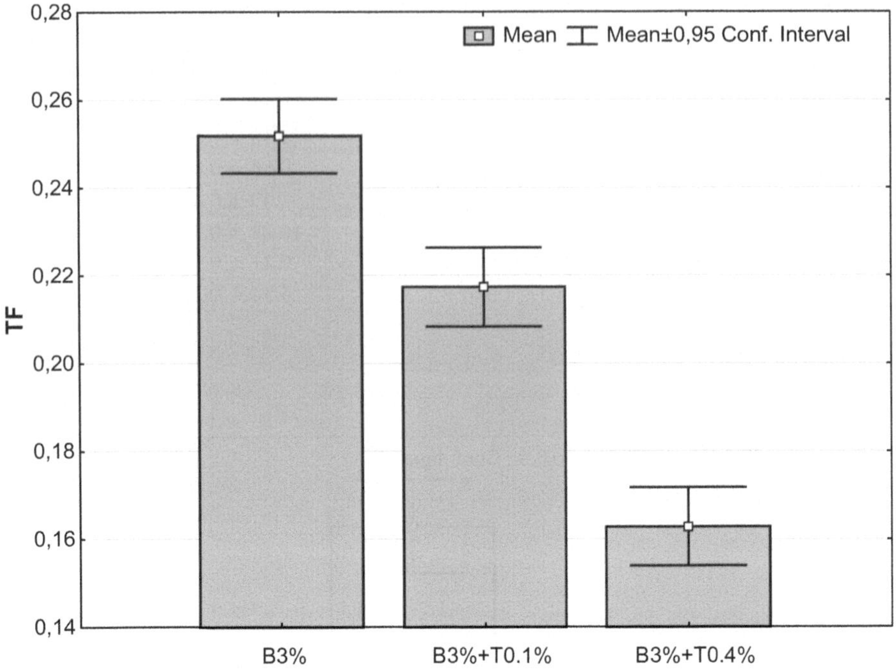

Fig. 5 Influence of bokashi and its compounds with bergmeal on transfer factor of ^{137}Cs into plants

Table 7 Influence of bergmeal on accumulation of ^{137}Cs in aboveground organs of plants

Option	Activity concentration of ^{137}Cs in soil, $10^3 \cdot Bq \cdot kg^{-1}$	Activity concentration of ^{137}Cs in plants, $10^3 \cdot Bq \cdot kg^{-1}$
Control	37.85 ± 0.99	8.98 ± 0.36
Bergmeal 0.1%	41.05 ± 0.64	5.87 ± 0.41
Bergmeal 0.4%	38.86 ± 0.64	5.83 ± 0.41

Transfer factor for ^{137}Cs into plants shows the reliable decrease at significance value of $p < 0.05$ after application of 3% of bokashi from the processing of waste of wheat with the addition of 0.1% and 0.4% of bergmeal into the soil (Fig. 5).

Application into the soil of bergmeal in concentration of 0.1% and 0.4% of the mass of the soil mixture reduces accumulation of ^{137}Cs in plants by 1.3 times concerning control values (Table 7). The carried-out statistical analysis of this dependence shows (Fig. 6) that the difference between the averages is statistically significant ($p < 0.05$).

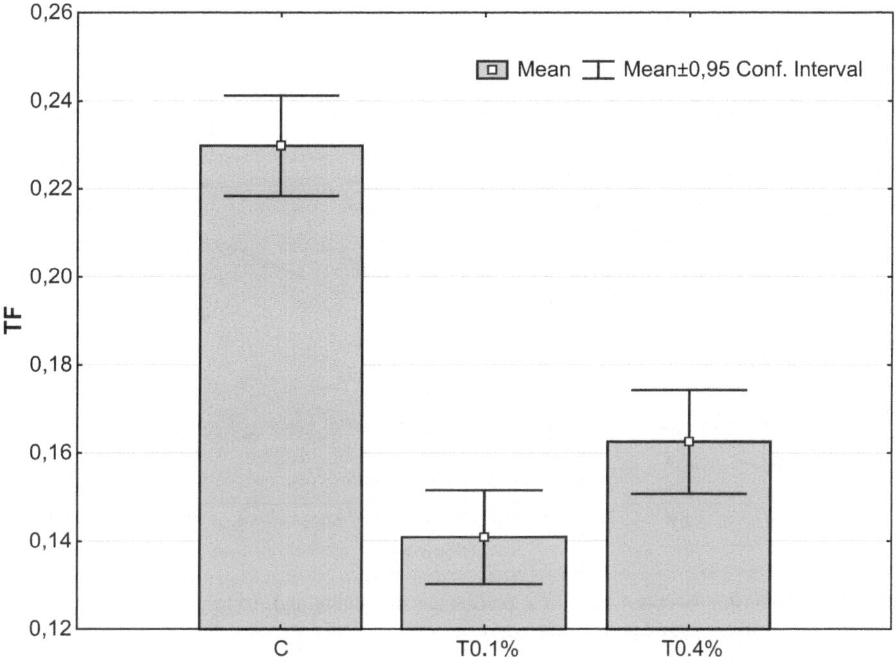

Fig. 6 Influence of application into the soil of bergmeal in concentration of 0.1 and 0.4% on transfer of ^{137}Cs in plants

So, analysis of the data shows that application into the soil with the high content of organic substance of 1% of biochar as well as a three-component additive on the basis of 1% of biochar in combination with 3% of bokashi and 0.1% of bergmeal reduces accumulation of ^{137}Cs in plants more than by 10%. Use of two-component and three-component additives on the basis of biochar in combination with bokashi and bergmeal in organic soils in other options of the experiment hasn't given the expected effect. Application of bergmeal into peat soil led to the decreasing in the accumulation of ^{137}Cs in plants.

4.4 *Influence of Biochar on Distribution of ^{137}Cs in Physical and Chemical Forms and Its Availability to Plants*

Processes of radionuclides interaction with soil mainly determine their migration in epy vertical profile, as well as their biological availability to plants. A wide variety of soils is the reason for significant differences in behavior of radionuclides in soils and their accumulation in plants. At cultivation of crops in different soil types at the same level of radioactive pollution, the intake of radionuclides can differ tens of times.

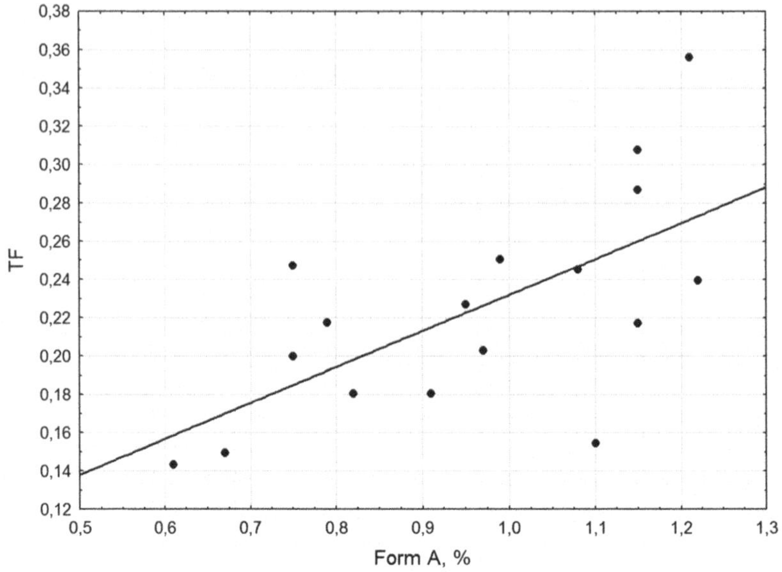

Fig. 7 Relationship between the ^{137}Cs percentage in soluble and exchange forms (A) and its transfer to the aboveground parts of mangold

The biological mobility of radionuclides mainly depends on their stock in soil in a soluble and exchange form. It can be assessed by extraction of the radionuclides from the soil with solutions of neutral salts. Confirmation of this fact is the linear relationship between the percentage of ^{137}Cs in the soluble and exchange forms and the transfer factor in aboveground parts of mangold, observed in the experiment (Fig. 7).

The analysis of the influence of various compounds of biochar from the wood of hard-wooded broadleaved breeds and products of enzymatic microbiological processing of organic waste (bokashi) and mineral sorbent (bergmeal) on redistribution of cesium-137 in the soil by physical-chemical forms with various biological availability, and transfer of radionuclides to biomass of plants is made. Data on the distribution of ^{137}Cs in the peat soil in physical-chemical forms are presented in Table 8.

On the 10th days after application into the soil of biochar and bokashi the increase the fraction of soluble ^{137}Cs is observed. Addition to the soil of bergmeal in the doses of 0.1 and 0.4% of the mass hasn't exerted a positive impact on the change of the fraction of soluble ^{137}Cs.

Application into the soil of 3% of biochar or 3% of bokashi leads to the increase the fraction of soluble ^{137}Cs (form A) concerning control for 135 and 152% respectively. The maximum increase in the percentage of soluble ^{137}Cs (for 253% concerning control) is observed at a joint application of 3% of biochar and 1% of bokashi into the soil.

Table 8 Distribution of ^{137}Cs in the soil between physical-chemical forms

Option	A before planting, %	A after harvesting of plants, %	Rate of form A decreasing $Bq{\cdot}kg^{-1}{\cdot}day^{-1}$	B, %	C, %	D, %
C	1.01	1.22	−1.9	4.48	90.66	3.64
BC3% + B1%	1.73	1.08	4.1	5.61	84.92	8.39
BC3% + B3%	1.80	0.99	5.3	4.45	71.49	23.08
BC1% + B1%	1.22	0.95	2.0	4.25	67.19	27.60
BC1% + B3%	1.33	1.15	1.1	3.96	72.18	22.72
BC1%	1.47	0.91	4.2	4.40	69.12	25.57
BC3%	1.36	0.79	3.9	4.20	69.41	25.60
B1%	1.34	1.15	1.4	4.40	77.47	16.99
B3%	1.53	0.75	7.0	3.47	64.29	31.49
Br3%	1.21	0.97	2.1	4.73	94.40	0.00
Bg3%	1.24	1.21	0.2	4.27	71.15	23.37
B3% + T0.1%	1.29	1.15	1.1	4.77	70.67	23.41
B3% + T0.4%	1.71	1.10	4.9	3.27	57.28	38.35
BC1% + B3% + T0.1%	1.30	0.82	3.9	3.34	77.20	18.64
BC1% + B3% + T0.4%	1.21	0.75	3.5	3.03	62.24	33.98
T0.1%	1.07	0.61	3.8	2.88	61.54	34.97
T0.4%	1.01	0.67	2.8	2.70	71.72	24.91

On the 65th day after application of additives into the soil, the extraction of the physical-chemical forms of ^{137}Cs in the earlier described technique was carried out. The obtained data before and after the experiment were compared with the results of control. After 65 days of carrying out the experiment the decrease in the percentage of soluble ^{137}Cs in the soil is observed. Application of 3% of biochar into the soil reduces the percentage of soluble ^{137}Cs by 1.7 times; application of 0.1% of bergmeal into the soil – by 1.8 times, and application of 3% of bokashi into the soil reduces the fraction of soluble ^{137}Cs twice in comparison with the value before the experiment. The greatest decrease the fraction of soluble ^{137}Cs by 2.4 times in comparison with the value before the experiment is observed at a joint application of 3% of biochar and 1% of bokashi into the soil. At the same time in control option, the increase the fraction of the soluble ^{137}Cs form is observed, which is caused by a set of processes of aerobic mineralization of the organic substance.

During application of the studied additives into the soil, the change of a stock of the available ^{137}Cs form is observed. A rather high rate of speed of transfer of ^{137}Cs into an insoluble form (5.3 $Bq{\cdot}kg^{-1}{\cdot}day^{-1}$) is observed at a joint application of 3% of biochar and 3% of bokashi into the soil. The maximum value of the decrease in the stock of the available ^{137}Cs form (7.04 $Bq{\cdot}kg^{-1}{\cdot}day^{-1}$) appears after application of 3% of bokashi into the soil.

At high speed of change of the available ^{137}Cs form, the application of 3% of bokashi into the soil leads to the increase in accumulation of ^{137}Cs in the biomass of plants. Biochar in a dose of 3% of the weight of soil mixture at high speed of change of the available ^{137}Cs form reduces its accumulation in the plants.

Application of 3% of biochar into the soil together with 1% of bokashi at the maximum speed of change of an available form of cesium reduces accumulation of ^{137}Cs in the biomass of plants.

Analyzing the received results allows us to distinguish additives whose application in the soil decreases the percentage of bioavailable cesium and increases its stock in an unextracted form. Application of 0.1% of bergmeal into the soil (option 16) makes the greatest impact on the decrease the fraction of mobile cesium and the increase in the un-extracted cesium percentage concerning control. Among two- and three-components soil additives the highest effectiveness concerning control has been shown by the mixture consisting of 3% of bokashi and 0.4% of bergmeal (option 13). Also, it should be noted the additive including mix of 1% of biochar, 3% of a bokashi and 0.4% of bergmeal (option 15).

As a result of carrying out the laboratory vegetative experiment on the investigation of influence of complex soil ameliorants on the basis of biochar and products of microbiological processing of organic waste on biological availability of cesium-137 on organic soils it has been pointed out that the main component causing the decrease in transfer of ^{137}Cs from the soil into plants is biochar in a dose of 3% of the weight of soil mixture. The largest rate of the decrease in the bio-available ^{137}Cs form is observed when applying two components into the soil in the ratio of 3% of biochar and 1% of bokashi. Application of biochar in combination with bokashi and bergmeal into the soil reduces the fraction of the soluble ^{137}Cs form and increases the proportion of the inaccessible (un-extracted) cesium form, but it increases the transfer of cesium from the soil into plants by 1.4–1.7 times concerning control. Which would be explained by the use in the experiment of peat (organic) soils, characterized by very high transfer factor for ^{137}Cs into plants. The separate or combined use of organic fertilizers (bokashi) leads to the increase in the percentage of fixed ^{137}Cs, decrease in its biological availability is observed.

5 Conclusions

1. It is shown that application of biochar and some complex ameliorants on its basis into the soil decreases the biological availability of ^{137}Cs and its transfer to the aboveground biomass of plants without a significant effect on their growth and development. But ameliorative properties of biochar against radioisotopes of Cs accumulation by plants are not strongly noticeable on soils with the high content of organic substances (peat).
2. Two mechanisms of impact of biochar on biological availability of cesium are revealed. The first mechanism (fast) is connected with the increase in the concentration of ions of potassium when applying biochar into the soil which causes switching from active transport of the macro element to the passive one through the ion channels having lower affinity to cesium. In soils, which are poor in available potassium, application of biochar can lower the transfer coefficient of ^{137}Cs much more. The second mechanism (slow) is related with sorption of

cesium on biochar. In first days after application of biochar into the soil biological availability of cesium can increase, and then it begins to decrease. In two months after adding of biochar and complex soil-improving additives on its basis into the soil, the content of soluble end exchange forms decreases by 15–25% with reference to control. The biologically active organic fertilizer– bokashi changes the speed of sorption of radionuclide on biochar.

References

Ageets VY (2001) The system of radioecological measures in agrosphere of Belarus. Institute of Radiology, Minsk

Ageets VY, Averin VS, Zuchenko YM, Timofeev SF, Tsygvintsev PN, Avtushko MI, Zhdanovich VP, Podolyak AG, Nenashev RA, Gvozdik AF, Naumchik AV, Tsarenok AA, Mashkov IA, Teshkovskij AV, Timofeev AS, Tsurankov EN, Staraseko EG (2003) Recommendations for the safe living and keeping of personal subsidiary plots in the conditions of radioactive contamination of the territory. Institute of Radiology, Minsk

Azarenko YV (2012) The influence of a new polymer-sorbent on the behavior of cesium-137 in soil. In: Proceedings of the international scientific and practical conference of young scientists, graduate students, undergraduates and students Innovations in technologies of cultivation of agricultural crops. Belarussian State Agricultural Academy, Gorki, pp 3–5

Bakunov NA, Archipov NP (1994) Binding of ^{137}Cs in organic soils with minerals and clays fixing ^{137}Cs. In: Proceedings of the international conference Radioecology of peat soils. St. Petersburg State Agricultural University, Saint Petersburg, pp 3–4

Beesley L, Moreno-Jiménez E, Gomez-Eyles JL, Harris E, Robinson B, Sizmur T (2011) A review of biochars, potential role in the remediation, revegetation and restoration of contaminated soils. Environ Pollut 159:3269–3282

Brennan JK, Bandosz TJ, Thomson KT, Gubbins KE (2001) Water in porous carbons. Collid Surf A Physicochem Eng Aspect 187–188:539–568

Bridle TR, Pritchard D (2004) Energy and nutrient recovery from sewage sludge via pyrolysis. Water Sci Technol 50:169–175

Burdakov VA, Majakov EA, Torubarova AA, Kalinin NF, Gelis VM, Mil'utin VV, Penzin RA (1994) The method for reducing the transition of cesium radionuclides from soil to plants. Russian Federation Patent 2013913

Chan KY, Van Zwieten L, Meszaros I, Downie A, Joseph S (2007) Agronomic values of green waste biochar as a soil amendment. Austral J Soil Res 45:629–634

Cheng CH, Lehmann J, Thies JE, Burton SD (2008) Stability of black carbon in soils across a climatic radient. J Geophy Res 113:20–27

Emmerich FG, Luengo CA (1996) Babassu charcoal: a sulfur less renewable thermo-reducing feedstock for steelmaking. Biomass Bioenergy 10:41–44

Fukuyama K, Kasahara Y, Kasahara N, Oya A, Nishikawa K (2001) Small-angle Xray scattering study of the pore structure of carbon fibers prepared from a polymer blend of phenolic resin and polystyrene. Carbon 39:287–290

Glaser B, Haumaier L, Guggenberger G, Zech W (2001) The Terra Preta phenomenon: a model for sustainable agriculture in the humid tropics. Naturwissenschaften 88:37–41

Hamilton TF, Martinelli RE, Kehl SR, Hayes MHB, Smith IJ, Peters SKG, Tamblin MW, Schmitt CL, Hawk D (2016) A preliminary assessment on the use of biochar as a soil additive for reducing soil-to-plant uptake of cesium isotopes in radioactively contaminated environments. J Radioanalyt Nucl Chem 307:2015–2020

Harris P (1999) On charcoal. Interdiscip Sci Rev 24:301–306

Israel YA, Bogdevich IM (2009) Atlas of modern and predictive aspects of the consequences of the Chernobyl accident in the affected territories of Russia and Belarus. Priroda, Moscow
Kishimoto S, Sugiura G (1980) Introduction to charcoal making on Sunday. Sougou Kagaku Shuppan, Tokyo
Kishimoto S, Sugiura G (1985) Charcoal as a soil conditioner. In: Symposium on forest products research, international achievements for the future, vol 5, pp 12–23
Krugliakov SV, Anisimov VS, Anisimova LN, Aleksahin RM (2008) The specific sorption capacity of soils and mineral sorbents for ^{137}Cs. Pedology 6:693–703
Kunovskij VM, Maister AA, Perepel'atnikova LV (1996) Impact of sapropel on accumulation of cesium-137 by potato tubers, In: Proceedings of II International Scientific Conference 'The problems of agricultural radioecology - ten years after the accident at the Chernobyl nuclear power plant', State Agricultural Academy of Ukraine, Zitomir, pp 24–25
Martínez ML, Torres MM, Guzmán CA, Maestri DM (2006) Preparation and characteristics of activated carbon from olive stones and walnut shells. Indus Crops Prod 23:23–28
Morley J (1927) Following through with grass seeds. Nat Greenkeep 1:15
Parr JF (2006) Effect of fire on phytolith coloration. Geoarch Int J 21:171–185
Perepel'atnikova LV, Prister BC, Omel'anenko NP (1993) Assessment of the effectiveness of the application of sapropels in the Ukrainian Polissye for crop production according to the level of radioactive contamination. Probl Agric Radiol Inst Agric Radiol Kiev 3:139–143
Petersen JB, Neves E, Heckenberger MJ (2001) Gift from the past: Terra Preta and prehistoric Amerindian occupation in Amazonia. In: McEwan C, Barreto C, Neves E (eds) Unknown Amazonia. British Museum Press, London, pp 86–105
Ratnikov AN, Popova GI, Petrov KV, Aleksakhin RM, Zhigareva TL, Vasil'ev AV (1996) Fundamental principles for obtaining agricultural produce of standard purity in those parts of the Russian Federation which were radioactively contaminated as a result of the Chernobyl accident, and the effectiveness of the measures being taken. In: One decade after Chernobyl: summing up the consequences of the accident. IAEA, Vienna, pp 233–236
Retan GA (1915) Charcoal as a means of solving some nursery problems. Forest Quart 13:25–30
Rondon MA, Lehmann J, Ramirez J, Hurtado M (2007) Biological nitrogen fixation by common beans (*Phaseolus vulgaris* L.) increases with bio-char additions. Biol Fert Soil 43:699–708
Santiago A, Santiago L (1989) Charcoal chips as a practical substrate for container horticulture in the humid tropics. Acta Hortic 238:141–147
Schmidt MW, Noack AG (2000) Black carbon in soils and sediments: analysis, distribution, implications, and current challenges. Glob Biogeochem Cycl 14:777–793
Schnitzer MI, Monreal CM, Facey GA, Fransham PB (2007) The conversion of chicken manure to bio oil by fast pyrolysis I. Analyses of chicken manure, bio oils and char by C-13 and H-1 NMR and FTIR spectrophotometry. Environ Sci Health 42:71–77
Shindo H (1991) Elementary composition, humus composition, and decomposition in soil of charred grassland plants. Soil Sci Plant Nut 37:651–657
Swiatkowski A, Pakula B, Biniak S, Walczyk M (2004) Influence of the surface chemistry of modified activated carbon on its electrochemical behaviour in the presence of lead(II) ions. Carbon 42:3057–3069
Trimble WH (1851) On charring wood. Plough, the Loom and the Anvil 3:513–516
Trompowsky PM, Benites VD, Madari BE, Pimenta AS, Hockaday WC, Hatcher PG (2005) Characterization of humic like substances obtained by chemical oxidation of eucalyptus charcoal. Org Geochem 36:1480–1489
Tryon EH (1948) Effect of charcoal on certain physical, chemical, and biological properties of forest soils. Ecol Monogr 18:81–115
Ure AM, Davidson CM (2002) Chemical speciation in the environment, 2nd edn. Wiley, New York
Woods WI, Teixeira W, Lehmann J, Steiner C, Winkler A (2009) Where to from here? A Tribute to Wim. Sombroek, Springer, Terra Preta Nova, pp 473–486
Wornat MJ, Hurt RH, Yang NC, Headley T (1995) Structural and compositional transformations of biomass hars during combustion. Comb Flame 100:131–143

Zhang X, Wang H, He L, Lu K, Sarmah A, Li J, Bolan NS, Pei J, Huang H (2013) Using biochar for remediation of soils contaminated with heavy metals and organic pollutants. Environ Sci Pollut Res 20:8472–8483

Zhu YG (2000) Plant uptake of radiocaesium: a review of mechanisms, regulation and application. J Exp Bot 51:1635–1645

Remediation of Agricultural Lands Contaminated by ^{90}Sr

Natalia I. Sanzharova, Irina V. Geshel, Rena A. Mikailova, Dmitry V. Krylenkin, and Ekaterina V. Gordienko

1 Introduction

Nuclear weapon tests, as well as accidents at nuclear facilities led to major contamination of agricultural lands. During the period of global radioactive fallout in middle latitudes of the Northern hemisphere the fallout density of ^{90}Sr was estimated as approx. $2.1 \cdot 10^4$ Bq·m^{-2} (Vasilenko and Vasilenko 2002). The area of East-Ural radioactive trace contaminated because of the accident at Southern Urals («Mayak" PA 1957) amounted to 106 thousand hectares. Moreover, 54% of total area was agricultural lands, including over 29% of arable lands (Ilyin and Gubanov 2001). The main dose forming radionuclide was ^{90}Sr. Contamination densities of abandoned lands ranged 230–300 kBq·m^{-2}. After the Chernobyl accident approximately 150 thousand km^{-2} of the territories of former USSR were contaminated with ^{137}Cs, which contamination density was over 37 kBq·m^{-2} (Israel et al. 1987). Fallout of ^{90}Sr was essentially high in the near zone of ChNPP and also in some areas of adjacent regions of Belarus and Ukraine. In 6 contaminated regions of Belarus the area of agricultural lands with contamination density of ^{90}Sr over 5.55 kBq·m^{-2} figured up to 375.0 thousand ha, and with contamination density over 111 kBq·m^{-2} – 0.07 thousand ha (Israel and Bogdevich 2009). On the territory of 7 regions of Ukraine areas with ^{90}Sr contamination density over 5.55 kBq·m^{-2} figured up to 307.8 thousand ha and with contamination density over 111 kBq·m^{-2} – 1.4 thousand ha (Nadtochiy 2003).

Under the influence of natural factors ^{90}Sr fallen out on the soil surface is included into migration processes, in particular along food chains, this fact determines its entrance into human diet. Biological hazard of ^{90}Sr is determined by several factors: high fission yield, long half-life (~30 years), intensive mobility in

N. I. Sanzharova (✉) · I. V. Geshel · R. A. Mikailova · D. V. Krylenkin · E. V. Gordienko
Russian Institute of Radiology and Agroecology, Kievskoe shosse, 109 km, Obninsk 249032, Russian Federation

D. K. Gupta, A. Voronina (eds.), *Remediation Measures for Radioactively Contaminated Areas*, https://doi.org/10.1007/978-3-319-73398-2_7

natural environments. Migration rates of this radionuclide depend on the fallout state, soil properties, biological mobility of the radionuclide, weather conditions, cultivation technologies of crops, etc. Fixation and distribution of ^{90}Sr in soil are mostly determined by the behaviour mechanism of its isotopic carrier-stable strontium, as well as by chemical analogue of non-isotopic carrier– stable calcium (Ca^{2+}) (Nishita and Taylor 1964; Klechkovsky et al. 1973).

To limit the entrance of radionuclides into human diet a complex of remediation measures was developed. Mechanical and physicochemical methods of remediation of agricultural lands take an important place in this complex. The purpose of these methods is to decrease the radionuclide mobility in soil and, as follows, decrease of its migration rates in the soil-plant system. Efficiency is usually estimated via the reduction factor i.e. reduction of ^{90}Sr uptake by plants after application of a remediation measure. Due to that fact that remediation measures in agriculture are mainly based on current crop cultivation technologies, they also influence on soil properties and soil fertility raise.

All soil-based remedial techniques, reducing entrance of ^{90}Sr into agricultural crops, are based on the regularities of their interaction with soils, agrochemical indicators, soil texture and mineralogical composition, and specific and varietal peculiarities of plants, conditions of their mineral nutrition and other factors.

2 The Behaviour of ^{90}Sr in Soils as a Basis for the Development of Remediation Methods

Timofeyev-Resovsky related strontium to the 2nd group of radionuclides according to its behaviour type in the soil-plant system (Timofeyev-Resovsky and Titlyanova 1966). Behaviour type is exchangeable, and the main mechanism of its fixation in soil is ion exchange.

The radionuclide can be present in soil either as part of liquid or solid phases. In solutions radionuclide can be in the form of cation, be a part of complex compounds and colloid particles. In solid phase this radionuclide exists in exchange-sorbed state. Three groups of ions of ^{90}Sr were defined: 1- those in soil solution; 2- exchangeable ions on the surface of mineral and organic particles; 3- ions constituent in practically insoluble compounds (Timofeyeva 1960; Pavlotskaya 1981). Major part of ^{90}Sr (60–90%) in soil are in an exchangeable form. The dynamic equilibrium exists between different forms of this radionuclide (Pavlotskaya 1974).

Organic matter of soil makes an influence on behaviour of ^{90}Sr. Sr-90 is bound mainly with low-molecular fulvic acids, this explains its high migration ability and availability to plants. ^{90}Sr can be a component of complex compounds of humic acids with fine minerals (Pavlotskaya 1981). Destruction of organic matter is conductive to increase of ^{90}Sr mobility due to its transfer to exchangeable form (Taylor 1968; Pavlotskaya 1974).

Isomorphous substitution in minerals, containing calcium and magnesium [(calcite and limestone ($CaCO_3$), gypsum ($CaSO_4{\cdot}2H_2O$), dolomite (CaMg $(CO_3)_2$)], has significant importance for ^{90}Sr (Spitsin and Gromov 1959). Ion exchange is the predominant mechanism of ^{90}Sr absorption by solid phase that is why depends on the presence of competing cations in the solution. The following range of ions influencing the sorption of ^{90}Sr by solid phase of soils was established: $Al^{3+} > Fe^{3+} > Ba^{2+} > Ca^{2+} > Mg^{2+} > K^+ > NH_4^+ > Na^+$ (Pavlotskaya 1974; Alexakhin and Korneyev 1992). At sorption of ^{90}Sr ion-exchangeable substitution of Ca or another cation can occur on the surface of crystal lattice, as well as formation of hardly soluble strontium phosphates (Havely and Tzur 1964; Kudelya 1969; Nikolayeva 1970).

The content of mineral part of soils makes a significant influence on the state and fixation of ^{90}Sr in soil adsorption complex. ^{90}Sr is more firmly fixed in soils with large content of silt particles (Yudintseva and Gulyakin 1968). Clay minerals of soils can absorb up to 99% of the radionuclide (Yudintseva and Gulyakin 1968). ^{90}Sr is rather absorbed by such minerals as ascanite, bentonite, vermiculite, phlogopite and gumbrine, than hydromuscovite and hydrogoethite. Minerals of montmorillonite group absorb from 92–99.9% of ^{90}Sr, minerals of kaolinite group absorb 40–68%, glimmers absorb from 71–87% of radionuclide, and hydro-glimmers absorb 80–88%. Minerals of calcite, fieldspar, quartz and gypsum groups absorb from 10–50% of ^{90}Sr (Spitsin and Gromov 1959). 44 and 21% of ^{90}Sr are correspondingly absorbed on calcite at absence and presence of calcium in the solution with pH 8–9.

Ion-exchangeable sorption places for the radionuclide are divided into 3 main types on selectivity rate. Regular Exchange Sites (RES) are located on the surface of soil solid phase; Frayed Edge Sites (FES) are located between layers of crystal lattice in the area of their splayed edges; and High Affinity Sites (HAS) (Bolt et al. 1963; Cremers et al. 1988).

Quantitatively the selectivity of exchange sites for a particular radionuclide (R) is defined by the constant of ion-exchange equilibrium of this radionuclide with one of the competing ions (M), and is called selectivity coefficient (Ks):

$$Ks = \frac{[R]_{ex} \times [M]_w}{[M]_{ex} \times [R]_w} \qquad (1)$$

where $[R]_{ex}$ and $[M]_{ex}$ are the content of exchangeable forms of the radionuclide and competing cation in solid phase; $[R]_w$ and $[M]_w$ are their concentrations in water extract.

Radioactive strontium is featured by simple and almost full exchange on RES of clay particles (Valcke 1992).

Extraction of ^{90}Sr into soil solution increases even in the case if the extraction of Ca is increased because Sr and Ca are contained in the solution in a particular ratio. However, distribution of ^{90}Sr between soil solution and absorbing complex differs from the same distribution of Ca. The ratios $^{90}Sr/Ca$ in soil solution varies from 0.49–0.78 of the ratio of these ions in soil. This is due to the firmer sorption of ^{90}Sr in comparison with Ca (Kokotov et al. 1960). Selectivity coefficient of exchange of

Table 1 Mobility factors of ^{90}Sr in the system soil - soil solution for different soil types

Soil type	Exchangeable form of ^{90}Sr, %	$K_S(^{90}Sr/Ca)$
Sod-podzolic	93 ± 0.07	2.5 ± 0.5
Light-grey forest soil	76 ± 0.09	1.5 ± 0.2
Leached chernozem	73 ± 0.10	2.1 ± 0.3
Podzolic chernozem	82 ± 0.11	1.3 ± 0.1
Usual chernozem	68 ± 0.01	1.8 ± 0.1
Modal chernozem	69 ± 0.02	1.9 ± 0.2
Meadow chernozem	64 ± 0.02	1.4 ± 0.1
Chestnut	76 ± 0.09	1.7 ± 0.5
Lowland peat	79 ± 0.01	1.1 ± 0.3

Adopted from Sysoeva (2004)

cations of strontium and calcium Ks(Sr/Ca) (Eq. 1) for ^{90}Sr varies from 1 to 3 for the majority of soils (Konoplev and Bulgakov 2000).

It was established that depending on soil properties the content of exchangeable form of ^{90}Sr varies from 64–93%; selectivity coefficient changes by 2.3 times (Table 1).

The distribution coefficient Kd, which is defined as the ratio of equilibrium concentration of radionuclide in solid phase to its concentration in the solution (Sheppard 1985) is normally used as an indicator of radionuclide distribution between solid and liquid phases of soils. A simple model based on Kd is based on the hypothesis that radionuclide in solid phase is in equilibrium with the radionuclide in solution, which makes the exchange process possible.

Kd(Sr) can be predicted by the ratio of Ca and Mg in exchange complex in soil solid phase ($mol{\cdot}kg^{-1}$) to the sum of concentrations of Ca and Mg in soil solution ($mmol{\cdot}kg^{-1}$) (Rauret and Firsakova 1996; Hilton and Comans 2001), with multiplication by trace selectivity coefficient of Sr to Ca and Sr to Mg (Kc (Sr/Ca-Mg)). This corresponds to the following equation:

$$\mathrm{Kd(RN)} = \mathrm{Kd(AN)} \times \mathrm{Kc}\left(\frac{\mathrm{RN}}{\mathrm{AN}}\right) \tag{2}$$

where Kd(RN) is calculated by linear increase of Kd(AN) on the coefficient, which is equal to selectivity coefficient RN-AN in sorption sites Kc(RN/AN).

Kd(Sr) can be predicted by the ratio of Ca and Mg in exchange complex in soil solid phase ($mol{\cdot}kg^{-1}$) to the sum of concentrations of Ca and Mg in soil solution ($mmol{\cdot}kg^{-1}$) (Rauret Firsakova 1996; Hilton and Comans 2001), with multiplication by trace selectivity coefficient of Sr to Ca and Sr to Mg (Kc (Sr/Ca-Mg)). This corresponds to the following equation:

$$\mathrm{Kd(Sr)} = \frac{\mathrm{Kc}\left(\frac{\mathrm{Sr}}{\mathrm{Ca}} - \mathrm{Mg}\right) \times (\mathrm{Ca_{exch}} + \mathrm{Mg_{exch}})}{\mathrm{Ca_{ss}} + \mathrm{Mg_{ss}}} \tag{3}$$

As far as Ks (Sr/Ca-Mg) is close to 1 (Valcke 1992), to get the simplified model in most of the cases the similar trace selectivity coefficients of Sr to Ca and Sr to Mg can be assumed. This way, Eq. 4 can be simplified:

$$Kd(Sr) = \frac{Ca_{exch} + Mg_{exch}}{Ca_{ss} + Mg_{ss}} \qquad (4)$$

If there are no data on exchangeable cations, the ratio of cation exchange capacity (CEC, $mol \cdot kg^{-1}$) to the sum of concentrations of Ca and Mg in soil solution can be used. Such approach gives satisfactory results at estimation of Kd(Sr), especially in case of soils with saturated exchange complex.

Analysis of the mechanism of behaviour of ^{90}Sr in soil educed the following peculiarities:

- High mobility of ^{90}Sr due to the exchangeable behaviour type;
- Influence of isotopic carrier - stable strontium, as well as by chemical analogue of non-isotopic carrier - stable calcium, as well as other competing ions;
- Influence of organic matter of soils;
- High sorption capacity of clay minerals for ^{90}Sr;
- Formation of insoluble compounds, in particular, strontium phosphates.

The educed mechanisms of behaviour of ^{90}Sr in soils became the basis of development of soil-based remediation methods.

3 Mechanical Methods of Remediation of Agricultural Lands Contaminated with ^{90}Sr

Development of methods of remediation of agricultural lands contaminated with ^{90}Sr was started after the accident at Southern Urals («Mayak" PA) (Klechkovsky 1973). This experience was used and further developed after the Chernobyl accident (The guide 1988; Alexakhin 1991; Guidelines for agricultural 1994).

Due to the large number of investigations, the effectiveness of various different remedial options was estimated, as well as the factors influencing this process. The purpose of application of remedial methods was to decrease the radionuclide mobility in soil and, as sequence, to decrease of its transfer to plants. Soil-based remedial options can be divided into two groups: techniques connected with tillage operations and techniques, which are based on change of physicochemical properties of soils.

Effectiveness of different techniques depends on many factors, one of which is time passed after the accident. During the first period after radioactive fallout the majority of radionuclides lies on the surface (or in the upper layer) of soils. To decrease the pollution direct decontamination, related with the removal of upper contaminated soil layer, can be applied. After the Kyshtym accident («Mayak" PA) the two main methods of decontamination of arable soils were developed (Klechkovsky 1973; Ternovsky and Teverovsky 1984; Klechkovsky 1973; Romanov 1993).

Table 2 Reduction factors for different types of mechanical methods of remediation

Remediation method	Depth of treatment, cm	Reduce factor
Soil removal	5	50–90
Ordinary ploughing (first application)	20–25	2.0 (1.5–3.0)
Skim and burial ploughing	50–70	2.0 (max 7.0)
Rotary cultivation, disking (first application)	10–12	1.5
Rotary cultivation, disking+ ordinary ploughing	20–25	3.0–4.0

Adopted from Alexakhin (1991) and Ilyin and Gubanov (2001)

The first option includes decontamination by removal of the topsoil. Its application decreased transfer of ^{90}Sr to crops by 50–90 times. The method was recommended for application at small area sites, in particular, at cultivation of vegetable crops. After the Kyshtym accident, the contaminated soil was removed only from the area of 50 ha at experimental agricultural enterprise.

The second option included burial of upper (most contaminated) layer of the soil in the subsoil horizons. It was developed for application at large areas of arable lands (Teplyakov et al. 1997). Special-purpose equipment was developed – modernized ploughs and plough-shifters of soil horizons. The modernized plough buried the contaminated soil layer to a depth of 30–40 cm (reducing ^{90}Sr concentration in arable layer by 80%) and the plough-shifters of soil horizons to a depth of 30–70 cm (reducing by 10–50 fold the ^{90}Sr content in arable layer) (Ilyin and Gubanov 2004) (Table 2). Application of such decontamination method in-situ ensured the decrease of transfer of ^{90}Sr to the yield by 2 times in average (7 times maximum). In 1960–1961 deep ploughing was implemented in the area of approximately 6550 ha.

Disadvantages and limitations of application of decontamination methods should also be mentioned. Thus, in case of the topsoil removal, a large amount of radioactive materials may be generated and their burial may cause substantial problems because of the high cost. The difficulties in solving this problem were faced with remediation of contaminated lands in Japan after the Fukushima Dai-ichi accident. The burial of contaminated layer can be effective for heavy in aggregate-size distribution soil with high thickness of humus-accumulated horizon. For soils with thickness of humus-accumulated horizon less than 20 cm the displacement of fertile upper layer to deeper level leads to the loss of fertility. The validation of this option after the Chernobyl accident showed that it was inappropriate for light-textured sandy soils and sandy loams. The cost of both these methods is high, for this reason statement of reasons of their application is required.

In agricultural practice application of remediation measures depends also on types of agricultural lands – arable and forage lands (natural and improved grasslands and pastures). For arable lands tillage operations (ploughing) are an essential element of cultivation technology and should be applied annually. At first application of ploughing after radioactive fallout the transfer of ^{90}Sr to agricultural crops decreases by ~2 times. Its next consequent application slightly influences the reduction of radionuclide transfer to crops.

At contamination of soil surface the method of burial of soil the contaminated layer by means of double-deck plough (skim and burial ploughing) was studied and tested. This technique ensures the reduction of accumulation of ^{90}Sr in the yield up to 10 times (Alexakhin 1991). The main limitation of using this method is that it cannot apply for light-textured soils with thin humus-accumulated horizon.

Tillage operations (rotary cultivation and ploughing) for pastures are implemented as a part of amelioration (radical improvement) of meadows and usually are performed once per 5 years. In the zone of the Chernobyl accident a large number of studies on the effectiveness of application of tillage operations (rotary cultivation, ploughing) on different types of meadows and on arable lands was performed. Rotary cultivation destructs sod and processes soil at depth of 10–12 cm. Ploughing is performed at a depth up to 20–25 cm. In practice soil treatment on meadows can be fulfilled with implementation of both techniques. This technique has its maximum effectiveness at the very first application at surficial contamination of land cover (Sanzharova et al. 1996; Firsakova et al. 2002). Reduction of accumulation of ^{90}Sr in grass ranged 1.5–3.0 times. At its next application the effectiveness of this measure decreases in average by factor of 2 (Firsakova et al. 2002).

These methods of remediation were based on mechanical redistribution of ^{90}Sr in the soil profile. Mixing of contaminated upper layer with mineral under layers, from one hand, leads to the reduction of specific concentration of the radionuclide per weight unit of soil, but, from the other hand, the number of exchange sites increases, and this leads to radionuclide fixation in soil adsorption complex (Sanzharova et al. 1996; Vidal et al. 2001).

4 Physicochemical Remediation Methods for Agricultural Lands Contaminated with ^{90}Sr

After the Kyshtym and Chernobyl accidents the various agrochemical measures were taken to reduce ^{90}Sr transfer to plant; soil liming and application of organic and mineral fertilizers were widely used (Lard radiation accidents: consequences and protective countermeasures (Ilyin and Gubanov 2001). These improved the physicochemical properties of soil, which resulted in increased soil fertility, increased crop yield and decreased radionuclide uptake to the crop yield from 1.2 to 3.0 times (Bogdevitch et al. 2002).

Numerous investigations have shown that the decrease of accumulation of ^{90}Sr in crops in case of application of fertilizers and ameliorants can be caused due to the following reasons:

- change of the availability of the radionuclide for plant because of their transfer to fixed state or formation of hardly soluble compounds (Havely and Tzur 1964; Kudelya 1969; Shaw 1993);

- increase of concentration of exchangeable cations, primarily, calcium cations (Nishita and Taylor 1964; Klechkovsky et al. 1973; Kozhevnikova and Teplyakov 1989);
- strengthening of antagonism between radionuclide ions and ions of applied salts in soil solution at root uptake (Timofeyev-Resovsky and Titlyanova 1966; Nisbet et al. 1994);
- improvement of conditions of plant nutrition and related to that increase of biomass, and therefore the "dilution" of radionuclide per weight unit of plants (Yudintseva and Gulyakin 1968).

4.1 Application of Mineral Fertilizers

Usually, in agricultural practices mineral fertilizers are applied to soil in a complex including nitrogen, potassium and phosphate fertilizers. Doses of fertilizers depend on the need of crops in nutrition elements and on their content in soil. On radioactively contaminated territories mineral fertilizers can be applied as remediation method. In this case it is recommended to change the doses of fertilizers and ratio of nutrition elements which they contain (i.e. N, K and P). The use of fertilizers to reduce plant root uptake of ^{90}Sr are based on decreasing of application rate of P (N:P:K 1:2:1.5). Nitrogen fertilizers are applied in doses, which can ensure the highest yield increase. Phosphate and potassium fertilizers are applied in doses, which are higher than the need of plants in these nutrition elements (Recommendations 2003). Thus separate types of fertilizers make a different influence on transfer of ^{90}Sr to agricultural plants (Bogdevich et al. 1996; Ageyets 2001).

4.1.1 Nitrogen Fertilizers

Data on effect of nitrogen fertilizers on accumulation of ^{90}Sr by plants are contradictory. Nitrogen fertilizers, especially physiologically acidic, can heighten accumulation of ^{90}Sr and majority of other radionuclides by 2 or more times (Prister et al. 1991; Bogdevich et al. 1996; Ageyets 2001). When plants use nitrogen of physiologically acidic or physiologically alkalinous fertilizers in soil the content of ^{90}Sr in a form easy for plant fixation correspondingly increases or decreases (Egorov et al. 1969).

A number of studies noted the possibility of reduction of ^{90}Sr accumulation by plants at application of nitrogen fertilizers (Gulyakin and Yudintseva 1962) or the absence of the effect on absorption of ^{90}Sr (Nisbet et al. 1994). The trend of an impact of nitrogen fertilizers to a great extent depends on the form of their application, as well as on the level of soil fertility (on the content of nitrogen in soils).

Special laboratory investigations on the effect of different types of nitrogen fertilizers showed that their application leads to the increase of concentration of cations of Ca^{2+} in soil solution, changes in characteristics of sod-podzolic soil and, as follows, increase of available forms of ^{90}Sr (Geshel et al. 2013). By the raise of

Table 3 Effect of different types and doses of nitrogen fertilizers on the value of distribution coefficients Kd, ml·g^{-1}

Type of fertilizers	Dose of nitrogen fertilizers, mg·kg^{-1}				
	Control	25	50	100	200
	Distribution coefficient, ml g^{-1}				
N_{aa} (NH_4NO_3)	313	254	182	103	7.2
N_c ($NaNO_3$)		125	183	175	39.9
N_a ($(NH_4)_2SO_4$)		113	47	20.9	24.4
	F_v (concentration ration) of ^{90}Sr in germs (Bq·g^{-1} plant/ Bq·g^{-1}soil)				
N_{aa} (NH_4NO_3)	10.88 ± 1.69	13.31 ± 3.10	17.77 ± 0.51	18.85 ± 1.0	22.75 ± 2.34
N_c ($NaNO_3$)		10.28 ± 1.61	9.66 ± 1.02	12.03 ± 1.0	12.78 ± 1.96
N_a ($(NH_4)_2SO_4$)		15.08 ± 2.33	15.54 ± 1.04	20.99 ± 0.91	16.56 ± 1.32

Adopted from Geshel et al. (2013)

fertilizer doses distribution coefficients Kd and soil adsorption capability decreased, and accumulation of ^{90}Sr in germs increased (Table 3).

An important issue in agricultural practices is the determination of optimal doses of fertilizers because the lack of available nitrogen in soil leads to the decrease of yield and heightened doses intensify the accumulation of ^{90}Sr in plants. For contaminated agricultural lands the optimal and maximum possible doses for nitrogen fertilizers were determined in order to form the planned yield on sod-podzolic soils in case of their contamination with ^{90}Sr (Rules 2002).

4.1.2 Potassium Fertilizers

Usually potassium fertilizers make a weak effect on ^{90}Sr transfer into plants (Yudintseva and Gulyakin 1968). But during the experiments of the Institute for Soil Science and Agrochemistry (Belarus) it was established that with increase of the dose of potassium fertilizers from 60–180 kg·ha^{-1} the content of ^{90}Sr decreases by 1.5–1.8 times for different crops (Table 4).

Probable mechanism of the influence of increasing doses of fertilizers on the heightening of the mobility of ^{90}Sr in soil is connected with the change of ratio of ions in soil solution of sod-podzolic soil. At increase of the number of K^+ ions concentrations of Ca^{2+} and Mg^{2+} grow by 1.5–2.0 times and this causes competitive impact on behaviour of ions of Sr^{2+} (Suslina 2004).

Beneficial effect of potassium fertilizers is determined both the antagonism of ions and increase of crop yield, especially on sod-podzolic sandy soils and sandy loams with low fertility. For example, at application of heightened doses of potassium fertilizers (up to 180 kg·ha^{-1}) accumulation of ^{90}Sr in tubers of different varieties of potatoes decrease on 33–57% at the simultaneous growth of yield on 20–50 hwt per ha (Putyatin 2008).

Table 4 Effect of different doses of potassium and phosphoric fertilizers on accumulation ^{90}Sr by crops from sod-podzolic soil

Alternative	Crops		
	Pea herbage	**Pea grain**	**Oats grain**
	Concentration of ^{90}Sr in crops, Bq·kg^{-1}		
Control	20	48	38
P_{60}	17	39	32
$P_{60}K_{60}$	15	33	28
$P_{60}K_{120}$	14	31	25
$P_{60}K_{180}$	13	27	22
	^{90}Sr uptake by grasses, T_{ag}		
	Perennial grasses	**Annual grasses**	
$N_{60}P_{60}K_{60}$	38.6±4.3	26.7±2.1	
$N_{60}P_{90}K_{120}$	21.7±2.4	17.1±1.9	
$N_{60}P_{180}K_{240}$	6.1±2.2	7.5±1.3	

Adapted from Putyatin (2008)

To prevent the over dosage of potassium fertilizers and depreciation of products quality the restrictions of the level of the limit soil saturation with exchangeable potassium depending on the cation exchange capacity were established: for sandy soils – 3.5%, for sandy loams – 4% and for loamy soils – no greater than 5% (Rules 2002).

4.1.3 Phosphate Fertilizers

One of the recommended techniques for decreasing the transfer of ^{90}Sr to plants is application of phosphate fertilizers, which can ensure the reduction of ^{90}Sr accumulation by 2–4 times (Yudintseva and Gulyakin 1968; Maraschino and Fedorov 1978). After the Chernobyl accident the experiments on application of heightened doses of phosphate-potassium fertilizers were performed on forage lands in Belarus; this ensured the reduction of transfer of ^{90}Sr to the grass stand by 1.6–6.3 times (Table 4). Effectiveness of phosphate fertilizers significantly depends on the content of phosphorus in soil. An important requirement to the conditions of their application is provision of balanced nutrition for crops with the consideration of the presence of labile phosphates in soil (Ageyets 2001).

4.1.4 Complex Application of Mineral Fertilizers (NPK)

Balanced application of various types of mineral fertilizers has an important significance at crop cultivation on lands contaminated with ^{90}Sr. It was established that vary minimum transfer of ^{90}Sr to plants was observed for soils with optimal parameters of soil fertility.

The effect of mineral fertilizers on entrance of ^{90}Sr into agricultural plants is well investigated (Yudintseva and Gulyakin 1968; Alexakhin and Korneyev 1992; Putyatin 2008).

It should be noted that the effectiveness of application of mineral fertilizers at soil contamination with ^{90}Sr significantly depends on the factors of soil fertility on low-fertility soils event the standard doses of fertilizers can reduce the accumulation of ^{90}Sr in plants.

At field experiments on sod-podzolic soils standard doses of nitrogen, potassium and nitrogenous potassium fertilizers had a weak influence on accumulation of ^{90}Sr in oats grain; this could be explained by the increase of hydrolytic acidity of soil. At application of phosphate fertilizers and their combinations with other types of fertilizers content of ^{90}Sr in oats grain decreased by 1.6–2.5 times.

4.2 Application of Organic Fertilizers

In most cases application of organic fertilizers reduces the transfer of ^{137}Cs to plants by 1.5–3.0 times (Gulyakin and Yudintseva 1973; Moiseyev et al. 1988). The maximum effect presents on poor with humus light-textured soils; accumulation of ^{90}Sr in crop yield reduces by 4.6–5.6 times (Vodovozova and Pogodin 1969; Gulyakin and Yudintseva 1973; Gulyakin et al. 1978; Marakushin and Fedorov 1978). Humus acids form complexes with ^{90}Sr or humates, which reduce that availability of strontium by 2–4 times. Ivanov et al. (1976) showed that soil rich in humus absorb ^{90}Sr in exchangeable form, but its bonding strength here is higher than in case of soil with low content of organic matter.

A number of studies showed that organic matter, from one hand, reduces the migration ability and improves the fixation of the radionuclide in soil, and, from the other hand, it heightens the mobility and, thereby, it increases the export of the radionuclide from soil to plant. The increase of the content of humus in sod-podzolic soils leads to the growth of ^{90}Sr accumulation by plants.

4.3 Liming

Liming is one of the most effective remediation methods on contaminated agricultural lands. It reduces the transfer of ^{90}Sr to crops by 2–4 times (Yudintseva and Gulyakin 1968; Gulyakin and Yudintseva 1973; Bondar et al. 1990). Lime ($CaCO_3$), dolomite powder, calcareous tuffs, carbonic calcium gypsum (lake lime) are used as ameliorants. The paper of Korneyev et al. (1975) shows that application of lime to sod-podzolic soil with low content of exchangeable Ca reduced the transfer of ^{90}Sr to grass stand by 1.4–2.7 times. Liming is beneficial because of its influence on agrochemical soil properties: soil solution acidity decreases, content of exchangeable Ca increases. Lime is a long-acting ameliorant that is why its impact to soil fertility, crop yield capacity and effectiveness of applied organic and mineral fertilizers can exert during 4–5 years.

During the experiments of (Kozhevnikova and Teplyakov 1989) on the impact of calcium-containing wastes of industrial enterprises (gypsum, lime, chalk, cement kiln dust) to transfer of radiostrontium to plants from soil under the conditions of simulative field experiment it was established that calcium-containing substances improved the reduction of transfer of ^{90}Sr to crop yield be 1.5–3.5 times. Investigation results showed that there is no significant distinction in the effectiveness of the decrease of ^{90}Sr transfer to agricultural plants when liming with dolomite powder, chalk, quicklime and its combinations with phosphogypsum in equivalent doses (Putyatin 2008).

4.4 Complex Application of Mineral Fertilizers, Organic fertilizers and Ameliorants

During field experiments on sod-podzolic soil dose heightening for nitrogen and phosphate fertilizers along with the application of dolomite powder reduced the content of ^{90}Sr in grain and straw of cereal crops by 1.1–2.8 times. Maximum effect was in the alternative with heightened doses of phosphate fertilizers ($N_{100}P_{240}K_{80}$) with dolomite background (Table 5). It should be mentioned that application of dolomite powder in a dose 100 t ha^{-1} maximally reduced the accumulation of ^{90}Sr in plants – the reduction factor was 2.1–3.6 times.

Yudintseva and Gulyakin (1968) showed that the maximum effect on the reduction of the accumulation of ^{90}Sr by plants was observed in case of combined application of lime and organic matter. Under the impact of lime and organic matter the content of ^{90}Sr in plants of oats and peas decreased by 3–4 times (Yudintseva and Gulyakin 1968). The studies performed after the Chernobyl accident showed that application of lime and heightened doses of organic fertilizers to sod-podzolic soil reduced the accumulation of ^{90}Sr in corn green material in average on 60% (Firsakova et al. 2002).

Table 5 Content of ^{90}Sr in straw and grain of cereal crops, Bq kg^{-1}

Alternative	Oats		Barley		Rye	
	Straw	Grain	Straw	Grain	Straw	Grain
$N_{100}P_{80}K_{80}$	1184	222	1058	211	455	37
Dolomite– 100 t ha^{-1} (background)	548	88	307	56	148	11
Background + $N_{100}P_{80}K_{80}$	381	80	259	44	115	7
Background + $N_{100}P_{160}K_{80}$	348	63	263	26	93	7
Background + $N_{100}P_{240}K_{80}$	244	30	241	26	70	7

Adopted from Decrease (1989)

4.5 *Action Mechanism of Combined Application of Fertilizers and Ameliorants*

To define the mechanisms of action of mineral fertilizers and ameliorants the special experiments were conducted on the territory of Bryansk region. During multifactorial field experiment studying the impact of mineral fertilizers, manure, liming and their combinations to biological mobility of ^{90}Sr there were selected distinctive alternatives, which allow defining accurate regularities in the behaviour of ^{90}Sr under the action of one or another type of ameliorants (Sysoeva 2004).

To evaluate the biological mobility of ^{90}Sr in soil-plant system various factors are used:

- Accumulation coefficient (AC), as integral indicator of ^{90}Sr transfer from soil to plant;
- Forms of ^{90}Sr in soils;
- Selectivity coefficient of the exchange of a pair ^{90}Sr – Ca (K_S ^{90}Sr/Ca).

4.5.1 Effect of Mineral Fertilizers and Ameliorants on Physicochemical Properties of Soil

The effectiveness of applied fertilizers is determined by a degree of their influence on physicochemical properties of soils. Data analysis shows that due to application of dolomite powder in doses 6 and 9 t ha^{-1} the content of exchangeable Ca increased by 1.4–1.6 times in any experiment alternative. Soil liming improved the reduction of its acidity (Table 6). Due to application of organic fertilizers the tendency to growth of humus content in soil was noted, but changes of this factor were slightly reflected.

Table 6 Main agrochemical indicators of soils on experimental sites

				Exchangeable cations			
				K	Ca	Mg	CEC
№	Alternative	pH_{KCl}	Humus, %	mg-eq 100 g^{-1}			
1	Control	5.08	1.70	0.46	2.84	0.56	3.86
2	$N_{30}K_{30}P_{20}$ + manure 30 t ha^{-1} + dolomite powder 9 t ha^{-1}	6.49	2.08	0.44	4.34	1.00	5.78
3	$N_{30}K_{90}P_{20}$ + manure 30 t ha^{-1} + dolomite powder 9 t ha^{-1}	6.46	2.07	0.47	4.07	0.48	5.02
4	$N_{60}K_{60}$ + dolomite powder 6 t ha^{-1}	6.34	1.87	0.44	4.11	1.15	5.70
5	N_{60} + dolomite powder 6 t ha^{-1}	6.46	2.00	0.48	4.00	0.74	5.22
6	K_{60}	5.93	2.06	0.46	3.93	0.76	5.15
7	K_{60} + dolomite powder 6 t ha^{-1}	6.11	2.05	0.53	4.49	0.81	5.83
8	Dolomite powder 6 t ha^{-1}	6.07	2.01	0.52	4.09	0.51	5.12

Adopted from Sysoeva (2004)

Oats is characterized by a comparatively high need to the level of potassium nutrition. On this basis close values of exchangeable K in different experimental alternatives can be explained. It seems probable that almost all applied K was absorbed by the plant during its vegetation period.

Application of mineral fertilizers and ameliorants led to the change of cation exchange capacity of sod-podzolic soil. CEC maximum increase by 1.5 is typical for the alternative, when full mineral fertilizer is applied with the background of manure and liming; for the alternative with the application of nitrogen and potassium fertilizers with the background of liming and for the alternative with application of only potassium fertilizers (K60) and dolomite powder (6 t ha^{-1}).

4.5.2 Effect of Mineral Fertilizers and Ameliorants on Biological Mobility of ^{90}Sr

The maximum effect on the decrease of ^{90}Sr accumulation coefficient in oats grain was made by the application of full complex of mineral fertilizers (N30P20K30) with the background of manure application (30 t ha^{-1}) and liming (9 t ha^{-1}), and also application of potassium fertilizers (K60) with the liming background (6 t ha^{-1}). For these alternatives AC reduced correspondingly by 1.7–1.9 in comparison with control.

Correlation analysis recognized the reliable direct correlation between AC and content of exchangeable Ca^{2+} ($r = -0.862$; $p < 0.01$) and soil cation exchange capacity ($r = -0.911$), and also it didn't recognize and correlations between AC and soil pH ($r = 0.601$), humus content ($r = 0.697$) and content of exchangeable K^+ ($r = -0.07$).

Analysis of received data showed that application of mineral fertilizers and ameliorants influenced the content change of exchangeable form of ^{90}Sr in sod-podzolic soil (Table 7). Minimum value of the fraction of exchangeable form was observed for the alternative, when full mineral fertilizer (N30P20K90) was applied with the background of liming (9 t ha^{-1}) and application of manure (30 t ha^{-1}). Fraction of the exchangeable form decrease by 1.4 in comparison with control. For alternatives, when there was application of potassium fertilizers (K60) with liming background (6 t ha^{-1}) and application of full mineral fertilizer (N30P20K30) with the background of manure (30 t ha^{-1}) and lime (9 t ha^{-1}) application, the fraction of exchangeable form decreased by 1.2 times. For the rest experiment alternatives, the content of strontium exchangeable form varied insignificantly and was close to the values of control.

Correlation analysis of data obtained shows that values of AC and fraction of exchangeable form has a very weak correlation ($r = 0,231$).

Table 7 Changes of the indicators of ^{90}Sr biological mobility in soil-plant system due to application of mineral fertilizers and ameliorants

Alternative	Exchangeable form fraction ^{90}Sr ($\alpha_{обм}$)	F_v (Bq kg^{-1})/(Bq kg^{-1})	$A \cdot 10^{-2}$
Control	0.86	0.85	1.21
$N_{30}P_{20}K_{30}$ + manure 30 t ha^{-1} + dolomite powder 9 t ha^{-1}	0.71	0.49	0.65
$N_{30}P_{20}K_{90}$ + manure 30 t ha^{-1} + dolomite powder 9 t ha^{-1}	0.60	0.64	0.59
$N_{60}K_{60}$ + dolomite powder 6 t ha^{-1}	0.81	0.55	0.79
N_{60} + dolomite powder 6 t ha^{-1}	0.75	0.68	0.75
K_{60}	0.88	0.53	0.90
K_{60} + dolomite powder 6 t ha^{-1}	0.69	0.44	0.61
Dolomite powder 6 t ha^{-1}	0.83	0.68	0.81

Adopted from Sysoeva (2004)

4.5.3 Mathematical Model as an Analysis Tool for Description of Processes of Influence of Mineral Fertilizers and Ameliorants on Biological Mobility of ^{90}Sr

From the position of soil chemistry to describe the processes occurring in soil-plant system under the conditions of field experiment ^{90}Sr bioavailability parameter (A) was used. It is a composed soil characteristic.

For this purpose, there were experimentally measured soil characteristics, which appear in an equation, in particular the fraction of exchangeable form of ^{90}Sr (a_{exch}) and content of exchangeable Ca^{2+}. Exchange selectivity coefficient of $^{90}Sr/Ca$ for sod-podzolic soil of the experiment, according to the data obtained, was equal to 2.5.

Data analysis shows that values of A parameter varied depending on experimental alternatives (Table 7). Maximum values of A parameter were calculated for control alternative and the alternative with application of nitrogen fertilizers (N60) and liming background (6 t ha^{-1}). Minimum values of A parameter were calculated for the alternatives with application of full complex of mineral fertilizers (N30P20K30) with the background of manure application (30 t ha^{-1}) and liming (9 t ha^{-1}), and with application of potassium fertilizers (K60) with the liming background (6 t ha^{-1}).

According to the hypothesis that accumulation coefficient of ^{90}Sr in plant is a linear function of the bioavailability parameter, obtained data on AC(^{90}Sr) in oats grain were compared to the bioavailability parameter (A) for ^{90}Sr, which was calculated for sod-podzolic soil after fertilizers application. Regression data analysis showed a linear correlation between AC and parameter A ($r = 0.701$; $p < 0.05$) (Fig. 1).

Conversion factor B, depending on plant biological peculiarities, was calculated in relation of AC to A. The result of calculation displayed that B factor in field experiment for oats grain was valued as 79 ± 15.

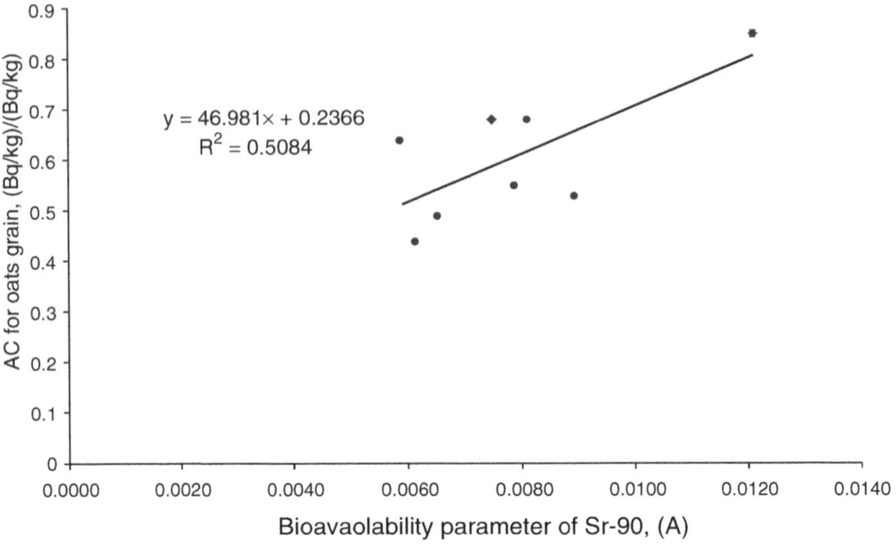

Fig. 1 Dependence of AC for oats grain on ^{90}Sr bioavailability parameter (A)

The offered bioavailability parameter (A) for ^{90}Sr, being a composed soil characteristic, can be used as a tool for analysis tool for interpretation of the observed regulations in the behaviour of ^{90}Sr in migration chain "soil-plant".

Good compliance of theoretical and experimental dependencies testifies to the offered method to be used for forecasting the AC of ^{90}Sr in agricultural plants basing on changes of soil properties due to fertilizers application.

5 Use of Sorbents

Application of sorbents for reduction of migration mobility of ^{90}Sr is based on mechanisms of its behaviour and physicochemical state in soils. The major part of ^{90}Sr (60–90%) exists in exchangeable forms and cation exchange is prevailing mechanism of ^{90}Sr absorption by soil solid phase. Different groups of minerals are characterized by different sorption capacity in relation to strontium ions (Table 8). The highest sorption capacity belongs to the minerals of montmorillonite group, which have high dispersion ability and absorb up to 99% of ^{90}Sr (Spitsin and Gromov 1959; Yudintseva and Gulyakin 1968). Sorption of ^{90}Sr by ascanite, bentonite and vermiculite results in the highest distribution coefficients, which correlate with the fraction of sorbed ^{90}Sr. It should be emphasized that ascanite and bentonite feature in the highest sorption strength.

Layer minerals with trioctahedral structure of crystal lattice, for example, montmorillonite clays of ascanite and gumbrine and glimmers of phlogopite, hydrophlogopite and vermiculite, are the most recommended for use as sorbents

Table 8 Sorption properties of minerals and montmorillonite clays and their effect on ^{90}Sr uptake by plants

Minerals and clays	Sorption of ^{90}Sr by minerals and montmorillonite clays: Distribution coefficient	Fraction of sorbed ^{90}Sr, % of deposited	Driven out by 0.1 N $CaCl_2$, % of absorbed	Concentration of ^{90}Sr in oats straw, n 10^3 Bq g^{-1} of dry plant weight
Ascanite	525.0	99.05	13.90	27.1
Bentonite	351.8	98.59	28.69	7.4
Gumbrine	130.1	96.29	34.19	12.0
Kaolin	101.8	95.31	73.58	6.4
Vermiculite	215.1	97.73	63.79	27.5
Hydro-phlogopite	103.5	95.39	52.90	15.9
Damourite	21.9	81.47	88.95	18.0
Muscovite	50.6	91.00	–	29.8
Biotite	77.2	93.91	70.57	27.7
Phlogopite	195.7	97.51	69.71	16.5
Hydrogoethite	3.4	40.26	100.0	16.6
Vivianite	47.0	90.36	47.27	30.7

Adopted from Yudintseva and Gulyakin (1968)

(Yudintseva and Gulyakin 1968). The performed studies showed that hydrophlogopite significantly reduces transfer of ^{90}Sr to plants. Admixing hydrophlogopite into sand results in ^{90}Sr accumulation decrease in oats straw by 2–3 times in comparison with control alternative. Under the impact of gumbrine the content of ^{90}Sr in oats straw decreased approximately by 5 times, in grain - by 4 times, and adding of ascanite influenced the decrease by 3.5 and 2.0 times correspondingly (Table 8). Vermiculite and phlogopite reduce ^{90}Sr accumulation considerably less than montmorillonite clays. Dosage of applied minerals and clays made a significant effect on export of ^{90}Sr with plant yield.

Besides the natural minerals the intensive development of artificial sorbents with high sorption capacity of ^{90}Sr is performed. Russian Institute of Radiology and Agroecology developed artificial sorbents (KC-10CX and KC-20CX) based on natural minerals of tripolite and diatomite.

Method of receiving of complex sorbent lies in the processing of refined natural materials on the basis of opal-cristobalite rocks with the solution of phosphoric acid. Before solution processing the original material is hydrolysed in hot alkaline solution containing potassium aluminate $K_3[Al(OH)_6]$. After the end of hydrolysing the suspension is conditioned in hermetic reactor at temperature and under the pressure. Artificial sorbents feature in higher cation exchange capacity and are enriched in phosphorus and potassium (Table 9).

To study sorption processes ^{90}Sr was deposited to sorbents as a solution of Sr $(NO_3)_2$. To study the mobility of ^{90}Sr the method of sequential extractions was applied: H_2O 1:5 (water soluble form); 1 N CH_3COONH_4 1:10 (exchangeable

Table 9 Sorbent physicochemical properties

Sample	pH_{KCl}	Forms of labile potassium, K_2O, %		Total phosphorus	CEC, mmol·eq.
		Water soluble	Assailable	P_2O_5, %	100 g^{-1}
Tripolite	3.84	–	–	0.96	25.00
Diatomite	6.53	–	–	1.44	16.00
KC-10CX	8.95	6.11	22.13	3.84	126.50
KC-20CX	8.52	7.25	13.49	7.19	162.50

Adopted from Krylenkin et al. (2012)

Table 10 Sorbents influence on the mobility of ^{90}Sr in sod-podzolic soil and its accumulation in barley vegetative mass

Sorbent	Forms of ^{90}Sr in soil, %[a]				Sorbents influence on ^{90}Sr accumulation	
	Water soluble	Exchangeable	Labile	Acid-soluble	^{90}Sr content in vegetative mass, F_v	Reduction factor
Control	1.45	55.45	21.82	5.30	44.5±0.9	–
Diatomite	0.15	54.09	20.45	4.70	30.3±1.8	1.47
Tripolite	0.66	56.82	21.05	5.61	43.2±0.6	1.03
KC-10CX	0.48	56.67	23.79	5.00	33.7±2.4	1.32
KC-20CX	1.77	55.91	22.27	5.18	22.6±0.4	1.97

Adopted from Krylenkin et al. (2013)

[a]Application of natural and artificial sorbents based on 3% CEC increase.

form); 1 N HCl 1:10 (labile form); 6 N HCl 1:10 (acid-soluble form) (Pavlotskaya 1974). Effectiveness of sorbent application for reduction of radionuclide mobility depends on CEC. Increase of CEC after sorbent application prevents ^{90}Sr transfer from soil to plants. Cation exchange capacity of artificial sorbents is 5–10 times higher than CEC of natural sorbents. Heightened potassium content in artificial sorbents is due to the fact that potassium is a part of hydrolysing solution, used in the technology of sorbent production (Krylenkin et al. 2012). Large amount of potassium, incorporated with sorbents, improves the growth of potassium capacity and displacement of Ca^{2+} and Mg^{2+} ions from SAC. These ions ensure the competitive interaction with ions of $^{90}Sr^{2+}$ reducing its transfer to plants. Artificial sorbents absorb ^{90}Sr in a stronger fixing labile form comparing to natural sorbents. Application of artificial sorbents to sod-podzolic sandy-loam soil leads to the decrease of ^{90}Sr availability because of the reduction of its content in water soluble form and increase of the content of its labile and firmly fixed forms (Table 10).

At this rate, artificial sorbents reduce radionuclide transfer to plants by 1.32–1.97 times. Whereas, natural sorbents reduce radionuclide content by 1.03–1.47 times (Krylenkin et al. 2013).

6 Combined of Soil-Based Remediation Methods of Grasslands

Studies on the effectiveness of combined application of mechanical and physico-chemical remediation methods are of separate interest. Such situation takes place at remediation of meadows, which are used in agriculture as pastures and haylands.

Much research has been carried out on the effectiveness of reducing root uptake by grasses of natural meadows from various amelioration techniques. The techniques have been placed into two categories: radical and surface improvement (Belli and Tikhomirov 1996; Sanzharova et al. 1996). Surface improvement includes application of soil mechanical treatment techniques (disking of the mat, rotary cultivation) and enhanced mineral fertilization. Radical improvement consists of similar options to those of surface improvement with additional soil ploughing. In addition, many combinations of ameliorants (lime, organic fertilizers and mineral fertilizers) have been evaluated experimentally. Rotary cultivation, disking and ploughing are countermeasures that can easily be applied as these form part of the normal agricultural practice. If meadow is ploughed as part of radical improvement, the use of the land is restricted for several years until the grass has re-established. Data on the effect of ploughing on subsequent root uptake by grass indicate that ploughing reduces root uptake by a factor of 2–2.5 for mineral soils and up to a factor of 3.0 for organic soils. Liming is one of the main countermeasures that have been used to reduce caesium contamination levels in grasses. The liming dosage depends on the acidity of the soil. Reductions in plant uptake are of up to a factor of 1.5–2.0. Mineral fertilizers have been used extensively in the CIS countries to reduce uptake to plants from soil and many experimental studies have been carried out to evaluate the optimum ratios of nitrates, phosphates and potassium to give the best reduction in root uptake. A specific feature of application of mineral fertilizers in conditions of radioactive contamination of a territory is a necessity of changing of ratio of main elements of nutrition.

To study the effectiveness of application of combined methods of meadow remediation experiments on the territories of Belarus, Russia and Ukraine were performed (Rauret et al. 1999; Vidal et al. 2001).

Two experimental sites were selected in the three republics. They were as follows: Russian sites: Rudnuy (180 km from the Chernobyl NPP); Belarussian sites: Dublin (42 km from the NPP) and Sawichi (32 km from the NPP); Ukrainian sites: Mateyki (70 km from the NPP) and Christinovka (75 km from the NPP).

Experiments were designed with consideration of the following treatments: ploughing, liming and NPK fertilizers, which had previously shown to be the most efficient in decreasing radionuclide transfer in meadows. The experimental design included two types of soil-meadow combination (peat soil in wet meadow and sod-podzolic soil in dry meadow), two tillage techniques (disking and ploughing-D + Pl), and two doses of liming and NPK fertiliser. The dose of fertilising (NPK) and liming (L) depended on the usual practice by each country/institute. Plots were reseeded with a mixture of three species: *Phleum pratense*, *Bromus inermis* and

Festuca pratensis. For every treatment, 4 randomly located plots were used as replicates in the all the experimental sites.

Soil samples taken during field experiments were analysed in order to define the changes in soil properties resulted from the application of agro technical measures. Moreover, ^{90}Sr accumulation in grass stand at different ways of mechanical soil treatment was determined.

6.1 *Effect of Agricultural Practices on RSr Transfer*

The changes in RSr soil-plant transfer after the field treatments were evaluated by calculating the aggregated transfer factor (T_{ag}, expressed as $m^2\ kg^{-1}$) as a mean value of 4 replicates.

Figure 2 shows a summary the results for RSr (the 1st harvest).

For RSr, no significant effect of agricultural practices on transfer was observed in the first year, not even after the addition of lime. However, time was also a key factor, since in the second year the effect of disking+ploughing (and reseeding) made the transfer to decrease from 2 to 4 times. However, no further decrease was noticed with the other treatments, with the exception of Dublin site in 1998 after the addition of lime.

Agricultural practices (disking+ploughing, combined with reseeding, and extra doses of NPK fertilizers and liming) are a suitable restoration strategy to decrease radiostrontium transfer. Disking+ploughing (with reseeding) was the most effective single treatment, whereas extra doses of NPK fertilizers led to further minor effects.

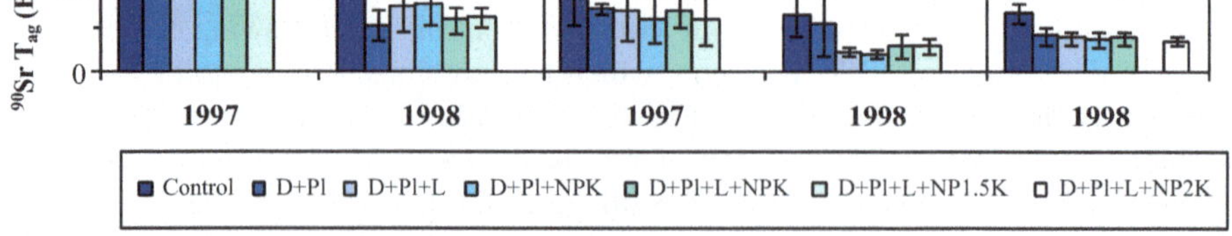

Fig. 2 Effect of agricultural practices on radiostrontium transfer in dry meadows. D + Pl: Disking+Ploughing; L: Liming; NPK: Addition of fertilizers. Aggregated transfer factors are mean values of 4 field replicates; bar errors indicate one standard deviation

7 Conclusions

The problem of agricultural production on the territories contaminated with ^{90}Sr is primarily related to the accidents on the Southern Urals («Mayak" PA) and Chernobyl NPP. The radionuclide has a range of peculiarities, which are required to consider while the assessment of its danger and ways of mitigation of negative contamination consequences. At first instance high migration ability of ^{90}Sr should be noted because it usually presents in soil in exchangeable form. Due to this fact the developed remediation techniques should be aimed on the fixation of the radionuclide in soils as the result of its sorbing and fixing, and, as follows, on the decrease of its transfer to crops. Development of remediation techniques is based, from one side, on understanding of radionuclide behaviour mechanisms in agricultural ecosystems, and, from the other side, on agricultural technologies, including soil fertility and biological peculiarities of crops. Agricultural techniques of soil treatment (rotary cultivation, ploughing) ensure redistribution of radionuclides in soil profile and increase of sorption surface by means of soil mineral layers. Agrochemical methods (application of fertilizers and ameliorants) improve the decrease of soil solution acidity, increase cation exchange capacity and soil fertility. These techniques, being applied either individually or in complex, change soil parameters and, as a consequence, influence on radionuclide mobility in soils and their availability to plants. The possibility of application of various natural and artificial sorbents for reduction of ^{90}Sr mobility in soils and decrease of its assimilation by plants is shown. Analysis of collected information shows that agriculture has a wide range of remediation methods, which ensure production meeting sanitary and hygienic standards.

References

Ageyets VY (2001) System of radiological countermeasures in agricultural sphere of Belarus. Minsk

Alexakhin RM (1991) Rekomendacii po vedeniyu selskogo khozyajstva v usloviyakh radioaktivnogo zagryazneniya territorii v rezultate avarii na Chernobylskoj AES na period 1991–1995 gg. (Recommendations for farming under the conditions of radioactive contamination of the territory resulted from the Chernobyl accident for the period 1991–1995) Moscow

Alexakhin RM, Korneyev NA (1992) Selskokhozyajstvennaya radioekologiya (Agricultural radioecology). Ekologiya, Moscow

Belli M, Tikhomirov F (1996) Behaviour of radionuclides in natural and semi-natural environments. Final report. Part Y11.5. EUR 16531 EN, pp 147

Bogdevich IM, Ageyets VY, Firsakova SK (1996) Fundamentals of agricultural industry. In: Konoplya YF, Rolevich IV (eds) Ecological, medicobiologic and socio-economic consequences of the Chernobyl accident in Belarus Minsk, pp 52–102

Bogdevitch I, Sanzharova N, Prister D, Tarasiuk S (2002) Countermeasures on natural and agricultural areas after Chernobyl accident. In: Kolejka J (ed) Role of GIS in lifting the cloud off Chernobyl. Kluwer Academic Publishers, Dordrecht, pp 147–158

Bolt GH, Sumner ME, Kamphorst AA (1963) Study of equilibria between three categories of potassium in an illitic soil. Soil Sci Soc Amer Proc 27:294–299

Bondar PF, Zaika VV, Dutov AI (1990) Influence of fertilizers and ameliorants on accumulation of radiocaesium in agricultural crops on limed soils. Abstracts of the 3rd All union conference on agricultural radiobiology, vol 1, Obninsk, pp 82

Cremers A, Elsen A, De Preter P, Maes A (1988) Quantitative analysis of radiocesium retention in soils. Nature 6187:247–249

Decrease of content of radioactive substances in plant products (1989) Recommendations (Snizhenie soderzhaniya radioaktivnykh veschestv v produkcii rastenievodstva). Agropromizdat, Moscow

Egorov AV, Klechkovsky VM, Sokolova LN (1969) On the effect of conditions of nitrogen nutrition on transfer of strontium-90 from soil to plants. Atomizdat, Moscow

Firsakova SK, Timofeyev SF, Shumilin VA, Timofeyev AS, Podolyak AG (2002) Accumulation of ^{90}Sr by field crops under the conditions of radioactive contamination of agricultural lands. Radiatsion Biol Radioekol 3:345–351

Geshel IV, Sanzharova NI, Krylenkin DV (2013) ^{90}Sr transfer to plants from sod-podzolic light-loam soil in case of nitrogen fertilizer application. Agrokhem vestnik 1:37–39

Guidelines for agricultural countermeasures following an accidental release of radionuclides (1994) IAEA, Vienna

Gulyakin IV, Yudintseva YV (1962) Radioactive fission products in soil and plants. Gosatomizdat, Moscow

Gulyakin IV, Yudintseva YV (1973) Agricultural radiobiology. Kolos, Moscow

Gulyakin IV, Yudintseva YV, Mamontovav LA, Levina EM (1978) Influence of lime, peat and peat ash on accumulation of strontium and caesium isotopes in crop yield. Agrokhimiya 5:109–115

Havely E, Tzur J (1964) Precipitation of Sr by $CaCO_3$ in calcareous and measurement of cation-exchange capacity. Soil Sci 1:66–67

Hilton J, Comans R (2001) Chemical forms of radionuclides and their quantification in environmental samples. In: Van der Stricht E, Kirchmann R (eds) Radioecology, radioactivity and ecosystems. Fortemps, Liège

Ilyin LA, Gubanov VA (2001) Krupnye radiacionnye avarii: posledstviya i zaschitnye mery (Major radiation accidents: consequences and protective measures). IzdAT, Moscow

Ilyin LA, Gubanov VA (2004) Lard radiation accidents: consequences and protective countermeasures. IzdAT Publisher, Moscow

Israel YA, Bogdevich IM (2009) Atlas sovremennykh i prognoznykh aspektov posledstvij avarii na Chernobylskoj AES na postradavshikh territoriyakh Rossii i Belarusi (ASPA Rossiya-Belarus) (Atlas of current and forecast effects of the Chernobyl accident on affected parts of the Russian Federation and Belarus (ACFE Russia-Belarus)). Foundation «Infosfera» – NIA Priroda, Moscow-Minsk

Israel YA, Petrov VN, Avdyushin SI, Gasilina NK, Rovincki FY, Vetrov VA, Vakulovski SM (1987) Radioactive contamination of natural environments in the Chernobyl nuclear power plant accident zone. Meteorol Hydrol 2:5–18

Ivanov SN, Shagalova ED, Shifrina SS (1976) Physicochemical mode of ^{90}Sr in sod-podzolic soils of Belarus. Pochvovedeniye 12:110–116

Klechkovsky VM (1973) Rekomendacii po vedeniyu selskogo i lesnogo khozyajstva pri radioaktivnom zagryaznenii vneshnej sredy (Recommendations for farmer and forestry at radioactive contamination of environment). Kolos, Moscow

Klechkovsky VM, Fedorov YA, Arkhipov NP, Romanov GN, Alexakhin RM, Fevraleva LT (1973) Mechanism of soil and aerial entrance of radioactive strontium to plants. Pochvovedeniye 5:38–47

Kokotov YA, Rusakov FF, Urbanyuk AP (1960) Influence of pH and salinity on sorption of ^{90}Sr by particular soils. In: Proceedings of the conference on radiation hygiene

Konoplev AV, Bulgakov AA (2000) Exchange distribution coefficient of ^{90}Sr and ^{137}Cs in the soil-water system. Atom Ener 2:152–158

Korneyev NA, Firsakova SK, Malysheva MR (1975) Transfer of strontium-90 to meadow grasses from different types of soils of non-chernozem belt. Pochvovedeniye 11:53–59

Kozhevnikova RN, Teplyakov IG (1989) Influence of substances containing calcium on transfer of strontium-90 to crop yield from leached chernozem. Agrokhimiya 2:91–94

Krylenkin DV, Sanzharova NI, Andreyeva NV, Geshel IV, Belova NV (2012) Dynamics of sorption of ^{90}Sr by minerals and sorbents. In: Proceedings of the IVth international scientific conference on "Sorbents as quality factor of life and health". Belgorod, pp 226–231

Krylenkin DV, Sanzharova NI, Geshel IV, Andreyeva NV (2013) Accumulation of ^{90}Sr by barley from sod-podzolic soil in case of natural and artificial sorbents application. Agrochem Herald 6:20–22

Kudelya AD (1969) On interaction trace quantities of strontium with phosphate ions in soils. Collected works on agrochemical physics. Kolos, Leningrad 18:134–141

Marakushin AV, Fedorov YA (1978) Accumulation of strontium-90 in field crops at different cultivation conditions. Agrokhimiya 8:125–128

Moiseyev IT, Tikhomirov FA, Martyushov VZ, Rerikh LA (1988) On the assessment of influence of mineral fertilizers on the dynamics of exchangeable ^{137}Cs in soils and its availability to vegetable crops. Agrokhimiya 5:86–92

Nadtochiya PP (2003) Opyt preodoleniya posledstvij chernobylskoj katastrofy (selskoye i lesnoe khozyajstvo) (Experience on overcoming the consequences of the Chernobyl disaster (agriculture and forestry)). Svit, Kiev

Nikolayeva AV (1970) Khimiya dolgozhivuschikh oskolochnykh elementov (Chemistry of long-lived fragmentation elements). Atomizdat, Moscow

Nisbet AF, Mocanu N, Shaw S (1994) Laboratory investigation into the potential effectiveness of soil–based countermeasures for soils contaminated with radiocaesium and radiostrontium. Sci Total Environ 149:145–154

Nishita H, Taylor P (1964) Influence of stable Sr and Ca on ^{90}Sr and ^{45}Ca in soil and clay minerals. Soil Sci 98:181–186

Pavlotskaya FI (1974) Migraciya radioaktivnykh produktov globalnykh vypadenij v pochvakh (Migration of radioactive products of global fallout in soil). Atomizdat, Moscow

Pavlotskaya FI (1981) Occurrence forms and migration of radioactive products of global fallout in soils. Dissertation, Institute of geochemistry and analytic chemistry, Moscow

Prister BS, Loschilov NA, Nemets OF, Poyarkov VA (1991) The fundamentals of agricultural radiology. Urozhay, Kiev

Putyatin YV (2008) Minimization of ^{137}Cs and ^{90}Sr radionuclides transfer to crops. Institute for Soil Science and Agrochemistry, Minsk

Rauret G, Firsakova S (1996) The transfer of radionuclides through the terrestrial environment to agricultural products, including the evaluation of agrochemical practices, EUR 16528 EN. - European Commission, Luxembourg

Rauret G, Vidal M, Rigol A, Camps M (1999) The reduction of ^{137}Cs and ^{90}Sr uptake by grasses in natural meadows–REDUP. IC15-CT96-0212. Final report, 1 February 1997–31 July 1999

Recommendations on agro industrial production under the conditions of radioactive contamination of the territories of the Republic of Belarus (Rekomendacii po vedeniyu agropromyshlennogo proizvodstva v usloviyakh radioaktivnogo zagryazneniya zemel Respubliki Belarus) (2003) Minsk

Romanov GN (1993) Likvidaciya posledstvij radiacionnykh avarij (Dealing with the consequences of radiation accidents). IzdAT, Moscow

Rules of on agro-industrial production under the conditions of radioactive contamination of territories of the Republic of Belarus for 2002–2005 (Pravila vedeniya agropromyshlennogo proizvodstva v usloviyakh radioaktivnogo zagryazneniya zemel RB na 2002–2005 gg.). Ministry of Agriculture and Food of the Republic of Belarus) (2002) Minsk

Sanzharova NI, Fesenko SV, Kotik VA, Spiridonov SI (1996) Behavior of radionuclides in meadow and efficiency of countermeasures. Rad Protec Dosim 64:43–48

Shaw G (1993) Blockade by fertilizers of caesium and strontium uptake into crops: effects on the root uptake process. Sci Total Environ 137:119–133

Sheppard MJ (1985) Radionuclide partitioning coefficients in soils and plants and their correlation. Health Phys 1:106–111
Spitsin VI, Gromov VV (1959) Absorption of radiostrontium by some soil minerals and grounds. Pochvovedeniye 12:45–50
Suslina LG (2004) Investigations on influence of acidity, potassium and ammonia nitrogen on sorption of ^{137}Cs by different soils and its adsorption by barley. PhD thesis, RIARAE, Obninsk
Sysoeva AA (2004) Experimental study and modelling of processes determining the mobility of ^{90} Sr and ^{137}Cs in soil-plant system. PhD thesis, RIARAE, Obninsk
Taylor AW (1968) Strontium retention in acid soils of the North Carolina coastal plain. Soil Sci 4:440–450
Teplyakov IG, Romanov GN, Spirin DA (1997) Returning of the EURT lands to agricultural use. Iss Rad Saf 3:33–41
Ternovsky IA, Teverovsky YN (1984) Izuchenie radiologicheskikh, radiacionno-gigienicheskikh i socialno-khozyaistvennykh posledstvij massivnogo radioaktivnogo zagryazneniya bolshikh ploschadej (1958–1984 gg.) (Study of radiological, radiation-hygienic and socio-economic consequences of massive radioactive contamination of large areas (1958–1984)), vol 4, 2166 Agricultural use of contaminated territories: Research report. Funds of «Mayak" PA
The guide to farming under the conditions of radioactive contamination of the part of territory of RSFSR, Ukrainian SSR and Belarusian SSR for the period 1988–1990 (Rukovodstvo po vedeniyu selskogo khozyajstva v usloviyakh radioaktivnogo zagryazneniya chasti territorii RSFSR, Ukrainskoj SSR i Belorusskoj SSR na period 1988–1990 gg. (1988) Industrial Press. Industrial complex AgroNIITEIITO, Moscow
Timofeyeva NA (1960) On the migration of radiostrontium in biogeocoenoses. Reports of AS USSR 2:488–491
Timofeyev-Resovsky NV, Titlyanova AA (1966) Povedeniye radioaktivnykh izotopov v sisteme pochva-rastvor (Behaviour of radioactive isotopes in soil-solution system. In: Soil radioactivity and methods of its determination), Nauka, Moscow pp 65–72
Valcke E (1992) The behavior dynamics of radiocaesium and radiostrontium in soil rich in organic matter, PhD thesis, Katholieke Universiteit Leuven, pp 135
Vasilenko IY, Vasilenko OI (2002) Radioactive strontium. Ener Econom Engg Ecol 4:26–32
Vidal M, Camps M, Grebenshikova N, Sanzharova N, Ivanov Y, Vandecasteele C, Shand C, Rigol A, Firsakova S, Fesenko S, Levchuk S, Cheshire M, Sauras T, Rauret G (2001) Soil- and plant-based countermeasures to reduce ^{137}Cs and ^{90}Sr uptake by grasses in natural meadows: the REDUP project. J Environ Radioact 56:139–156
Vodovozova IG, Pogodin RI (1969) Influence of organic matter on transfer of radioactive isotopes to plants. In: Abstracts of the conference "Radioactive isotopes in soil and freshwater systems". Sverdlovsk pp 29
Yudintseva YV, Gulyakin IV (1968) Agrokhimiya radioaktivnykh izotopov stronciya i ceziya (Agro chemistry of radioactive isotopes of strontium and caesium). Atomizdat, Moscow

Rehabilitation of Radioactively Contaminated Soil: Use of Bioremediation/Phytoremediation Techniques

Joana Lourenço, Sonia Mendo, and Ruth Pereira

1 Introduction

Just before the end of the Second World War, during the race for the construction of nuclear weapons by the U.S. and Germany, a new source of anthropogenic environmental contamination has arisen: the radioactive contamination. This new type of contamination reached all environmental compartments. Although, the soil cover has in its composition radionuclides that are naturally present, usually referred to as NORMs, Naturally Occurring Radioactive Materials (like ^{238}U, ^{235}U, ^{234}U, ^{226}Ra, ^{210}Po, ^{222}Rn, ^{210}Pb, for example), originating from the formation of our planet, their concentration may be enhanced due to anthropogenic activities like mining and milling activities, particularly in the case of uranium extraction and processing (IAEA 2005). This caused the radioactive contamination of soils and foodstuffs, contributing for the exposure of animals and humans to ionizing radiation (IAEA 2005). The construction and deployment of nuclear weapons, the use of nuclear energy for the production of electricity, the development of a complete nuclear fuel cycle, and also the occurrence of nuclear disasters, released a new type of radionuclides, the artificial radionuclides, produced due to the fission/fusion reactions that occur during those processes. These artificial or technogenic radionuclides were massively released during nuclear weapons tests and nuclear disasters, such as, the Chernobyl, Fukushima and Kyshtym disasters, contaminating very large surface areas (Aleksakhin 2009; Evrard et al. 2014). The Chernobyl nuclear power plant

J. Lourenço (✉) · S. Mendo
Department of Biology & CESAM, University of Aveiro, Aveiro, Portugal
e-mail: joanalourenco@ua.pt; smendo@ua.pt

R. Pereira
Department of Biology & GreenUPorto, Faculty of Sciences of the University of Porto, Porto, Portugal
e-mail: ruth.pereira@fc.up.pt

D. K. Gupta, A. Voronina (eds.), *Remediation Measures for Radioactively Contaminated Areas*, https://doi.org/10.1007/978-3-319-73398-2_8

disaster, the most serious nuclear disaster in history, contaminated an area that reached 195,000 Km^2 with a level of ^{137}Cs above 37KBq/m^2 (Aleksakhin 2009). The testing of thermonuclear bombs, caused the global dispersion of radioactive contamination, since the fine grained aerosol sized particles, formed during the explosion, may reach the stratosphere, remain there for several months and be dispersed globally by strong winds (Lourenço et al. 2016).

The contamination of soils with radioactive elements and their integration and migration through the food chains, means that animals, plants and humans will be exposed to the radiation (Aleksakhin 2009). Depending on its physicochemical characteristics (e.g. pH, organic matter content), which will influence radionuclides mobility, that may disturb the soil ecosystem, destabilizing and compromising its functions and services, making it inappropriate, for example, for agricultural activities (Smičiklas and Šljivić-Ivanović 2016). The rehabilitation of radioactively contaminated soils is, therefore, of paramount importance. Soils can be remediated using mainly physical, chemical and biological remediation techniques like bio- and phytoremediation. Different approaches can be used solely or simultaneously, however, the choice of the methods to be used will depend on contaminant's mobility, the distribution pattern in different fractions of the soil and future land use (Smičiklas and Šljivić-Ivanović 2016). In this chapter, we will begin by identifying and describing the main sources of soil radioactive contamination, as well as the radionuclides that most contribute to the contamination in each scenario. Then the mostly studied bio- and phytoremediation techniques for the rehabilitation of such contaminated soils will be described and discussed.

2 Radioactive Contamination–Site-Specific Problematics

2.1 Nuclear Fuel Cycle

2.1.1 Uranium Mining and Milling

The mining industry has long been associated with a significant environmental impact (IAEA 2009). Mining and processing of uranium and thorium bearing materials, produce a diversity of waste materials that contain a number of radioactive and non-radioactive hazardous constituents, namely metals and radionuclides (IAEA 2005, 2010; Carvalho et al. 2007; Pereira et al. 2014). Uranium can be extracted using three different methods: open cast, underground and in situ leaching (ISL) (IAEA 2002a, 2005). Considering all the mining methods, the major environmental issues related to uranium mining and milling are: (a) the production of large amounts of waste rock and mill tailings, containing radionuclides with long half lives (>1000 years) and a great variety of metals (IAEA 2005, 2009; Winfield et al. 2006; Carvalho et al. 2007; (b) potential of waste rock and mill tailings to generate acid mine drainage (IAEA 2005, 2009; Winfield et al. 2006; Carvalho et al. 2007); (c) water and wind erosion, which affect mainly above-ground tailings impoundments and waste piles, spreading contamination to adjacent areas mostly to surface water and soil (IAEA 2002a; Winfield et al. 2006); (d) the failure of waste management facilities that can cause an uncontrolled release of waste material and contaminated water to the environment (IAEA 2002a; Winfield et al. 2006); (e) controlled release of contaminated water, mainly from mill sites located in areas with low evaporation and/or high precipitation potentially contaminating surface and groundwater and soil (IAEA 2002a; Winfield et al. 2006); (f) spills during the transport of tailings or mine wastes to an impoundment area (leaks in pipes or trucks not properly covered) (IAEA 2002a); (g) construction of buildings on tailings and mine waste and unauthorized removal of those wastes (IAEA 2002a); (h) radon emission, a radioactive gas that is produced by the decay of ^{238}U and ^{232}Th (^{222}Rn and ^{220}Rn respectively) (IAEA 2002a; Winfield et al. 2006; Carvalho et al. 2007); and (i) seepage of contaminated water from above and bellow ground waste management facilities into surface and groundwater (IAEA 2002a; Winfield et al. 2006).

After the Chernobyl accident in 1986 and after the end of the Cold War, there was a severe drop of uranium prices worldwide that lead to the closure of several uranium mines and mills (Merkel and Schipek 2011; Pereira et al. 2014; Lourenço et al. 2016). As a consequence, many of these sites were left abandoned with little or no remediation at all, creating the so called "legacy" sites (Waggitt 2008; IAEA 2009) and the same happened at the end of mines life (IAEA 2009). At the time uranium mines and mills began operation, there was no environmental protection legislation whatsoever in the majority of the countries where they were established (IAEA 2009). The main ongoing environmental problems associated with legacy sites are: (a) migration of radionuclides and metals into the environment due to wind and water erosion of unprotected deposits containing Technologically Enhanced levels of Naturally Occurring Radioactive Materials (85% of the amount present in the original ore) like ^{226}Ra, ^{222}Rn, ^{210}Po, ^{230}Th, ^{234}U, ^{210}Pb, ^{238}U and associated metals (Waggitt 2008; Salbu 2013; Matshusa and Makgae 2017); (b) radon flux or gamma dose rate from unprotected waste piles, as tailings contain a very high percentage of the radium present in the original ore (90%) (Matshusa and Makgae 2017). Radium 226 is one of the most problematic radionuclides due to its very long half-life (1600 years) during which it is continuously decaying and consequently producing radon gas (a radioactive gas that is carcinogenic), which in turn will decay and produce ^{210}Po, which is also a very hazardous radionuclide. Radium as also the potential of polluting surface and groundwater (Salbu 2013; Matshusa and Makgae 2017); (c) the transport of radionuclides into surface and groundwater, spreading contamination and posing serious problems to water management, shared water resources and potential soil contamination (Salbu 2013; Matshusa and Makgae 2017); (d) acid mine drainage, which will acidify soil and water and increase the concentration of several stable elements like arsenic and selenium and also of several metals like nickel, zinc and manganese (Waggitt 2008; Salbu 2013; Matshusa and Makgae 2017); (e) soil contamination from the disposal of sludge containing high concentration of metals and radionuclides from mine water treatment (Fig. 1).

Fig. 1 Soil contamination at uranium mines: deposition site for sludge from mine water treatment ponds

Uranium will be in demand for many decades to come due to the increase in the investment in nuclear power generation, in which is forecasted that China will have the main role (OECD-NEA 2014). Although the current mine production is already providing 90% of the requirements of power utilities, both the NEA (Nuclear Energy Agency) and the IEA (International Energy Agency) in a Technology Roadmap published in 2015 (NEA-IEA 2015), stated that investments in environmentally sustainable uranium mining should be developed to address expected long-term demand and to avoid the generation of new legacy sites.

2.1.2 Uranium Fuel Fabrication (Refining, Conversion and Enrichment)

The "front end" of the nuclear fuel cycle consists in the extraction of uranium and all of the facilities that are involved in the production of the nuclear fuel until its irradiation in a nuclear reactor (Turner 2013). Fuel fabrication for nuclear reactors consist mainly of uranium purification, conversion to tetrafluoride and/or hexafluoride (if uranium is to be enriched), conversion to the necessary chemical form (metal or oxide), shaping into the required form, loading the fuel pellets into tubes of Zircaloy or stainless steel, sealing the ends with welded caps and locating the fuel rods to form the reactors assemblies (IAEA 2002b). The fuel consists in reactive material containing uranium or plutonium and a metal casing to protect the material and prevent fission products from escaping (IAEA 2002b). Not all reactors require uranium enrichment like CANDU or fast breeding reactors (IAEA 2002b). Uranium fuel can be produced in two forms: metal fuel and oxide fuel. Metal fuel production is performed by converting uranium to UF4 and mixing it with chips of magnesium, not needing uranium enrichment. To produce oxide fuel, uranium is converted to UF6 and enriched to 3–5% uranium-235 (IAEA 2002b). Then UF6 is converted to uranium dioxide powder and formed into pellets (IAEA 2002b). MOX fuel fabrication implies the mechanical mixing of uranium and plutonium oxide powders or dilution of a master blend of uranium and plutonium oxide to the required enrichment (dry process) or by using a uranium-plutonium nitrate solution (wet process) (IAEA 2002b). The uranium used can be natural or depleted uranium (IAEA 2002b). The downstream production steps (pelletizing, sintering, rod fabrication and assembling) are similar to those used for uranium fuel production (IAEA 2002b).

The major environmental impacts that arise from nuclear fuel fabrication are the generation of low level radioactive wastes, air emissions and the risks of radiation exposure to workers and host communities (Winfield et al. 2006). Low-level radioactive wastes include contaminated material and contaminated equipment normally used during nuclear fuel production (IAEA 1996, 2002b; Winfield et al. 2006). The main atmospheric releases consist in uranium, low quantities of uranium and plutonium radionulides (^{235}U, ^{238}U, ^{238}Pu, ^{239}Pu, ^{240}Pu, ^{241}Pu, ^{241}Am (NRC 2012)) metals, dioxins and furans, hexachlorobenzene, ammonia, particulate matter, sulphuric acid, nitrogen oxides, hydrogen chloride, hydrogen fluoride and greenhouse gases as a result of transportation from mill to processing facilities and all stages of the fuel production (IAEA 1996, 2002b; Winfield et al. 2006). Water impacts include the release of uranium, low quantities of ^{226}Ra and ^{230}Th nitrates, ammonia and phosphorus. There is also the problem of facility land impacts and off-site ecosystem impacts of pollutants release (Winfield et al. 2006). Depending on the operating history of the plants, namely the occurrence of accidents, including spills and unintentional releases, cleanup and decommissioning procedures in this kind of facilities may include remediating soil/groundwater and waste management, such as disposing of contaminated materials (Turner 2013). Decommissioning issues related to the front end of the nuclear fuel cycle are mainly associated with naturally occurring radionuclides, such as uranium and radium and other hazards (chemicals) associated with the processing of uranium ore (Turner 2013). The size and complexity of these plants can lead to large costs for full decommissioning of all facilities that can reach billions of dollars (Turner 2013) (Fig. 2).

2.1.3 Nuclear Power Plants

Nuclear power represents 18% of the overall share of low carbon energy production being the largest low-carbon energy source in OECD countries in 2013 and second at global levels with an 11% share (NEA-IEA 2015). Currently, among the 30 countries that are operating nuclear power plants (NPPs), 13 are either constructing new units or completing previously suspended construction projects and 16 have plans or proposals for new reactors (IAEA 2017). Twenty-eight other countries are considering, planning or starting nuclear power and an additional 20 have expressed

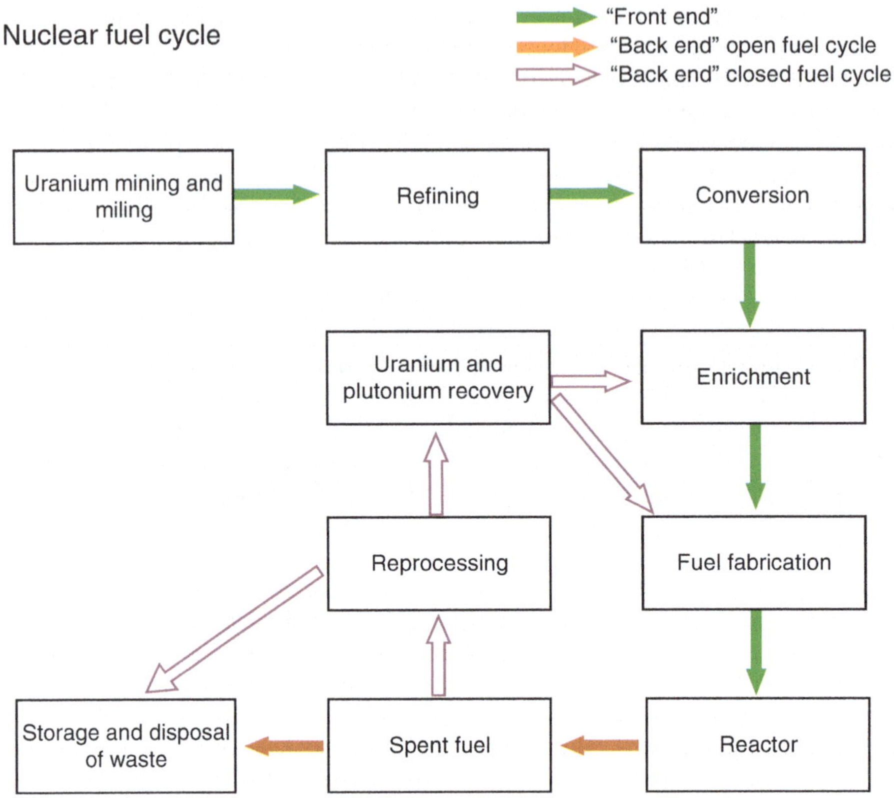

Fig. 2 The nuclear fuel cycle

interest in nuclear power (IAEA 2017). Although a significant decline is expected in North America and almost all Europe, the International Atomic Energy Agency (IAEA), states that the long term potential remains high, with a significant increase projected for central and eastern Asia, with an expected increase in power capacity of 43% by 2025 (IAEA 2017). There is also a significant interest in developing innovative designs and technologically advanced nuclear fuel cycles to improve safety and competitiveness of nuclear power plants (NEA-IEA 2015).

The use of nuclear reactions as a power source is a very controversial matter, which as been fueled recently by the Fukushima accident in 2011 (Seier and Zimmermann 2014). This recent accident, along with others like Chernobyl or Three Mile Island has been calling the attention for human health and environmental risks associated with nuclear power plants. However, the potential for environmental contamination does not exist only in the case of an accident, but also during the normal functioning of nuclear power plants. The main sources of environmental contamination are associated with air emissions, contributing for the contamination of soil and water, discharges of cooling waters into the sea, rivers and lakes and leakages to aquifers (Fairlie 2014). The main radionuclides released during air emissions are ^3H and ^{14}C (Winfield et al. 2006; Fairlie 2014). These emissions are relatively large (around 1.1E + 03 GBq of both ^3H and ^{14}C) and result in elevated nuclide concentrations in vegetation and foodstuffs near NPPs (Fairlie 2014). Moreover, the replacement of the fuel from the reactors causes spikes in air emissions, which occurs generally once a year (Fairlie 2014). During these spikes, 20–100 times higher levels of nuclides are emitted than in the rest of the year (Fairlie 2014). Other common nuclides found in air emissions are iodine (131, 132, 133, 134, 135), bromine (82), cobalt (58, 60), cesium (134, 137), chromium (51), manganese (54) and niobium (95), besides noble gases (NRC 2012). These radionuclides are likely to contaminate surface and groundwater, soil and foodstuffs, putting human populations at risk of developing genetic diseases like cancer (NRC 2012).

2.1.4 Waste Repositories and Nuclear Fuel Reprocessing

The management of spent nuclear fuel is one of the most important stages in the nuclear fuel cycle, beginning with the removal of the spent fuel from the reactors and ending with its final disposal in a waste repository site (IAEA 2012). There are two main options to manage the spent fuel: an open fuel cycle, where the spent fuel is immediately disposed of after it is removed from the reactor and a closed fuel cycle, where the fuel is sent to reprocessing plants for the recycling of U and Pu into new fuel assemblies (back end of the nuclear fuel cycle) (IAEA 2012). Nowadays, the spent fuel is being stored in reactor pools or in independent spent fuel storage installations using wet (spent fuel pools) or dry technology (metal casks and concrete silos or vaults) (IAEA 2012). For the instance, NPPs have to keep the spent fuel on site, but the amount of accumulated spent fuel is growing and fuel pools were never intended to serve as long term storage sites (IAEA 2012). Consequently, there is an increasing need for additional storage capacity, by establishing new independent spent fuel storage facilities (IAEA 2008, 2012).

Nuclear fuel reprocessing is a chemical treatment that allows to recover fissionable actinides from the spent fuel, namely uranium and plutonium isotopes and also to stabilize fission products into durable waste forms (Simpson and Law 2013). Reprocessing can also reduce the need for geologic repositories for spent fuel, by decreasing the volume of waste that has to be sent to such repositories (Simpson and Law 2013).

The function of a repository is to isolate the waste and retard the transport of radionuclides to the surface (SNCNW 2016). As long as the engineered barriers and the canisters remain intact no radionuclides will leak out. If a leakage is to occur from these repositories, it means that radionuclides have passed through the engineered barriers and were transported to the surface by groundwater (SNCNW 2016). They will appear first in the water that will also contaminate soils, and then in locally produced foodstuffs (SNCNW 2016). There could be also the emission of gaseous radionuclides, although probably in very small amounts, mainly of ^{14}C in the form of carbon dioxide and radon (SNCNW 2016). The type of radionuclides that reach the surface is affected by the mobility of the nuclides and to what extent they are water-soluble (SNCNW 2016). Therefore, even if a large fraction of the activity in the repository is due to long-lived radionuclides, it will be their fission products that will reach the surface (SNCNW 2016). The radionuclides that will reach the surface in the event of a leakage will depend on the how long they were isolated in the repository site due to their specific half-lives. For example, if the leakage occurred less than about 100 years after closure, the radionuclides that may occur are ^{14}C, ^{129}I, ^{137}Cs and ^{90}Sr (SNCNW 2016). If the leakage occurred about 1000 years after closure, the radionuclides that may occur are ^{14}C, ^{129}I, ^{79}Se and ^{108m} Ag. If a leakage is to occur after 10,000 years of closure, radioactive contamination would come mainly from ^{129}I and other long-lived fission products like ^{79}Se and ^{59} Ni. However, there is still a great contribution of ^{14}C (SNCNW 2016). After 100,000 years the radioactivity would come mainly from ^{226}Ra, since it is the decay product of ^{238}U, which is the main constituent of spent fuel. There would be also some contribution of ^{129}I and ^{237}Np (SNCNW 2016). Above all, radionuclides that dissolve in water and have a low tendency to bind to particles will be the ones to reach the surface.

2.2 *Nuclear Disasters*

Since the end of the World War II and also since the beginning of the commercial operation of nuclear power plants, there have been some major nuclear accidents. The ones that were rated higher in the INES scale, causing a higher environmental contamination and public exposure were the Chernobyl accident in 1986 (level 7, major accident), the Fukushima accident in 2011 (also level 7) and the Kyshtym accident in 1957 (level 6).

The Chernobyl accident occurred in April 1986 at a nuclear power plant, and the following reactor fire that lasted for 10 days, resulted in a massive and unprecedented release of radioactive material from the reactor core and serious consequences for people and the environment (IAEA 2001, 2006). The release of radioactive material included ^{131}I, ^{137}Cs and other cesium isotopes, ^{90}Sr and also plutonium isotopes (IAEA 2006). It caused the contamination of more that 200,000 km^2 in Europe with ^{137}Cs, and the majority was deposited in the most affected countries (Belarus, Russia and Ukraine) (IAEA 2001, 2006). The deposition of the radioactive material was highly affected by rain and therefore very heterogeneous (IAEA 2006). After the initial period that followed the accident, where mainly radionuclides with short half-lives were emitted to the atmosphere (meaning that many of them already decayed), ^{137}Cs became the radionuclide with the greatest radiological significance (being ^{90}Sr of less importance), particularly in the agricultural environment as well as in the forest environment (IAEA 2006; SNCNW 2016). At the long term (thousands of years), the radionuclides that will be important are ^{241} Am and plutonium isotopes, due to their very long half-lives (IAEA 2006).

The Fukushima Daiichi nuclear accident began with the Great East Japan Earthquake in March 2011, which was a seismic event of great magnitude that generated a massive tsunami beyond the resistance capacity of Fukushima Daiichi nuclear power plant (IAEA 2014). The earthquake and the tsunami caused a general failure in the power supply and backup generators needed to provide support for the cooling system of the reactor and spent fuel (IAEA 2014). This lead to explosions and severe core damage in three of the six nuclear reactors and release of significant amounts of radioactive material to the atmosphere and to the Pacific ocean, during several days (IAEA 2014). An area of about 500 km^2 show radiation dose levels above 20 mSv y^{-1} and about 1300 km^2 show radiation dose levels between 5 and 20 mSv y^{-1} (IAEA 2011). The accident caused the atmospheric and marine releases of high amounts of ^{131}I, ^{137}Cs and ^{134}Cs, which represented about 10–20% of the releases that occurred during the Chernobyl accident (IAEA 2014). Along with these radionuclides, smaller amounts of tritium were also released (IAEA 2014). As in Chernobyl, ^{137}Cs will be the radionuclide that will contribute the most to radiation doses in the future, since ^{134}Cs as a shorter half-life (IAEA 2014).

The Mayak plant was the first established in the former Soviet Union for the production and separation of ^{239}Pu as nuclear weapons material (Christensen et al. 1995). Since its early years of operation, the plant directly discharged high-level nuclear waste into nearby small lakes and into the Techa river (Christensen et al. 1995). A total of 100 PBq of liquid waste, containing a mixture of radionuclides (from which ^{90}Sr and ^{137}Cs represented 25% of the total activity), were directly discharged in the Techa river from 1948–1956 (Christensen et al. 1995). Since 1951, the radioactive waste started to be sent to Lake Karachay instead of being directly discharged in the Techa river (Christensen et al. 1995). The main radionuclides associated to these discharges were ^{90}Sr and ^{137}Cs, in a total quantity of 4440 PBq in the 1990s (Christensen et al. 1995). In 1957, a problem in the cooling system of the high-level nuclear waste storage tanks of the Mayak plant caused an explosion that spread contamination for 20,000 km/m^2 leading to the formation of the East Urals Radioactive Trace (EURT), due to a strong wind from the south west. The radionuclide causing higher concern in terms of dose is ^{90}Sr due to its longer half-life (Molchanova et al. 2014; SNCNW 2016). Later in 1967, due to a very hot summer and a dry winter, Lake Karachay's water completely evaporated, which caused the spreading about 20TBq of radioactive dust with the wind, contaminating parts of the same area as the EURT (Christensen et al. 1995; Molchanova et al. 2014). This time the primary contaminant was ^{137}Cs (Molchanova et al. 2014).

2.3 *Nuclear Test Sites*

The end of the Second World War launched the nuclear arms race and countries like the US, USSR, the United Kingdom, France and China became nuclear powers between 1945 and 1964 (Schenck and Youmans 2012; Prăvălie 2014). The detonation of nuclear weapons was mainly performed in the so called "test sites". As a consequence, in some sites several hundreds of tests were performed, resulting in a severe environmental contamination. Among nuclear test sites, the Marshall Islands

(USA), the Novaya Zemlya Soviet test site (USSR), the Nevada Test Site (USA) and the Semipalatinsk Test Site (USSR) were the most important and the ones that suffered higher environmental impact from several nuclear detonations performed in each of those sites. In the Marshall Islands, namely in the Bikini and Enewetak Atolls, 67 nuclear detonations were performed, namely some very powerful H-Bombs detonations, among which the Castle Bravo test is included. The area is severely contaminated having high levels of radioactivity, contamination of marine ecosystems and soils, rendering them improper for agriculture (Lachner et al. 2014; Prăvălie 2014). This contamination is mainly due to the release of ^{137}Cs (present in seawater, lagoons sediments and fish), ^{90}Sr (present in coral soils) and $^{239\text{-}240}Pu$ and ^{241}Am (both found mainly in coral sediments) (Prăvălie 2014). The Novaya Zemlya test site was the place where the former Soviet Union performed 20% of the country's nuclear detonations (130 tests). It was also the site where the most powerful nuclear device ever built (the Tsar Bomba) was detonated (Wendel et al. 2013; Prăvălie 2014). The nuclear tests performed in this site, namely the "Tsar Bomba" test (with a yield of 50Mt) released high amounts of radioactive debris giving a great contribution to the presence of artificial radionuclides in the stratosphere and allowing their global dispersion (Wendel et al. 2013; Prăvălie 2014). At the site, the radionuclides ^{137}Cs and $^{239\text{-}240}Pu$ are the main sources of contamination, mainly in seawaters (Prăvălie 2014). Forty four percent of all nuclear tests performed worldwide were carried out at the Nevada Test Site, in a total of 86 atmospheric detonations and approximately 800 underground detonations (Zhao et al. 2011; Stone et al. 2013; Prăvălie 2014;). The main environmental consequences are the spreading of fallout through rainfalls (UNSCEAR 1993), which contains among other radionuclides ^{14}C and ^{137}Cs, that contaminate soils and biota (Snyder et al. 2012; Stone et al. 2013; Prăvălie 2014). In this test site there is also a high risk of groundwater contamination with several radionuclides, mainly the $^{239\text{-}240}Pu$ isotopes due groundwater hydrodynamic processes that will increase their mobility, with the risk of reaching the surface (Prăvălie 2014). There was also the accidental release of radionuclides, mainly ^{131}I, to the atmosphere during underground tests, due to venting, which occurred at least 32 times (UNSCEAR 1993). The former USSR is responsible for conducting 35% of the tests conducted worldwide (715 tests) and 65% of those tests (456 nuclear explosions) were conducted at the Semipalatinsk test site. Among all those detonations, 110 were atmospheric, causing severe contamination of soil and water (Evseeva et al. 2012; Prăvălie 2014). The site is contaminated with ^{234}U, ^{235}U and ^{238}U, mainly found in local water bodies and also with ^{90}Sr, ^{137}Cs, $^{239\text{–}240}Pu$, and ^{241}Am, mainly present in soil and vegetation (Carlsen et al. 2001; Kadyrzhanov et al. 2005; Yamamoto et al. 2010; Prăvălie 2014). There is also the contribution of Chinese nuclear tests performed at Lop Nur, which further increased the radioactive environmental contamination at Semipalatinsk by adding more ^{137}Cs, ^{106}Ru, ^{103}Ru, ^{141}Ce, ^{144}Ce and ^{95}Zr, in the south part of the site (Prăvălie 2014).

Although several countries have already joined the Nuclear Non-Proliferation Treaty (NPT) with the main goal of limiting the production of weapons of mass destruction and avoid a potential nuclear war, there are countries that are still performing nuclear tests, nowadays.

2.4 Depleted Military Use Uranium

Depleted uranium (DU) is one of the by products of uranium enrichment. The "depleted" term comes from the fact it retains only about 60% of natural uranium radioactivity, due to a lower relative concentration of the isotopes ^{235}U and ^{234}U (Bleise et al. 2003; Jiang and Aschner 2009; Handley-Sidhu et al. 2010);. DU has a wide range of peaceful applications, however, compared to its military applications, the use of DU for civilian purposes is only marginal (Bem and Bou-Rabee 2004). DU as been used for armor-piercing ammunitions, since it self sharpens on impact due to its pyrophoricity, allowing for a superior penetrative ability (Bleise et al. 2003). Moreover, it is also used in protective armors for tanks (Bleise et al. 2003). Depleted uranium ammunitions were used in several conflicts, mainly in the Gulf War (1990–1991), Bosnia and Herzegovina (1994–1995), Kosovo (1999) and in the Iraq War (2003) (Burger 2012). During the military conflicts, DU can enter the environment through projectiles that become lodged into soil or through aerosols produced upon ammunition firing and their impact on hard targets (Handley-Sidhu et al. 2010). When ammunitions hit the targets 10–35% of their mass is transformed into aerosols (Bleise et al. 2003; Jiang and Aschner 2009; Handley-Sidhu et al. 2010). According to the United Nations Environmental Program (UNEP) reports (UNEP 2001, 2002, 2003a), although some contamination of soil and water was detected in Serbia and Montenegro, Bosnia Herzegovina and Kosovo, due to the spreading of aerosols produced upon ammunition impact on armored vehicles, such contamination did not extended more than 100 m from the target (UNEP 2001, 2002, 2003a). Also, results from the analysis on soil, air, water and milk samples showed values for uranium within the normal range (UNEP 2001, 2002, 2003a). However, as highlighted by the UNEP reports, there is the problem of the high number of ammunitions that missed the target and that are still berried in the ground (UNEP 2001, 2002, 2003a). Those projectiles can pose a risk of groundwater and surface water contamination, specially in areas of heavy DU firing (UNEP 2001, 2002, 2003a). Therefore, long-term studies will be needed to determine accurately the DU dispersion to the environment. In Iraq there is the problem of the abandon of DU-impacted tanks that still lay abandoned in former battle fields, with unrestricted access (UNEP 2007). Also, there is the problem of uncontrolled recycling of DU-impacted military scrap, which may impact people and the environment (UNEP 2007). Regarding the Gulf War, according to the report produced by IAEA (International Atomic Energy Agency), the DU ammunitions used do not pose a long-term hazard to people and the environment (IAEA 2003). However, the report UNEP (UNEP 2003b) highlights the need for additional research to clarify the uncertainties regarding the effect of DU ammunitions deployed during the war, in the environment and human health.

2.5 Use of Radioactive Sources for Medical Purposes

The use of radionuclides in medicine is very common and a very well established field (IAEA 2000). Radionuclides can be used in medicine for: a) diagnostic imaging procedures (e.g. ^{67}Ga, ^{111}In, ^{201}Tl, ^{123}I and ^{131}I; b) for the labeling of human blood components (e.g. ^{99m}Tc, ^{111}In, ^{51}Cr, ^{59}Fe and ^{125}I); c) bone density scanners (e.g. ^{241}Am, ^{153}Gd or ^{125}I); d) lung ventilation imaging (e.g. ^{81m}Kr, ^{133}Xe and ^{99m}Tc); e) direct implant or application to a patient and also for stenosis treatments (e.g. ^{106}Ru, ^{192}Ir, ^{137}Cs, ^{198}Au, ^{192}Ir and ^{89}Sr); f) teletherapy and gamma knife surgery (e.g. ^{137}Cs and ^{60}Co) (IAEA 2000). The use of radioactive substances, as also the use of chemical and organic substances for example, always generates wastes. The management of such wastes needs to begin with the purchase of isotopes, then isotope application, waste production and waste collection and segregation (IAEA 2000). The waste is then divided into long-lived/short lived isotopes (IAEA 2000). Long-lived isotopes will suffer a pretreatment and then a treatment on a centralized waste processing facility (IAEA 2000). The short-lived isotopes will be stored for decay and control, treated for disinfection, incineration, quality control and then released as municipal waste (IAEA 2000). All the waste management procedure needs to be fully controlled and accomplished with the maximum security, to make sure that wastes are treated properly to preserve humans and the environment (IAEA 2000). However, as in other areas that include the handling and use of radioactive sources, accidents can occur, with a great risk of humans and environmental contamination. That was the case of the worst accident ever recorded involving a sealed ^{137}Cs source, used for teletherapy. At the end of 1985, a private radiotherapy institute in Goiania, Brazil, moved to new premises, leaving behind a ^{137}Cs teletherapy unit without notifying the licensing authority as required (IAEA 1988). As a consequence, it was taken by persons who were unaware of the risks and unprepared to handle the source (IAEA 1988), leading to the rupture of the capsule containing highly soluble and readily dispersible caesium chloride salt. Four people were killed and 249 were found to have significant levels of radioactive material in or on their bodies (IAEA 1988). In terms of environmental contamination, countermeasures were necessary for soil and fruit within a 50 m radius from the primary contaminated site (IAEA 1988). The heavy rain that felled at the time of the accident further dispersed cesium into the environment instead of washing out, causing its deposition on roofs, being the main contributor to dose rates in houses of that area (IAEA 1988).

3 Bioremediation of Radioactively Contaminated Soils

Remediation of these sites to acceptable levels of radioactive contaminants is mandatory and is/should be the mission of the regulatory bodies. Various methods based on both active and passive treatment systems have been applied. Most of the strategies employed to clean up these contaminated sites, in situ or ex situ, are based on physicochemical approaches, which render the whole process very expensive (Kumar et al. 2007; Groudev et al. 2010). Even so, in situ chemical leaching of contaminated sites through lixiviation is less expensive and less risky than ex-situ management and it is still quite used (Groudev et al. 2010). Notwithstanding, there has been a growing interest in microorganisms, mainly bacteria, fungi, and plants to decontaminate these areas, including those contaminated by radionuclides and metallic pollutants (Gadd 2010; Prakash et al. 2013). Although this may be a promising technology, its implementation requires thorough understanding of factors governing the growth, metabolism, dynamics and functions of indigenous microbial communities as well as the characteristics of the contaminated sites (Tiwari and Singh 2014), since environmental factors/conditions will also control the composition and dynamics of the microbial community (Williams et al. 2013). The combination of genomics, transcriptomics, proteomics and metabolomics has provided crucial insights into microbial communities, their interactions and mechanisms for the bioremediation of many polluted environments (Prakash et al. 2013). Ultimately, genome wide transcriptome analysis of the microbial community will certainly provide clues on the pathways involved in the process and will contribute to advance on the application of microorganisms to improve this technology. Yet, we are still at a very early stage of understanding, on how the whole process is conducted by these microorganisms. Although some studies have been carried out, most of them were at a lab scale and, therefore, further studies are still required until we can see the full and effective application of bioremediation in real scenarios, particularly in extensive areas contaminated by mixed radioactive contaminants, namely radionuclides and metals.

3.1 Bioremediation: A Cost Effective Environmental Friendly Technology

Bioremediation is a simple and cost effective process that exploits the naturally occurring biodegradative processes, resulting from the activities of indigenous microbiota (bacteria and fungi) or plants (phytoremediation), to clean up contaminated sites. For reasons related to the drafting strategy of this chapter, phytoremediation will be will be reviewed in a different section. Bioremediation has a positive environmental impact and its application has been investigated as an environmental friendly natural process (Williams et al. 2013; Chung et al. 2014; Newsome et al. 2014). It is primarily used in organic pollutants; notwithstanding, other applications have been attempted, like, for instance, for the bioremediation of metals and radionuclides.

Microorganisms have developed efficient and versatile mechanisms that enable them to thrive even in the most extreme environments and that capacity can be exploited for our benefit. In bioremediation, the appropriate organisms, in the correct amounts and combinations and at the appropriate environmental conditions in which they are generally well adapted to survival, play a key role. As above referred, bacteria, fungi, and plants are the major players in this process (NABIR Primer 2003; Gadd 2010). Under aerobic conditions microorganisms use oxygen as an electron acceptor, alternatively, if oxygen is absent they can use other substrates namely, nitrates, sulfates, metals, and carbon dioxide. Accordingly, the availability of electron donors, nutrients, electron acceptors (Fe^{3+}, Mn^{4+}, NO_3^-, SO_4^{2-}, organic compounds), and site-specific environmental factors (pH, Eh, temperature, moisture) determine the type, rate, and extent of microbial activity, and hence, the transformation of radionuclides and metals at contaminated sites.

Microorganisms can solubilize metals by oxidation or reduction with the interference of microbial key enzymes. Metal reduction can also occur passively through binding to reactive sites on the surface or within microbial cells (Francis and Nancharaiah 2015). It is clear though that the composition of the microbial community and the environmental conditions of the site are crucial and will determine the success of the bioremediation process.

3.1.1 In Situ Versus Ex Situ Bioremediation

Different bioremediation methodologies are known and can be employed accordingly. Bioremediation can take place ex situ (such as land farming, biopiling, composting and bioreactor treatment), which involves the removal of the contaminated soil from the site, its transport and further treatment elsewhere and in situ, which involves the treatment of a contaminated site in-place (Elekwachi et al. 2014). The later is desirable as there is low disturbance of the site, avoids excavation and transport, and thus it is less costly. Additionally and probably more important, the exposure of the workers to the contaminants is reduced. In situ bioremediation can be intrinsic, when it relies on naturally occurring indigenous microbiota and it occurs without any stimulation or treatment or it can involve **biostimulation** (with the addition of nutrients, oxygen, etc., to stimulate the growth of the microorganisms involved in the bioremediation process) or **bioaugmentation**, which rely on the introduction of new organisms to perform the biotransformation/ biodegradation (Dixit et al. 2015). In ex-situ remediation, contaminated material is removed and stored or disposed at a low-intermediate level waste facility. Here, microorganisms and nutrients can also be added to accelerate and optimize/biostimulate the transformation of the contaminant. Ex-situ bioaugmentation is also a quite common process. Likewise, bioleaching, a process that is being adapted from the mining industry to the remediation of metals, can be used either in-situ or ex-situ. The process involves microorganisms that help removing metals from soils through solubilization, or by interactions with the metabolic products, or both (Francis and Nancharaiah 2015).

3.1.2 Active Remediation *versus* Environmental Restoration by Natural Attenuation

The degradation of metals, metalloids and radionuclides is inherently impossible. Thus alternative biologically-based technologies have been developed. Bioremediation is now regarded as the "golden" technology for the rehabilitation of polluted environments due to its low costs and environmental friendly methodology. However, its full application depends on many factors, the extent of which must be carefully analysed. In situ natural attenuation of large contaminated areas, thorough the stimulation of indigenous microbial communities, offers many advantages over ex situ treatments, which involve mechanical processes accompanied by the use of physical and chemical alternatives (Harms et al. 2011). As such, the potential of these biological treatments, which lead to the stabilization of the pollutants rather than their elimination, is being fully explored.

As for passive remediation schemes, also referred to as natural remediation/ attenuation, it depends on the natural processes of sorption, retardation (physical, chemical and biological) and radioactive decay. Here, the autochthonous microbial communities transform the chemical state of the radioactive contaminants and modify their solubility and therefore their mobility, while there is a low degree of disturbance of the natural soil (IAEA 1999). A major advantage of natural attenuation is the reduced generation of remediation wastes. Nonetheless, besides a slower cleanup, natural attenuation might not always result in a decrease of the contaminant (s) and may result in transformation products more toxic than the original contaminants (IAEA 1999). For instance, after Chernobyl's accident, the levels of radioactivity were found to be rising after an initial fall (IAEA 1999). Therefore, the combination of both methodologies might produce more desirable results.

3.1.3 Biological Remediation *versus* Physico-chemical Remediation: The Pros and the Cons

Physical remediation methods are those that do not change the physicochemical properties of the pollutants accumulated in the soil to be cleaned. As for chemical remediation, it involves the degradation of the pollutants accumulated or induction of changes to their physico-chemical properties by oxidation/ reduction, extraction, precipitation, pH stabilization. It includes ex situ and in situ methods (Das and Adholeya 2012). Physico-chemical processes such as chemical precipitation, coagulation and adsorption are tremendously costly, particularly if one considers the extent of the contaminated areas to be remediated. Besides, these processes have some disadvantages since large amounts of wastes are produced, that have to be further treated/ recycled. Thus, bioremediation is now regarded as the default method for the rehabilitation of polluted environments considering its cost efficiency and environmental friendliness. Furthermore, bioremediation can also be conducted on site, which can be advantageous as no excavation and transportation is required

(Das and Adholeya 2012). In addition, bioremediation strategies can be refined to improve the overall process, namely, i) bioaugmentation which can be seen as more invasive, since microorganisms are added to the environment to be decontaminated, and ii) biostimulation, where nutrients, electron donors and acceptors are added to stimulate the indigenous microbiota, to more efficiently treat the area (Hazen and Tabak 2005; Shukla et al. 2010). Nevertheless, the extent to which these advantages apply, depend on the degree of technical intervention required at the site itself. Also, the selection of the appropriate technology depends on i) the amenability of the pollutant to biological transformation, ii) the accessibility of the contaminant to microorganisms and iii) the optimization of overall process.

Due to their particular lifestyle and biochemical abilities, fungi have a wide range of biotechnological applications, to remove or stabilize organic and inorganic contaminants (Harms et al. 2011). Most fungi produce high amounts of extracellular enzymes that can be used as intermediates and carbon sources, which will stimulate growth and metabolism of the existing pollutant-degrading bacteria. Therefore, bacteria-fungi associations sound as ecologically relevant bioremediation approaches. Even so, further studies are required until we can see any outcome and possible application of this technology (Harms et al. 2011). As for bacterial-based bioremediation, it requires that the microorganism grow, fully-function and outcompete other bacteria in the presence of all the contaminants present at a specific site (Ruggiero et al. 2005). Accordingly, as referred above, various factors and conditions have to be observed for the success of the biological remediation as opposed to the physico-chemical process.

3.2 *Bioremediation of Radionuclides and Metals*

Radionuclides and metals cannot be biodegraded, but they must be removed or stabilized so that a particular soil/environment is restored. Bioremediation of metals is usually based on the complexation, oxidation-reduction (redox), microbial leaching, microbial surfactants (biosurfactants), volatilization, and bioaccumulation/complexation. In bioleaching, specific microorganisms such as *Thiobacillus ferrooxidans* and *T. thiooxidans* play a major role. For instance, *T. ferrooxidans* plays an important role in the extraction of uranium from ore (Francis and Nancharaiah 2015). Microorganisms who are able to remediate these elements have developed strategies for survival in these habitats, using chemicals for their growth and development, and for developing detoxifying mechanisms. Therefore, they are able to solubilize and immobilize a wide range of these elements. They do so by different mechanisms such as: oxidation-reduction reactions, synthesis of sequestering agents (such as siderophores), bioaccumulation, and biomineralization. Additionally, microorganisms can interact with radionuclides and metals through biotransformation, change their oxidation state by reduction or oxidation and thus increasing the mobility of the contaminant, allowing its easy removal from the environment (Alexander and Loehr 1992; Amachi et al. 2010; Francis and Nancharaiah 2015).

3.2.1 The Major Players: Bacteria and Fungi

Bacteria, in particular, due to their small size, high surface area to volume ratio, ease of cultivation and adaptation to harsh conditions, and their genetic arsenal are preferred candidates for bioremediation of contaminated sites.

Various studies are available in the literature reporting bioremediation of uranium and metals by pure cultures of different bacterial species (Prakash et al. 2013; Li et al. 2014). Still, we believe that bacterial consortiums could to be more effective than pure cultures, since microorganisms live in communities were the whole consortium play an important role in a given process. Furthermore, genetically engineered bacterial strains have also proved effective to enhance the bioremediation capabilities of selected organisms and few examples can be found in the literature (Dash and Das 2015; Das et al. 2016). As for fungi, their hyphae have a large surface area for absorbing nutrients as well as radionuclides and metals from the environment. Mycoremediation studies that have been performed at a lab scale showed that saprotrophic fungi can accumulate and absorb radionuclides (Dighton et al. 1991, 2008). It appears that fungi, including some mushroom species (basidiomycete fruiting body) can mineralize radionuclides and translocate them into their own biomass. Once incorporated, radionuclides are retained and become resistant to leaching. Accordingly, in bioremediation, microorganisms mineralize the organic contaminants to end-products such as carbon dioxide and water, or to metabolic intermediates which are used as primary substrates for cell growth (Dixit et al. 2015).

3.3 Microbial Transformation and Immobilization in the Context of Bioremediation of Radionuclide and Metal Contaminated Soils

Bioremediation can be performed by one of the following mechanisms: biosorption/ bioaccumulation, biomineralisation, biotransformation and bioleaching (Prakash et al. 2013; Dixit et al. 2015). Different bioremediation strategies are employed resulting in radionuclides and toxic metals immobilization: **Biosorption/ bioaccumulation,** is one of the most currently used and promising techniques to bioremediate radionuclide/ metal contaminated sites. It involves the passive immobilization/ adsorption of metals/ radionuclides to cell walls, exopolymers and other structural components of the microorganisms, trough electrostatic interactions and/or chemical reactions (Gadd 2010; Tiwari and Singh 2014; Francis and Nancharaiah 2015). Biosorption is pH dependent, as it was observed that in various fungal species biosorption of some metals was inhibited at pH 3.0 (Tiwari and Singh 2014). Also, it is important to highlight that there is not only one organism able to accumulate all the metals, thus it seems wise to admit that microbial consortia have a key role in this process (Tiwari and Singh 2014). The microbial bioaccumulation of various radionuclides and metals, including ^{137}Cs, ^{85}Sr and uranium, was described by various authors (Sasaki et al. 2013; Francis and Nancharaiah 2015; Seeprasert et al. 2016). Different types of biosorbents exist and many more are currently being developed. For example, Joshi and co-workers have isolated an *Aspergillus terreus* able to biosorb Pb, Cd, Cr and Ni (Joshi et al. 2011). Another fungal isolate *Aspergillus flavus* NA9 was also effective in the biosorption of Zn (Aftab et al. 2013). Also, various fungal species have been isolated from contaminated areas, like Chernobyl, and have been extensively investigated for their application in bioremediation, given their ability to accumulate and/or biotransform radionuclides present in the soil (Dighton et al. 2008). For instance, after the Chernobyl accident high levels of radionuclides were detected in mushrooms. *G. glutinosus* has been reported to absorb and concentrate radioactive ^{137}Cs (Dighton et al. 2008). Likewise, after the Fukushima accident, in 2011, large quantities of Cs and Sr were released and are still present in terrestrial ecosystems nearby. Cs and Sr adhere to organic matter and therefore, it is predictable that Cs and Sr are strongly retained in the top layer of organic soil (Yoshida et al. 2004), and this can be attributed, mainly, to microbial (fungi and bacteria) activity. In fact, Seeprasert et al. (2016), demonstrated that when comparing biotic and abiotic systems, soil fungi play an important role in the biosorption and retention of Cs and Sr, since the retention of both elements was greater in biotic systems than in abiotic systems, and that binding was irreversible in biotic systems (Seeprasert et al. 2016). As for bacteria, *Bacillus thuringiensis* OSM29 was efficient in the removal of Cd, Cr, Cu, Pb and Ni from soils (Oves et al. 2013). Accumulation of arsenic was observed in *Lysinibacillus* sp. B1CDA (Rahman et al. 2014). Another example of a natural biosorbent was employed in Bukhovo (Bulgaria) uranium mine, where an artificial peat bog with dispersed iron was used to adsorb and accumulate uranium from mining waters (IAEA 1999). Extracellular Polymeric Substances (EPS) are another type of biopolymers produced by the microorganisms that possess many adsorption sites for metals and, thus, might play an important role in metal sorption to bacterial cells and, therefore, can be potentially applied for soil remediation. EPS, possess many functional anionic groups (carboxyl, phosphoryl, sulfhydryl, phenolic and hydroxyl) that have cation exchange potential and hence can complex with metals (Ha et al. 2010). Zhang et al. (2014), found that a mixed bacterial community present in a soil conditioner sold in Ishinomaki, Miyagi, located near the area contaminated during the Fukushima nuclear power plant accident, was able to uptake and accumulate Cs^+. pH and temperature greatly affected the uptake process: the higher the temperature (4–37 °C) and the pH range, from 7–8, the higher the Cs^+ uptake. This accumulation was mediated by a K^+ transport system, since Cs^+ uptake was dose-dependently inhibited by potassium ion (K+). Similar results were obtained with *Rhodococcus erythropolis* CS98 and *Rhodococcus* sp., strain CS402 (Tomioka et al. 1995). The authors also suggest that the same approach might be effective to remove ^{137}Cs from contaminated soils. Still, and regarding to radioactive cesium, a study was conducted by Sasaki et al. (2013), were the authors collected the blue-green algae Cyanobacterium *Nostoc commune*, from different soils, in Japan, after the Fukushima accident, in 2011. They demonstrated that *N. commune* is able to accumulated high levels of cesium (415,000 Bq kg^{-1} dry weight ^{134}Cs and 607,000 Bq kg^{-1} dry weight ^{137}Cs) and observed that the concentration of cesium accumulated was higher in areas where soil radioactivity was high. It appears that the radioactive material is deposited in jelly-like clumps of polysaccharides, which allows easy removal of Cs from the environment. Laboratory studies with collected soil samples indicate that *N. commune* might be a suitable candidate for bioremediation of radioactively contaminated soil. Nevertheless, further and deeper studies are required until we can see this approach in practice. Also, based on the fact that proteins from halophilic bacteria have highly acidic protein surfaces (they are rich in acid amino acids) which can interact with metal ions, such as Sr^+ and Cs^+, Arai et al. (2015) argue that these binding sites can be identified by X-Ray crystallography, and the structures can be used as scaffolds for constructing artificial binding sites for metal ions, that could be used as environmentally friendly non-toxic protein-based specific metal-ion absorbents. These protein absorbents will then be heterologous expressed in parts of plants (roots, seeds and flowers), which will accumulate the metal. Nevertheless, and regardless of the considerable amount of information gathered, biosorption has not been applied industrially (Gadd 2009).

In **precipitation**, microorganisms are capable of dissolving metals and reducing or oxidizing transition metals. Bacteria, for instance, use organic compounds as food, and oxidize them in the process. During bioremediation, reduction occurs by indigenous or introduced bacteria that use an electron donor (a food source such as organic matter, sulfides, or ferrous iron) during the process of respiration and transfer electrons to an electron acceptor (a radionuclide, for example), resulting in a lower valence or oxidation state. For some radionuclides the result is that they precipitate out of solution in a more stable, less soluble form (NABIR Primer 2003). In the case of uranium, soluble U(VI) can be bacterially reduced to insoluble U(IV) (Anderson et al. 2003). As for technetium, it can be bacterially reduced from soluble Tc(VII) to the less soluble Tc(IV) (Lloyd et al. 2000). Sulfate-reducing bacteria play an important role in the precipitation of metals since under anaerobic conditions they are able to oxidize organic matter and reduce sulfate to sulfide that in turn reacts with metals to form water-insoluble, metal-sulfide compounds. An extensive review on microorganisms-radionuclide interactions was recently published and readers are referred to it (Newsome et al. 2014).

As for **biomineralisation**, bacteria are able to produce ligands (sulfides, phosphate, carbonates or hydroxides) that interact, immobilize and concentrate metal ions, eventually generating small minerals. For example, uranium biomineralisation was observed in Serratia species (previously classified as Citrobacter spp.) when cells are grown in the presence of glycerol phosphate (Pattanapipitpaisal et al. 2002). Cell's phosphatases are able to cleave the organic phosphate, releasing inorganic phosphate which precipitated with U(VI) as extracellular hydrogen uranyl phosphate minerals [HUO_2PO_4].

Some authors consider that biomineralisation is a toxicity resistance mechanism whereas others state that the process is detrimental to the cells, because it seems to block microorganism's metabolic activity (Benzerara et al. 2011). Evidence available, suggests that, at least in Serratia, uranyl phosphates were found deposited on the cell wall or 'tethered' within the lipopolysaccharide, thus preventing fouling of the cell surface (Macaskie et al. 2000). Various other studies report the isolation of bacteria able to bioremediate radionuclides and metals from soils, namely U, Sr, Th and Cs (Janssen et al. 2010; Young et al. 2010; Achal et al. 2012; Salem et al. 2012). Notwithstanding, although most of the studies show some application at a lab/pilot scale and under controlled conditions, none of them is able to show the application and validation of this technology in real scenarios. In addition, in many cases, bioremediation is combined with other techniques (physico- chemical treatment) and can last for a long time (many months to many years) (Das and Adholeya 2012) and, thus, considering the large areas that have to be remediated and the inherent difficulties of scale up, we believe that, despite all the benefits of this technology and the valuable contribution that we have assisted to, its efficacy might be questioned and we are far from seeing its full application in the near future.

3.4 Limitations of Bioremediation, the Final Remarks

Bioremediation of soils contaminated with radionuclides and metals are still far from being put into practice, at least in real scenarios. Over the years we have seen an improvement of this technology and its application/attempted application in different areas. Nonetheless, the majority of the studies reported have been carried out either at a laboratory scale, using soils from sampled contaminated sites, or soils contaminated in the laboratory. Also, much of its application is restricted to the decontamination of individual metals/radionuclides. As regards to poly-metal decontamination or mixtures of metals/radionuclides in real scenarios, to the best of our knowledge, no information is available.

4 Phytoremediation-Applications and Problems

Phytoremediation can be defined as a plant-based biotechnology to recover and/or manage contaminated soils, by removing (phytoextraction) or by stabilizing contaminants in the soil matrix. Phytostabilization, also known by explant remediation, reduces the leaching of contaminants to groundwater, their spread by wind and runoff by water erosion, through absorption and accumulation by roots, adsorption onto roots surface and/or precipitation within root's zone (McGrath et al. 2002; Dushenkov 2003; Gavrilescu et al. 2009; Vangronsveld et al. 2009; Lee 2013; Sharma et al. 2015). Other strategies, such as the concentration and removal of economic valuable contaminants (usually trace metals) by plants (phytomining) and volatilization of compounds through the soil-plant system (phytovolatilization) will not be the focus of this chapter as they cannot be applied to radionuclide contaminated soils (McGrath et al. 2002; Lee 2013).

For decades, phytoremediation has been referred to as a more cost-effective and environmental friendly approach to restore contaminated soils, as opposed to engineer-based methods, involving the transport and the physical and chemical immobilizations of huge amounts of soil (Entry et al. 1996; Ebbs et al. 1998), with serious physical disturbances (e.g. soil compaction and sealing) which frequently compromise the restoration of functions of the affected areas.

The interest of the academic community on phytoremediation of soils contaminate with radionuclides, arouse in the nineties, after the Chernobyl's nuclear power plant accident, in Ukraine in 1986, which brought the need of finding an approach to phytostabilize an extensive area of soils (about 130,000 km^2), contaminated by the radioactive fallout of the accident (Victorova et al. 2000; Vandenhove 2013). This objective was pursued by an EC-funded project PHYTOR aimed at assessing the ability of willow (*Salix vimialis)* to immobilize radiocesium and radiostrontium (Victorova et al. 2000). Soil surface excavation and physical removal was the method used to immediately deal with the most contaminated soils closer to the reactor (an approach also followed in the vicinity of the Fuskushima nuclear power plant, after the accident), however a more cost effective solution was necessary for the vast affected area. The results from this project were unsatisfactory and since then there are no examples of in situ application of phytoremediation to rehabilitate soils contaminated with both industrial and or natural radionuclides were reported. Therefore, the progresses made on this topic, through field and laboratory studies and, the main reasons for the limited application of phytoremediation to decontaminate soils polluted with radioactive wastes (Fesenko and Howard 2012) almost along 30 years, will be discussed, although some of the limitations are already known and were described by Lee (2013).

Radionuclides tend to accumulate in the first few centimeters of soil, associated with high concentrations of metals, when part of NORM, where they persist due to their long half-life and low mobility (Kato et al. 2012; Endo et al. 2013; Sugiura et al. 2016; X. Wang et al. 2017a), representing a serious hazard especially to soil biota and even to humans, as some of them are α and β emitters and interfere with calcium and potassium metabolism being also responsible for DNA damage (Lourenço et al. 2011, 2012, 2013a, b, 2017). Theoretically, the phytoextraction would be the best approach to deal with radionuclides rich soils, however it may represent the concentration but also the transference of contaminants to other areas, as well as an opportunity for food-webs transfer (Lee et al. 2009), since, plant species with higher bioconcentration factors (BCF) and roots to shoots transfer factors (TF) are required to attain a reasonable degree of success. Thus the best approach needs to be site-specific and the balance has to be made between: efforts of risk mitigation within the contaminated area, with the recovery of soil functions and uses, that can be potentially attained by phytoextraction and, risk mitigation outside the contaminated area by reducing radionuclides and/or metals mobility, by keeping them stabilized in the vicinity of plant roots through phytostabilization. In this case, although the landscape can be recovered, soil uses will be kept restrict within the contaminated area. Further, phytostabilization should preferentially involve perennial plant species, which frequently have a slow development, thus requiring more time for attaining meaningful results. However, this approach can have additional benefits by improving soil quality and soil carbon sequestration and several authors already demonstrated the ability of different tree species to accumulate radionuclides (*Acer rubrum*, *Liquidambar stryaciflua*, *Liriodendron tulipifera*, *Pinus halepensis*) (e.g. Pinder et al. 1984; Huang et al. 2016).

The effectiveness of phytoremediation is determined by a complex plethora of factors which are strictly interconnected, namely the soil physical and chemical properties that determine the bioavailability and the speciation of associated metals and radionuclides and the plant physiological traits that govern their ability to tolerate soils poor in nutrients and rich in toxic and genotoxic elements. Several studies have been published reporting the efforts made to find out hyperaccumulator species within the flora of contaminated sites (e.g. Yamashiro et al. 2013; Kaewtubtim et al. 2017; Skoko et al. 2017) or within fast-growing species (e.g. crop species) under laboratorial (both in soil and hydroponic conditions) (Ebbs et al. 1998; Eapen et al. 2006; Tomé et al. 2009; Smodiš et al. 2015; Singh et al. 2016; Yamashiro et al. 2013; X. Wang et al. 2017a, b), open top chambers and greenhouse (Lasat et al. 1998; Cook et al. 2009; Tang et al. 2011) and field experiments (Dushenkov et al. 1999). As far as radionuclides are considered, transfer factors (TF) have been used as an indicator of the phytoremediation potential of terrestrial plants for contaminated soils and expressed, as described in Eq. 1 (Kaewtubtim et al. 2017):

$$\mathrm{TF} = \mathrm{Cr}/\mathrm{Cs} \tag{1}$$

C_r –radionuclide activity in the plant organ in Bq kg^{-1} (or in mg kg^{-1})
C_s –radionuclides activity in the soil Bq kg^{-1} (or in mg kg^{-1})

As previously described TF, are also used by some authors to describe the rate of translocation from the roots to the shoots of plants, while concentration ratio (CR) has also been used to replace BCF. With this explanation, the terminology used by each author will be respected.

In fact, the selection of the appropriate species is one of the most challenging tasks, as plants need to be hypertolerant to toxic elements and frequently to poor and degraded soils, hyperaccumulators and, depending on site specific situations and management targets, they also need to have a fast growth rate, high biomass production and resistance to diseases and pests (Watanabe 1997). The great majority of studies have focused on the most hazardous radionuclides (^{137}Cs, ^{90}Sr, ^{214}Am, ^{226} Ra, ^{238}U), although very frequently and for understandable reasons surrogate stable radionuclides have been used on the experiments (e.g. Wiesel et al. 2015; X. Wang et al. 2017a). Several promising results were obtained, but a compilation of available data demonstrates, that it will be difficult to find a species with similar accumulation abilities, for different radionuclides and metals (Gouthu et al. 1997; Kaewtubtim et al. 2017; X. Wang et al. 2017a, b), thus an in situ phytoextraction strategy, for many contaminated soils, will always require a multi-species approach. The selection of different species, with different niches, even if they are poor accumulators or excluders (with low roots to shoots translocation), but hypertolerant to contamination, will also contribute to the establishment of hyperaccumulator plants and also to restore soil functions (Pinheiro et al. 2013; Kaewtubtim et al. 2017) (Fig. 3).

Fig. 3 Laboratory experiments to determine the phytoextraction potential of *Fraxinus angustifolia*

Wang et al. (2017a) tested the ability of *Sorghum bicolor* to growth and accumulate the stable radioisotope ^{133}Cs (used as surrogate of ^{137}Cs) by exposing plants for 100 days in soil and in hydroponic conditions, to 100 and 400 mg kg^{-1} and to 50–1000 μmol L^{-1}, respectively. The authors recorded BCF and TF values higher than 1 (both in soil and in hydroponics), demonstrating the ability of two varieties of this species to phytoextract ^{133}Cs from soil, at the lowest concentration tested in soil. At the highest concentration (400 mg kg^{-1}) ^{133}Cs was highly toxic to plants causing a decrease in plants height and dry height higher than 80%. In another study, Wang et al. (2017b) observed that *S. bicolor* was not so efficient to deal with the stable radioisotope ^{88}Sr (used as a surrogate of the radioactive isotope ^{90}Sr) under a pot experiment that lasted for 140 days. The authors observed that ^{88}Sr was not toxic to *S. bicolor* at concentrations between 50 and 400 mg kg^{-1} in soil and it was able to promote the growth of plants both in terms of height and dry mass of stems. The ^{88}Sr accumulation was higher in stems and the TF values of *S. bicolor* varied between 0.92 and 1.29 at the concentrations tested, however the BCF values did not surpassed 0.39, demonstrating that despite the ability of this species to translocate the isotope from the roots to the stems, it was not able to uptake it from the soil. Further, the BCF value decreased as the concentration of strontium increased, most probably due to the toxic effects of this element.

Although tests were not made in soils, Eapen et al. (2006) reported the ability of *Calotropsis gigantean* R.Br.), a high biomass and fast growth terrestrial shrub plant to remove both ^{90}Sr and ^{137}Cs from solution, in few hours/days. The plant removed 99% and 41% of ^{90}Sr and ^{137}Cs, respectively, after an incubation period of 168 h, during which both radioisotopes were supplied together. Further, the same species was able to remove 99% of the radioactivity of a Low Level Nuclear Waste (LLNW), after 15 days of exposure, by accumulating radioactive elements mainly in the roots. In parallel with the preferential accumulation in the roots, this plant species also produces latex, which may contribute for reducing herbivory, aspects that gives this species a high potential for phytoremediation of soils contaminated with radionuclides. At contaminated sites, Sugiura et al. (2016) compared the radiocesium accumulation (^{137}Cs and ^{133}Cs) in five deciduous woody tree species (*Chengiopanax sciadophylloids*, *Acer crataegifolium*, *Quercus serrata*, *Fraxinus sieboldiana* and *Ilex macropoda*) from Kawamata Fukushima prefecture, 37 km northeast the Fukushima Dai-ichi Nuclear power plant, and found a higher ability only for *C. sciadophylloids* to accumulate radiocesium, especially in the leaves, demonstrating the ability of the plants to translocate radionuclides to the leaves, as a possible strategy of detoxification, which will make radionuclides return to the soils. The authors, justified such differences with important plant characteristics, such as the ability of *C. sciadophylloids* trees to spread their roots horizontally on surface layers of soil, where mainly ^{137}Cs tends to be immobilized. However, the authors also suggested other factors to explain the differences between species as the expression of Cs transporters, which would be the same of potassium, as well as the possible colonization of roots by mycorrhizal fungi. Yamashiro et al. (2013) analyzed 99 wild plants species growing in paddy and upland fields in Iatate-mura area, in the Fukushima Prefecture, aimed in understanding how to handle weeds growing in the area, as well as candidates for the phytoextraction of radiocesium (^{134}Cs and ^{137}Cs) from affected soils. The highest TF were recorded for plants from paddy fields (0.703), when plant species commonly growing in both fields were analyzed, but for all plants species never surpassed 1. The differences between fields, point to the relevance of soil properties on plants uptake of radiocesium (e.g. pH, exchangeable K^+, organic matter, cation exchange capacity), however such differences were not analysed by the authors. Similarly, Skoko et al. (2017) in a disposal site of a coal-burning thermoelectric unit, at the Kastela Bay, Croacia, recorded a high activity of some naturally occurring radionuclides (^{238}U, ^{235}U, ^{232}Th, ^{226}Ra, ^{210}Pb and ^{40}K) in coal ash slag (CAS), when compared with a control site, but this was translated in a significant high activity in the tissues of Mediterranean plants growing on site, but only for ^{226}Ra. The weathering of CAS (as demonstrated by a pH lower than the expected but still alkaline, and by the low electrical conductivity) and the potential low availability of radionuclides was likely to be the reason for the low concentration ratios (CR) recorded for plants found at the CAS disposal site. However, no statistical significant relationship was found between CR values and CAS physico-chemical properties.

In fact, the role of different medium/soil properties in the accumulation of radionuclides by plants has been discussed by different authors and, once again, data suggests that the best phytoextraction protocol will be species, radionuclides and soil properties dependent. For example, Tomé et al. (2009) observed that

Helianthus annuus was more efficient in transferring ^{238}U from solution to the roots, than *Brassica juncea*, at all pH tested (5.0, 6.7 and 8.0), and the highest transfer factors for each species were attained at different pH values (8 for *H. annuus* and 6.7 for *B. juncea*). The same tendency was observed for ^{226}Ra and for *B. juncea*, but regarding *H. annuus*, this radioelement was more accumulated at acidic pHs. Despite the effect of pH, the authors also consider that plant physiological traits have their role, since the two radioisotopes have different speciation at pH 6.7, thus *B. juncea* must have a different uptake behavior. The production of specific exudates by plants that may form complexes of both radionuclides may favor their uptake by some plant species (Tomé et al. 2009). Another study from Ebbs et al. (1998) conducted under hydroponic conditions, also related the translocation of uranium from roots to shoots of Pisum sativum with the pH, assuming that the acidic pH (5.0) has favored this process, due to the dominance of uranyl cations in solution. Nevertheless, and similarly to the previously described study, the uptake/adsorption of uranium by the roots of this plant species was always higher at pH 6 and 8. These authors also reported a highest ability of *Phaseolus acutifolius* and *Beta vulgaris* to accumulate uranium in their shoots in comparison with many other monocotyledonous and dicotyledonous crop species. Smodiš et al. (2015), find out a positive and a significant correlation between soil cation exchange capacity and ^{238}U and ^{226}Ra activity in Chinese cabbage (*B. rapa* L. subsp. *pekinensis* (Lour.)), and significant and negative correlations with available K_2O, organic matter, organic carbon, total nitrogen and C/N ratio, when plants were exposed to soils contaminated with uranium mining tailings. Nevertheless, these authors reported concentrations ratios for ^{238}U, 30 times lower than those usually reported in the literature for the same species, attributing these results to site-specific properties of soils, thus highlighting the need of understanding the soil-to-plant transfer of natural radionuclides in real contaminated soils. The highest CR values were recorded for ^{226}Ra.

Through field experiments, carried out in the Northwest Border of the Chernobyl, 10 km south of the nuclear power plant, and by screening several cultivars, mainly varieties of the *Amaranth* species, *Brassica juncea* (L.) Czern. (Indian mustard), *Zea mays* L. (corn), *Pisum sativum* L. (peas), *Helianthus annuus* L. (sunflower), *H. tuberosus* (Jerusalem artichoke), *H. tuberosus* L. X *H. annuus*, Dushenkov et al. (1999) find out that *A. retroflexus* cvs aureus and PT-95 were the only varieties able to combine a bioaccumulation coefficient higher than one, with a high biomass production and with accumulation of ^{137}Cs, especially in the roots. However, these authors, pointed out to a major limitation for the phytoextraction of ^{137}Cs from the soils in the region affected by Chernobyl nuclear power plant accident, namely the accumulation of the majority of ^{137}Cs radionuclides in the top 5 cm of soil, where they were progressively incorporated in the crystalline structure of minerals becoming unavailable for uptake by plants. In order to overcome this limitation, several soil amendment solutions have been tested to increase the availability of radionuclides by changing soil properties or by providing chelating and extracting agents to increase their uptake by plants (Dushenkov et al. 1999; P. Wang et al. 2017) and also to promote their translocation from the roots to shoots (Singh et al. 2016). Some studies providing both results from hydroponic and soil experiments

(Singh et al. 2016), for the same plant species, highlight the limitations of the previous experiments, since it is clear that many factors are not being taken into account, including the lower bioavailability of radionuclides in soils (even in soils spiked with radionuclides solutions), which in part protects plants from the toxicity of these elements, allowing high biomass yields, but also contributing for lower CR and TF values. The great majority of laboratorial studies have also focused on the efficiency of plants to extract individual radionuclides, when in natural conditions they never occur isolated, thus potential synergistic or antagonist effects on the health of plants must be expected as well as competitive interactions during the uptake process by plants. Huang et al. (2016) testing seedlings of the variant *Sedum alfredii* Hance, a perennial herb from the region of Asia, to extract uranium and thorium (Th) from solution, concluded that at low concentrations Th stimulated growth of the plant (through an hormetic effect) and subsequently the uptake of U, but as the concentration of U and Th increases the U uptake by the plants decreases, due to the translocation of U from the cell walls to cell organelles increasing the cellular damages, probably caused by Th.

4.1 Chemically Assisted Phytoextraction

It is well known that to increase the solubility and the subsequent uptake of metals, plants release root exudates (rich in sugars, polysaccharids, aminoacids, peptides, proteins and organic acids), decreasing the pH in the rizhosphere (Sharma et al. 2015; Wiszniewska et al. 2016) and it is based on this knowledge that several artificial and natural occurring organic acids have been tested to help phytoextraction of radionuclides from soils. Chelating agents prevent the precipitation and the adsorption of radionuclides to soil, increasing their bioavailability for uptake by plants, but also help to mitigate their toxic effects directly by sequestration in plant cell vacuoles or by stimulating the growth of arbuscular mycorrhizal fungi (Sharma et al. 2015). Singh et al. (2016), after growing a subtropical grass *Vetiveria zizanioides* in soil spiked with ^{239}Pu for 30 days, observed a very low efficiency in the reduction of activity of this radionuclide in the soil (only 7%), which increased for 15 and 28% with the addition of citric acid and of the synthetic chelate diethylenetriaminepenataacetic acid (DTPA), respectively. Even more relevant was the role of DTPA in increasing the translocation of ^{239}Pu from the roots to the shoots of *V. zizanioides*, both in hydroponic and soil conditions (80.4 and 55.4%, respectively). The use of natural chelating agents is preferable, as they are not toxic and can be easily degraded by soil microbial community. However, several studies like the one from Singh et al. (2016) also demonstrated that they are less efficient than the artificial chelators, and their use as substrate by soil microorganisms can be one of the reasons for that. Meers et al. (2005) for example, described very short half-life of ethylenediamine disuccinate (EDDS), an aminopolycarboxylic acid produced by microorganisms (3.8 and 7.5 days in a soil experiment, depending on the applied doses), suggesting that, to be effective, natural chelators will require frequent applications in the field, what may have other unpredictable consequences on soil properties. Nevertheless, their short half-lives are a guarantee that post-harvest mobilization of contaminants will not continue, with risks to soil biota, surface waters and aquifers. Testing a soil with natural high levels of radionuclides (^{226}Ra) from Ramsar (Iran), Nezami et al. (2016) also confirmed the low efficiency of oxalic and citric acid in increasing the extraction of ^{226}Ra from the soil by *Zea mays*. Despite the significant reduction of soil pH promoted by both organic acids (added at a concentration $\geq$ 10 mM to the irrigation water, for 10 days), the maximum uptake of plants growing in a natural loam soil for 17 days, was of 6.3 and 6%, respectively when citric and oxalic acids were added individually to soil irrigation water, at a concentration of 100 mM. In this natural soils only 4.21% of ^{226}Ra was in the exchangeable fraction of soil, bound to carbonates, Fe/Mn oxides or to organic matter, being the greatest percentage unavailable, and weak organic acids added to soil, were not able to change this scenario, limiting the phytoextraction of this radionuclide.

The selection of the best chelating agents becomes even more complex, as their efficiency can be determined by the interaction of different soil factors and plant physiological traits, thus requiring the tests for each specific situation. Tomé et al. (2009) observed very different and highly variable behaviors of *H. annuus* and *B. juncea* in response to the addition of ethylene-diamine-tetraacetic acid (EDTA) and citric acid, in the presence of phosphates and at different pH, in terms of plants ability to uptake ^{238}U and ^{226}Ra. But despite that, at the end the authors concluded that both chelators did not contribute for a best performance of species in terms of phytoextraction of both radionuclides from the soil.

An alternative strategy could be to enhance soil properties or plant conditions (e.g. soil fertility, soil moisture and CO_2) to increase plant biomass and subsequently the uptake of radionuclides (Cook et al. 2009; Tomé et al. 2009; Tang et al. 2011). Fertilization, as the addition of nitrogen in poor soils, is expected to increase the plant uptake of radionuclides by stimulating plants and roots growth (Entry et al. 1996). However, depending on the radionuclide, the presence of some soil major nutrients, as phosphates may decrease its uptake by plants. The addition of phosphates through the nutrients solution reduced the uptake of ^{238}U by *H. annus*, but not by *B. juncea*. The authors suggested that probably the phosphates caused the precipitation of ^{238}U, rendering it less available, at least for one of the species, but differences in the mechanisms of uptake between species were likely the reason for the highest TF recorded for *B. juncea*. In the absence of phosphates *H. annus* was a better phytoextractor, being able to translocate more radionuclides from the roots to the shoots. Cook et al. (2009) also observed that under high cesium concentrations in soil (added as CsCl), soil fertilization and an enhanced soil moisture (35% and 70% of maximum soil water holding capacity) increased the biomass of four species of grasses, but despite that the concentrations of Cs in their shoots did not changed significantly. The role of some major soils elements, as competitors for common mechanisms of transport (e.g. as between potassium K^+ and Cs) requires a careful evaluation of soil properties at each contaminated site. As potassium is a macronutrient for plants, when K^+ increases, the uptake of Cs decreases (Zhu and Smolders 2000).

Ebbs et al. (1998) also call the attention for the long-term impacts of chelators, as they can have beneficial effects in the short-term, but in the long term they can render the contaminants even more unavailable in soils. These authors observed a highly significant effect of citric acid in the accumulation of U in the shoots of *Pisum sativum*, in detriment of N-(hydroxyethyl)-ethylenediaminetriacetic acid (HEDTA) trisodium salt, however the author's highlighted that the biodegradation or the photodegradation of U-citric acids complexes in the soil, could give rise to a new species of uranium (U trioxide) which will be less available than the original forms in the soils. Furthermore, the addition of chelating agents to soils will also contribute for the highest mobility of contaminants but also of nutrients which may be lost by percolation, making soils less fertile and with poor conditions for the establishment of plants, in the medium, long-term. Tang et al. (2011) with experiments carried out in open top chambers, that allowed the increase of CO_2 provided to plants also observed the beneficial effects of elevated CO_2 levels in the growth and yield of *Phytolacca Americana* and *Amaranthus cruentus* but also on Cs uptake from soil.

4.2 Phytoremedition Assisted by Arbuscular Mycorrhizal Fungi

The symbiosis between the roots of plants with arbuscular mycorrhizal fungi (AMF) has been explored both to improve plants phytoextraction and/or phytostabilization performance, but also to protect plants from the toxic effects of radionuclides as well as their resistance to pathogenic agents (e.g. Huang et al. 2016). AMF fungi increase the extension of plant roots, helping them to explore a large volume of soil to obtain more nutrients and water, while the fungi benefits by receiving sugars from plants (Entry et al. 1996). AMF fungi are able to deal with metals and radionuclides by precipitation in polyphosphate granules, adsorption in the cell walls and chelation through the secretion of ligands (Sharma et al. 2015). Thus hypothetically inoculation of roots with AMF can increase the ability of plants to acquire nutrients while simultaneously extract radionuclides. Despite that, the results of studies focusing the effect of roots inoculation with AMF fungi, in terms of changes in the uptake of radionuclides provided inconsistent results. Wiesel et al. (2015) advanced the inhibition of the colonization or of the symbiosis between plants and AMF, caused by the radionuclide, as a possible explanation. These authors recorded that even at very low concentrations Cs (3.13 μg k^{-1} in the form of stable $^{133}CsCO_3$) reduced the frequency and the intensity of mycorrhiza colonization (%), as well as the arbuscule abundance in roots (%) of *Medicago truncatula.* However, the colonization with *R. intraradices* did not influence the content of Cs both in the roots and in the shoots of plants growing in different Cs concentrations up to 78.1 μg kg^{-1}. Vinichuk et al. (2013) also evaluated the effect of the inoculation with the AMF *Glomus mosseae* in terms of ^{137}Cs uptake by four plant species (cucumber, ryegrass, sunflower and quinoa) in different types of soil (silty clay, loamy and loamy sand soils). The effect of AMF inoculation in plants biomass was different depending on the type of soil, but without a clear pattern between soils. Similarly the ^{137}Cs uptake by plants was determined by different soil types, with high CR values in plants infected with the AMF in loamy soils probably due to the most favorable texture of this soil. CR values were four times higher in loamy sand soils. AMF had a highest effect in the ability of *H. annuus* to uptake the radionuclide, however cucumber plants were the most effective regardless of soil type and treatment with AMF. Apparently, and contrary to the low success of previous studies, Huang et al. (2016) observed that the infection of *Pinus halepense* with the AMF species, *Glomus mosseae* and *G. versiforme* (with a 52 and 61% of success, respectively) increased significantly the CR factors for roots and leaves, as well as the transfer factor for Cs. Although *P. halepense* was per se a potential Cs phytoextractor, the colonization with both AMF species increased the production of ascorbate (AsA), by increasing the activity of enzymes like monodehydroascorbate (MDHAR), dehydroascorbate reductase (DHAR) that oxidize dehydroascorbate (DHA) in AsA like other enzymes from the ascorbate-gluthatione cycle AsA-GSH. The AMF infection was thus able to protect *P. halepense* from the toxicity of Cs while increasing in parallel their ability to uptake and transfer Cs to leaves. Although, this apparently increased the interest of this species for phytostabilization purposes, it was not discussed the possibility of this high translocation to leaves also be an additional mechanism of detoxification, and how unfortunately this will contribute to transfer Cs back to the soil. However, Chen et al. (2005b), by using an interesting experimental system, proved that AMF may contribute to fix radionuclides in the roots of plants. The system formed by two compartments, one of them separated by nets with nylon mesh with two different sizes (20 or 650 μm) and containing the uranium contaminated soil to allow the colonization only by the hyphae of the AMF *G. intraradices* or by both the roots of *Medicago truncatula* and the hyphae, concluded that U uptake through the route pathway can be more easily translocated to the shoots, than U taken up through the AMF pathway, as U is probably immobilized by fungal structures. Once attained, such symbiosis would be extremely interesting for the stabilization of radionuclides in the roots of plants. In another study conducted with two mutants of barley plants (*Hordeum vulgare*) Chen et al. (2005a, 2008) confirmed that AMF once again decreased the translocation from the roots to the shoots of plants, but also report that soil fertilization with P, may both stimulate or impair U accumulation in the roots of plants, depending on the level added to soil (Chen et al. 2005a) but also can influence the colonization by AMF (2008). In this later study by Dupré de Boulois et al. (2008), the authors also confirmed once more that the responsiveness and the success of infection with AMF is also dependent of roots morphology and surrounding environment.

5 Conclusion

The number of sites contaminated with radioactive wastes is tremendous, and they are responsible for the chronic exposure of biota and humans, thus requiring appropriate risk mitigation strategies. The conventional strategies are prohibitive due to the logistic required and the costs involved. Thus more environmental friendly approaches, able to recover the areas, at list for some uses have been announced as the most promising solution. As far as the bioremediation of radioactively contaminated sites is concerned, despite all the benefits advocated for the application of this technology, it seems that it is beneficial for small sites decontamination, under controlled conditions and requiring a long-term control. It is clear that its application is not suitable for areas such as, those of nuclear accidents (e.g. Chernobyl and Fukushima), due to the large distribution of Cs-137, for example. In fact, this technology, based solely on microorganism's activity, has not been employed in real scenarios, so far. Yet, it has been employed in combination with other methods. Although some research has been done *hitherto*, further research is still needed until we can see a practical application of this technology in the field. Furthermore, this technology is currently still hampered by an incomplete understanding of the true role of the different microorganisms involved, their genetics and genome-level characteristics, the metabolic pathways involved and their kinetics. The result of this is an inability to model and predict the outcome of these processes, and hence a huge difficulty in developing and employing natural bioremediation processes at a field scale. Additionally, for better/improved bioremediation very stable conditions of the environment should be observed and maintained, which becomes even more difficult when the whole process is done in the open and subjected to changing weather conditions. Thus, without neglecting the potential application of this technology we truly believe, that bioremediation can be successfully employed in small-medium mesocosm systems, where one can establish a link between field studies and highly controlled laboratory experiments, but nothing more.

As for the use of phytoremediation, despite the apparent few progresses that were made in the upscaling of phytoremediation of sites contaminated with radioactive wastes, much information was collected by several research studies, that needs to be systematized and used to develop models integrating site specific conditions (e.g. soil properties, main radionuclides, climate, previous management measures to mitigate radionuclides mobility in soils) plant species physiological traits and potential AMF symbionts, as variables, to support the design of phytoremediation approaches for each site in particular, as well as to better identify data gaps. Furthermore, investment has to be made in field studies in detriment of controlled laboratorial conditions, as most likely a great part of the promising results recorded in the laboratory will not be confirmed in situ, under more complex conditions. And, even more concerning, and as opposed to what has been argued by many authors phytoextraction of radionuclides from contaminated areas (even for those with low concentrations of radionuclides) is probably not a good approach for extensive areas since, if applied, as highlighted by the International Atomic Energy Agency (Fesenko and Howard 2012), it will generate a huge amount of biomass rich in radionuclides that must be treated as radioactive material that has to be properly disposed and thus with additional costs and complex management. In this context, the focus should be mainly in phytostabilization taking benefit of plant-microorganisms symbiotic relationships.

Acknowledgements The Portuguese Foundation for Science and Technology (FCT), through National Funds (Ministry for Science and Education in Portugal), provided financial support to Joana Lourenço by means of a Post-Doc grant (SFRH/BPD/92554/2013). This work was also supported by the Strategic Funding UID/Multi/04423/2013 (CIIMAR) and UID/AMB/50017/2013 (CESAM) through national funds provided by FCT and European Regional Development Fund (ERDF), in the framework of the PT2020 Partnership Agreement.

References

Achal V, Pan X, Zhang D (2012) Bioremediation of strontium (Sr) contaminated aquifer quartz sand based on carbonate precipitation induced by Sr resistant *Halomonas* sp. Chemosphere 89:764–768

Aftab K, Akhtar K, Jabbar A, Bukhari IH, Noreen R (2013) Physico-chemical study for zinc removal and recovery onto native/chemically modified *Aspergillus flavus* NA9 from industrial effluent. Water Res 47:4238–4246

Aleksakhin RM (2009) Radioactive contamination as a type of soil degradation. Euras Soil Sci 42:1386–1396

Alexander M, Loehr RC (1992) Bioremediation review. Science 258:874

Amachi S, Minami K, Miyasaka I, Fukunaga S (2010) Ability of anaerobic microorganisms to associate with iodine: 125I tracer experiments using laboratory strains and enriched microbial communities from subsurface formation water. Chemosphere 79:349–354

Anderson RT, Vrionis HA, Ortiz-Bernad I, Resch CT, Long PE, Dayvault R, Karp K, Marutzky S, Metzler DR, Peacock A, White DC, Lowe M, Lovley DR (2003) Stimulating the in situ activity of Geobacter species to remove uranium from the groundwater of a uranium-contaminated aquifer. Appl Environ Microbiol 69:5884–5891

Arai S, Yonezawa Y, Okazaki N, Matsumoto F, Shibazaki C, Shimizu R, Yamada M, Adachi M, Tamada T, Kawamoto M, Tokunaga H, Ishibashi M, Blaber M, Tokunaga M, Kuroki R (2015) Structure of a highly acidic β-lactamase from the moderate halophile *Chromohalobacter* sp. 560 and the discovery of a Cs^+ – selective binding site. Acta Crystallogr Sect D Biol Crystallogr 71:541–554

Bem H, Bou-Rabee F (2004) Environmental and health consequences of depleted uranium use in the 1991 Gulf War. Environ Int 30:123–134

Benzerara K, Miot J, Morin G, Ona-Nguema G, Skouri-Panet F, Férard C (2011) Significance, mechanisms and environmental implications of microbial biomineralization. Comptes Rendus Geosci 343:160–167

Bleise A, Danesi PR, Burkart W (2003) Properties, use and health effects of depleted uranium (DU): a general overview. J Environ Radioact 64:93–112

Burger M (2012) The risks of depleted uranium contamination in post-conflict countries: Findings and lessons learned from UNEP field assessments. Assess Restoring Nat Resour Post-conflict Peacebuil 5:163

Carlsen TM, Peterson LE, Ulsh BA, Werner CA, Purvis KL, Sharber AC (2001) Radionuclide contamination at Kazakhstan's Semipalatinsk test site: Implications on human and ecological health. Hum Ecol Risk Assess An Int J 7:943–955

Carvalho F, Madruga M, Reis M, Alves J, Oliveira J, Gouveia J, Silva L (2007) Radioactivity in the environment around past radium and uranium mining sites of Portugal. J Environ Radioact 96:39–46

Chen B, Jakobsen I, Roos P, Zhu YG (2005a) Effects of the mycorrhizal fungus *Glomus intraradices* on uranium uptake and accumulation by *Medicago truncatula* L. from uranium-contaminated soil. Plant Soil 275:349–359

Chen B, Zhu YG, Zhang X, Jakobsen I (2005b) The influence of mycorrhiza on uranium and phosphorus uptake by barley plants from a field-contaminated soil. Environ Sci Pollut Res Int 12:325–331

Chen B, Roos P, Zhu YG, Jakobsen I (2008) Arbuscular mycorrhizas contribute to phytostabilization of uranium in uranium mining tailings. J Environ Radioact 99:801–810

Christensen GC, Malyshev SV, Salbu B, Romanov GN, Selnais TD, Oughton D, Glagolenko YV, Amundsen I, Liv Rudjord A, Bjerk TO, Lind B (1995) Radioactive contamination in the environment of the nuclear enterprise Mayak, PA. Results from the joint Russian-Norwegian field work in 1994. The nuclear enterprise Mayak, PA. In: International conference on environmental radioactivity in the Arctic, pp 21–25

Chung AP, Sousa T, Pereira A, Morais PV (2014) Microorganisms – tools for bioremediation of uranium contaminated environments. Proc Earth Planet Sci 8:53–58

Cook LL, Inouye RS, McGonigle TP (2009) Evaluation of four grasses for use in phytoremediation of Cs-contaminated arid land soil. Plant Soil 324:169–184

Das M, Adholeya A (2012) Role of microorganisms in remediation of contaminated soil. In: Satyanarayana T, Johri BN (eds) Microorganisms in environmental management: microbes and environment. Springer, Dordrecht, pp 81–111

Das S, Dash HR, Chakraborty J (2016) Genetic basis and importance of metal resistant genes in bacteria for bioremediation of contaminated environments with toxic metal pollutants. Appl Microbiol Biotechnol 100:2967–2984

Dash HR, Das S (2015) Bioremediation of inorganic mercury through volatilization and biosorption by transgenic *Bacillus cereus* BW-03(pPW-05). Int Biodeterior Biodeg 103:179–185

Dighton J, Clint GM, Poskitt J (1991) Uptake and accumulation of ^{137}Cs by upland grassland soil fungi: a potential pool of Cs immobilization. Mycol Res 95:1052–1056

Dighton J, Tugay T, Zhdanova N (2008) Fungi and ionizing radiation from radionuclides. FEMS Microbiol Lett 281:109–120

Dixit R, Wasiullah E, Malaviya D, Pandiyan K, Singh U, Sahu A, Shukla R, Singh B, Rai J, Sharma P, Lade H, Paul D (2015) Bioremediation of heavy metals from soil and aquatic environment: an overview of principles and criteria of fundamental processes. Sustainability 7:2189–2212

Dupré de Boulois H, Joner EJ, Leyval C, Jakobsen I, Chen BD, Roos P, Thiry Y, Rufyikiri G, Delvaux B, Declerck S (2008) Impact of arbuscular mycorrhizal fungi on uranium accumulation by plants. J Environ Radioact 99:775–784

Dushenkov S (2003) Trends in phytoremediation of radionuclides. Plant Soil 249:167–175

Dushenkov S, Mikheev A, Prokhnevsky A, Ruchko M, Boris S (1999) Phytoremediation of radiocesium-contaminated soil in the vicinity of Chernobyl, Ukraine. Environ Sci Technol 33:469–475

Eapen S, Singh S, Thorat V, Kaushik CP, Raj K, D'Souza SF (2006) Phytoremediation of radiostrontium (^{90}Sr) and radiocesium (^{137}Cs) using giant milky weed (*Calotropis gigantea* R.Br.) plants. Chemosphere 65:2071–2073

Ebbs SD, Brady DJ, Kochian LV (1998) Role of uranium speciation in the uptake and translocation of uranium by plants. J Exp Bot 49:1183–1190

Elekwachi C, Andresen J, Hodgman TC (2014) Global use of bioremediation technologies for decontamination of ecosystems. J Biorem Biodegrad 5:1–9

Endo S, Kajimoto T, Shizuma K (2013) Paddy-field contamination with ^{134}Cs and ^{137}Cs due to Fukushima Dai-ichi Nuclear Power Plant accident and soil-to-rice transfer coefficients. J Environ Radioact 116:59–64

Entry JA, Vance NC, Hamilton MA, Zabowski D, Watrud LS, Adriano DC (1996) Phytoremediation of soil contaminated with low concentrations of radionuclides. Water Air Soil Pollut 88:167–176

Evrard O, Chartin C, Onda Y, Lepage H, Cerdan O, Lefèvre I, Ayrault S (2014) Renewed soil erosion and remobilisation of radioactive sediment in Fukushima coastal rivers after the 2013 Typhoons. Sci Rep 4:4574

Evseeva T, Belykh E, Geras'kin S, Majstrenko T (2012) Estimation of radioactive contamination of soils from the "Balapan" and the "Experimental field" technical areas of the Semipalatinsk nuclear test site. J Environ Radioact 109:52–59

Fairlie I (2014) A hypothesis to explain childhood cancers near nuclear power plants. J Environ Radioact 133:10–17

Fesenko S, Howard BJ (2012) Guidelines for remediation strategies to reduce the radiological consequences of environmental contamination. International Atomic Energy Agency, Vienna

Francis AJ, Nancharaiah YV (2015) In situ and ex situ bioremediation of radionuclide-contaminated soils at nuclear and norm sites. In: Velzen Leo van BT (ed) Environmental remediation and restoration of contaminated nuclear and norm sites, in: Woodhead Publishing Series in Energy. Woodhead Publishing, Cambridge, pp 185–236

Gadd GM (2009) Biosorption: critical review of scientific rationale, environmental importance and significance for pollution treatment. J Chem Technol Biotechnol 84:13–28

Gadd GM (2010) Metals, minerals and microbes: geomicrobiology and bioremediation. Microbiology 156:609–643

Gavrilescu M, Pavel LV, Cretescu I (2009) Characterization and remediation of soils contaminated with uranium. J Hazard Mater 163:475–510

Gouthu S, Arie T, Ambe S, Yamaguchi I (1997) Screening of plant species for comparative uptake abilities of radioactive Co, Rb, Sr and Cs from soil. J Radioanal Nucl Chem 22:247–251

Groudev S, Spasova I, Nicolova M, Georgiev P (2010) In situ bioremediation of contaminated soils in uranium deposits. Hydrometallurgy 104:518–523

Ha J, Gélabert A, Spormann AM, Brown GE (2010) Role of extracellular polymeric substances in metal ion complexation on *Shewanella oneidensis*: Batch uptake, thermodynamic modeling, ATR-FTIR, and EXAFS study. Geochim Cosmochim Acta 74:1–15

Handley-Sidhu S, Keith-Roach MJ, Lloyd JR, Vaughan DJ (2010) A review of the environmental corrosion, fate and bioavailability of munitions grade depleted uranium. Sci Total Environ 408:5690–5700

Harms H, Schlosser D, Wick LY (2011) Untapped potential: exploiting fungi in bioremediation of hazardous chemicals. Nat Rev Micro 9:177–192

Hazen TC, Tabak HH (2005) Developments in bioremediation of soils and sediments polluted with metals and radionuclides: 2. Field research on bioremediation of metals and radionuclides. Rev Environ Sci Bio/Technol 4:157–183

Huang RH, Lu YM, Yang HL, Huang W, Chen K (2016) Effects of arbuscular mycorrhizal fungi on caesium accumulation and the ascorbate-glutathione cycle of *Sorghum halepense*. Sci Asia 42:323–331

IAEA (1988) The radiological accidents in Goiania, Vienna

IAEA (1996) Health and environmental aspects of nuclear fuel cycle facilities p /r\, nuclear fuel cycle and materials section. International Atomic Energy Agency, Vienna

IAEA (1999) Technologies for Remediation of Radioactively Contaminated Sites, Vienna

IAEA (2000) Management of radioactive waste from the use of radionuclides in medicine, Vienna

IAEA (2001) Present and future environmental impact of the Chernoby, Vienna

IAEA (2002a) Monitoring and surveillance of residues from the mining and milling of uranium and thorium, Vienna. IAEA

IAEA (2002b) Environmental aspects based on operational performance of nuclear fuel fabrication facilities, Nuclear Fuel Cycle and Materials Section, International Atomic Energy Agency, Vienna
IAEA (2003) Radiological Conditions in Areas of Kuwait with Residues of Depleted Uranium, Vienna
IAEA (2005) Environmental contamination from uranium production facilities and their remediation. In: Proceedings of an international work held on Lisbon, Portugal. 11–13 February 2004
IAEA (2006) Environmental Consequences of the Chernobyl Accident and their Remediation: Twenty Years of Experience, Vienna
IAEA (2008) Spent fuel reprocessing options. International Atomic Energy Agency, Vienna
IAEA (2009) Establishment of uranium mining and processing operations in the context of sustainable development. International Atomic Energy Agency, Vienna
IAEA (2010) Best practice in environmental management of uranium mining. International Atomic Energy Agency, Vienna
IAEA (2011) International mission on remediation of large contaminated areas off-site Fukushima Daiichi Nuclear Power Plant – final report, Vienna.
IAEA (2012) Efficient water management in water cooled reactors. In: IAEA nuclear energy series no. NP-T-2.6, Vienna.
IAEA (2014) Radiation protection after the Fukushima Daiichi Accident: promoting confidence and understanding, Vienna.
IAEA (2017) General distribution international status and prospects for nuclear power
Janssen PJ, Van Houdt R, Moors H, Monsieurs P, Morin N, Michaux A, Benotmane MA, Leys N, Vallaeys T, Lapidus A, Monchy S, Médigue C, Taghavi S, McCorkle S, Dunn J, van der Lelie D, Mergeay M (2010) The complete genome sequence of *Cupriavidus metallidurans* strain CH34, a master survivalist in harsh and anthropogenic environments. PLoS One 5:e10433
Jiang GCT, Aschner M (2009) Depleted uranium. In: Handbook of toxicology of chemical warfare agents. Academic, Amsterdam, pp 393–406
Joshi PK, Swarup A, Maheshwari S, Kumar R, Singh N (2011) Bioremediation of heavy metals in liquid media through fungi isolated from contaminated sources. Ind J Microbiol 51:482–487
Kadyrzhanov KK, Khazhekber S, Solodukhin VP, Lukashenko SN, Kazachevskiy IV, Rofer C, Poznyak VL, Knyazev BB, Knatova MK, Nazarenko LM, Yakushev EM (2005) Plutonium at the Semipalatinsk Nuclear Test Site (SNTS). J Radioanal Nucl Chem 263:229–234
Kaewtubtim P, Meeinkuirt W, Seepom S, Pichtel J (2017) Radionuclide ((226)Ra, (232)Th, (40)K) accumulation among plant species in mangrove ecosystems of Pattani Bay, Thailand. Mar Pollut Bull 115:391–400
Kato H, Onda Y, Teramage M (2012) Depth distribution of ^{137}Cs, ^{134}Cs, and ^{131}I in soil profile after Fukushima Dai-ichi Nuclear Power Plant accident. J Environ Radioact 111:59–64
Kumar R, Singh S, Singh OV (2007) Bioremediation of radionuclides: Emerging technologies. Omi A J Integr Biol 11:295–304
Lachner J, Christl M, Alfimov V, Hajdas I, Kubik PW, Schulze-König T, Wacker L, Synal HA (2014) ^{41}Ca, ^{14}C and ^{10}Be concentrations in coral sand from the Bikini atoll. J Environ Radioact 129:68–72
Lasat MM, Fuhrmann M, Ebbs SD, Cornish JE, Kochian LV (1998) Phytoremediation of a radiocesium-contaminated soil: Evaluation of cesium-137 bioaccumulation in the shoots of three plant species. J Environ Qual 27:165–169
Lee JH (2013) An overview of phytoremediation as a potentially promising technology for environmental pollution control. Biotechnol Bioprocess Eng 18:431–439
Lee SH, Kim EY, Hyun S, Kim JG (2009) Metal availability in heavy metal-contaminated open burning and open detonation soil: Assessment using soil enzymes, earthworms, and chemical extractions. J Hazard Mater 170:382–388
Li X, Ding C, Liao J, Lan T, Li F, Zhang D, Yang J, Yang Y, Luo S, Tang J, Liu N (2014) Biosorption of uranium on Bacillus sp. dwc-2: preliminary investigation on mechanism. J Environ Radioact 135:6–12

Lloyd JR, Sole VA, Van Praagh CV, Lovley DR (2000) Direct and Fe(II)-mediated reduction of technetium by Fe(III)-reducing bacteria. Appl Environ Microbiol 66:3743–3749.

Lourenço J, Pereira RO, Silva AC, Morgado JM, Carvalho FP, Oliveira JM, Malta MP, Paiva AA, Mendo SA, Gonçalves FJ (2011) Genotoxic endpoints in the earthworms sub-lethal assay to evaluate natural soils contaminated by metals and radionuclides. J Hazard Mater 186:788–795

Lourenço J, Pereira R, Silva A, Carvalho F, Oliveira J, Malta M, Paiva A, Gonçalves F, Mendo S (2012) Evaluation of the sensitivity of genotoxicity and cytotoxicity endpoints in earthworms exposed in situ to uranium mining wastes. Ecotoxicol Environ Saf 75:46–54

Lourenço J, Pereira R, Gonçalves F, Mendo S (2013a) Metal bioaccumulation, genotoxicity and gene expression in the European wood mouse (*Apodemus sylvaticus*) inhabiting an abandoned uranium mining area. Sci Total Environ 443:673–680

Lourenço J, Pereira R, Pinto F, Caetano T, Silva A, Carvalheiro T, Guimarães A, Gonçalves F, Paiva A, Mendo S (2013b) Biomonitoring a human population inhabiting nearby a deactivated uranium mine. Toxicology 305:89–98

Lourenço J, Mendo S, Pereira R (2016) Radioactively contaminated areas: Bioindicator species and biomarkers of effect in an early warning scheme for a preliminary risk assessment. J Hazard Mater 317:503–542

Lourenço J, Marques S, Carvalho FP, Oliveira J, Malta M, Santos M, Gonçalves F, Pereira R, Mendo S (2017) Uranium mining wastes: The use of the Fish Embryo Acute Toxicity Test (FET) test to evaluate toxicity and risk of environmental discharge. Sci Total Environ. 605–606:391–404

Macaskie LE, Bonthrone KM, Yong P, Goddard DT (2000) Enzymically mediated bioprecipitation of uranium by a Citrobacter sp.: a concerted role for exocellular lipopolysaccharide and associated phosphatase in biomineral formation. Microbiology 146:1855–1867

Matshusa K, Makgae M (2017) Prevention of future legacy sites in uranium mining and processing: The South African perspective. Ore Geol Rev 86:70–78

McGrath SP, Zhao J, Lombi E (2002) Phytoremediation of metals, metalloids, and radionuclides. Academic Press 1–56

Meers E, Ruttens A, Hopgood MJ, Samson D, Tack FMG (2005) Comparison of EDTA and EDDS as potential soil amendments for enhanced phytoextraction of heavy metals. Chemosphere 58:1011–1022

Merkel B, Schipek M (2011) The new uranium mining boom: challenge and lessons learned. Springer. Berlin

Molchanova I, Mikhailovskaya L, Antonov K, Pozolotina V, Antonova E (2014) Current assessment of integrated content of long-lived radionuclides in soils of the head part of the East Ural Radioactive Trace. J Environ Radioact 138:238–248

NABIR Primer (2003) Bioremediation of metals and radionuclides: what it is and how it works, 2nd edn. Lawrence Berkeley National Laboratory for U.S. DOE, Office of Biological and Environmental Research, Natural and Accelerated Bioremediation Research Program

NEA-IEA (2015) Technology roadmap: nuclear energy. OECD/IEA, Paris

Newsome L, Morris K, Lloyd JR (2014) The biogeochemistry and bioremediation of uranium and other priority radionuclides. Chem Geol 363:164–184

Nezami S, Malakouti MJ, Bahrami Samani A, Ghannadi Maragheh M (2016) Effect of low molecular weight organic acids on the uptake of ^{226}Ra by corn (*Zea mays* L.) in a region of high natural radioactivity in Ramsar-Iran. J Environ Radioact 164:145–150

NRC (2012) Effluent releases from nuclear power plants and fuel-cycle facilities, analysis of cancer risks in populations near nuclear facilities: Phase I. National Research Council, National Academies Press (US), Washington, DC

OECD-NEA (2014) Perceptions and realities in modern uranium mining – extended summary

Oves M, Khan MS, Zaidi A (2013) Biosorption of heavy metals by *Bacillus thuringiensis* strain OSM29 originating from industrial effluent contaminated north Indian soil. Saudi J Biol Sci 20:121–129

Pattanapipitpaisal P, Mabbett AN, Finlay JA, Beswick AJ, Paterson-Beedle M, Essa A, Wright J, Tolley MR, Badar U, Ahmed N, Hobman JL, Brown NL, Macaskie LE (2002) Reduction of Cr (VI) and bioaccumulation of chromium by gram positive and gram negative microorganisms not previously exposed to CR-Stress. Environ Technol 23:731–745

Pereira R, Barbosa S, Carvalho FP (2014) Uranium mining in Portugal: a review of the environmental legacies of the largest mines and environmental and human health impacts. Environ Geochem Health 36:285–301

Pinder JE, McLeod KW, Alberts JJ, Adriano DC, Corey JC (1984) Uptake of ^{244}Cm, ^{238}Pu and other radionuclides by trees inhabiting a contaminated flood plain. Health Phys 47:375–384

Pinheiro JC, Marques CR, Pinto G, Bouguerra S, Mendo S, Gomes NC, Gonçalves F, Rocha-Santos T, Duarte AC, Roembke J, Sousa JP, Ksibi M, Haddioui A, Pereira R (2013) The performance of *Fraxinus angustifolia* as a helper for metal phytoremediation programs and its relation to the endophytic bacterial communities. Geoderma 202:171–182

Prakash D, Gabani P, Chandel AK, Ronen Z, Singh OV (2013) Bioremediation: a genuine technology to remediate radionuclides from the environment. Microb Biotechnol 6:349–360

Prăvălie R (2014) Nuclear weapons tests and environmental consequences: A global perspective. Ambio 43:729–744

Rahman A, Nahar N, Nawani NN, Jass J, Desale P, Kapadnis BP, Hossain K, Saha AK, Ghosh S, Olsson B, Mandal A (2014) Isolation and characterization of a *Lysinibacillus* strain B1-CDA showing potential for bioremediation of arsenics from contaminated water. J Environ Sci Heal Part A 49:1349–1360

Ruggiero CE, Boukhalfa H, Forsythe JH, Lack JG, Hersman LE, Neu MP (2005) Actinide and metal toxicity to prospective bioremediation bacteria. Environ Microbiol 7:88–97

Salbu B (2013) Preface: uranium mining legacy issue in Central Asia. J Environ Radioact 123:1–2

Salem IB, Sghaier H, Trifi H, Heni S, Khwaldia K, Saidi M, Landoulsi A (2012) Isolation and characterization of a novel Micrococcus strain for bioremediation of strontium in radioactive residues. Afr J Microbiol Res 6:851–858

Sasaki H, Shirato S, Tahara T, Sato K, Takenaka H (2013) Accumulation of radioactive cesium released from Fukushima Daiichi Nuclear Power Plant in terrestrial cyanobacteria *Nostoc commune*. Microbes Environ 28:466–469

Schenck LM, Youmans RA (2012) From start to finish: A historical review of nuclear arms control treaties and starting over with the new start. Cardozo J Int Comp Law 20

Seeprasert P, Yoneda M, Shimada Y (2016) The influence of soil fungi on the sorption of cesium and strontium in the soil organic layer. Int J Environ Sci Dev 7

Seier M, Zimmermann T (2014) Environmental impacts of decommissioning nuclear power plants: methodical challenges, case study, and implications. Int J Life Cycle Ass 19:1919–1932

Sharma S, Singh B, Manchanda VK (2015) Phytoremediation: role of terrestrial plants and aquatic macrophytes in the remediation of radionuclides and heavy metal contaminated soil and water. Environ Sci Pollut Res 22:946–962

Shukla KP, Singh NK, Sharma S (2010) Bioremediation: developments, current practices and perspectives. Genet Eng Biotechnol J:1–20

Simpson MF, Law JD (2013) Nuclear fuel, reprocessing of. In: Nuclear energy. Springer, New York, pp 153–173

Singh S, Fulzele DP, Kaushik CP (2016) Potential of *Vetiveria zizanoides* L. Nash for phytoremediation of plutonium (^{239}Pu): Chelate assisted uptake and translocation. Ecotoxicol Environ Saf 132:140–144

Skoko B, Marović G, Babić D, Šoštarić M, Jukić M (2017) Plant uptake of ^{238}U, ^{235}U, ^{232}Th, ^{226}Ra, ^{210}Pb and ^{40}K from a coal ash and slag disposal site and control soil under field conditions: A preliminary study. J Environ Radioact 172:113–121

Smičiklas I, Šljivić-Ivanović M (2016) Radioactive contamination of the soil: Assessments of pollutants mobility with implication to remediation strategies. In: Larramendy ML, Soloneski S (eds) Environmental sciences, soil contamination-current consequences and further solutions. InTech, Rijeka, pp 253–276

Smodiš B, Černe M, Jaćimović R, Benedik L (2015) Transfer of uranium and radium to Chinese cabbage from soil containing elevated levels of natural radionuclides. J Radioanal Nucl Chem 306:685–694

SNCNW (2016) Nuclear waste state-of-the-art report 2016 risks, uncertainties and future challenges, Stockholm

Snyder DC, Delmore JE, Tranter T, Mann NR, Abbott ML, Olson JE (2012) Radioactive cesium isotope ratios as a tool for determining dispersal and re-dispersal mechanisms downwind from the Nevada Nuclear Security Site. J Environ Radioact 110:46–52

Stone MB, Stanford JB, Lyon JL, VanDerslice JA, Alder SC (2013) Childhood thyroid radioiodine exposure and subsequent infertility in the intermountain fallout cohort. Environ Health Persp 121:79

Sugiura Y, Kanasashi T, Ogata Y, Ozawa H, Takenaka C (2016) Radiocesium accumulation properties of *Chengiopanax sciadophylloides*. J Environ Radioact 151:250–257

Tang S, Liao S, Guo J, Song Z, Wang R, Zhou X (2011) Growth and cesium uptake responses of *Phytolacca americana* Linn. and *Amaranthus cruentus* L. grown on cesium contaminated soil to elevated CO_2 or inoculation with a plant growth promoting rhizobacterium Burkholderia sp. D54, or in combination. J Hazard Mater 198:188–197

Tiwari G, Singh SP (2014) Application of bioremediation on solid waste management: A review. J Biorem Biodegrad 5:248

Tomé V, Blanco Rodríguez P, Lozano JC (2009) The ability of *Helianthus annuus* L. and *Brassica juncea t*o uptake and translocate natural uranium and ^{226}Ra under different milieu conditions. Chemosphere 74:293–300

Tomioka N, Uchiyama H, Yagi O, Fujii T (1995) Kinetic studies on cesium transport in *Rhodococcus erythropolis* CS98 and *Rhodococcus* sp. strain CS402. Biosci Biotechnol. Biochem 59:2219–2222

Turner DR (2013) Nuclear facilities, decommissioning of. In: Tsoulfanidis N (ed) Nuclear energy: selected entries from the encyclopedia of sustainability science and technology. Springer, New York, pp 223–268

UNEP (2001) Depleted uranium in Kosovo: Postconflict environmental assessment. United Nations Environment Programme, Imprimerie Chirat, France

UNEP (2002) Depleted uranium in Serbia and Montenegro: Postconflict environmental assessment in the Federal Republic of Yugoslavia. United Nations Environment Programme, Imprimerie Chirat, France

UNEP (2003a) Depleted uranium in Bosnia and Herzegovina: Postconflict environmental assessment. United Nations Environment Programme, Imprimerie Chirat, France

UNEP (2003b) Environment in Iraq: UNEP progress report. United Nations Environment Programme, Geneva

UNEP (2007) UNEP. Technical report on capacity-building for the assessment of depleted uranium in Iraq. United Nations Environmental Program, Geneva

UNSCEAR (1993) Report to the general assembly, with scientific annexes-sources and effects of ionizing radiation. United Nations Publications, New York

Vandenhove H (2013) Phytoremediation options for radioactively contaminated sites evaluated. Ann Nucl Ener 62:596–606

Vangronsveld J, Herzig R, Weyens N, Boulet J, Adriaensen K, Ruttens A, Thewys T, Vassilev A, Meers E, Nehnevajova E, van der Lelie D, Mench M (2009) Phytoremediation of contaminated soils and groundwater: lessons from the field. Environ Sci Pollut Res 16:765–794

Victorova N, Voitesekhovitch O, Sorochinsky B, Vandenhove H, Konoplev A, Konopleva I (2000) Phytoremediation of Chernobyl contaminated land. Radiat Prot Dosim 92:59–64

Vinichuk M, Mårtensson A, Ericsson T, Rosén K (2013) Effect of arbuscular mycorrhizal (AM) fungi on ^{137}Cs uptake by plants grown on different soils. J Environ Radioact 115:151–156

Waggitt P (2008) Uranium mining legacies remediation and renaissance development: an international overview. In: Merkel BJ, Hasche-Berger A (eds) Uranium, mining and hydrogeology. Springer, Cham, pp 11–18

Wang P, Du L, Tan Z, Su R, Li T (2017) Effects of organic acids and sylvite on phytoextraction of ^{241}Am contaminated soil. Bull Environ Contam Toxicol 98:407–412

Wang X, Chen C, Wang J (2017a) Phytoremediation of strontium contaminated soil by *Sorghum bicolor* (L.) Moench and soil microbial community-level physiological profiles (CLPPs). Environ Sci Pollut Res 24:7668–7678

Wang X, Chen C, Wang J (2017b) Cs phytoremediation by *Sorghum bicolor* cultivated in soil and in hydroponic system. Int J Phytorem 19:402–412

Watanabe ME (1997) Phytoremediation on the brink of commercialization. Environ Sci Technol 31:182–186

Wendel CC, Oughton DH, Lind OC, Skipperud L, Fifield LK, Isaksson E, Tims SG, Salbu B (2013) Chronology of Pu isotopes and ^{236}U in an Arctic ice core. Sci Total Environ 461–462:734–741

Wiesel L, Dubchak S, Turnau K, Broadley MR, White PJ (2015) Caesium inhibits the colonization of *Medicago truncatula* by arbuscular mycorrhizal fungi. J Environ Radioact 141:57–61

Williams KH, Bargar JR, Lloyd JR, Lovley DR (2013) Bioremediation of uranium-contaminated groundwater: a systems approach to subsurface biogeochemistry. Curr Opin Biotechnol 24:489–497

Winfield M, Jamison A, Wong R, Czajkowski P (2006) Nuclear power in Canada: an examination of risks, impacts and sustainability. https://www.pembina.org/reports/Nuclear_web.pdf

Wiszniewska A, Hanus-Fajerska E, Muszynska E, Ciarkowska K (2016) Natural organic amendments for improved phytoremediation of polluted soils: a review of recent progress. Pedosphere 26:1–12

Yamamoto M, Tomita J, Sakaguchi A, Ohtsuka Y, Hoshi M, Apsalikov K (2010) Uranium isotopes in well water samples as drinking sources in some settlements around the Semipalatinsk Nuclear Test Site, Kazakhstan. J Radioanal Nucl Chem 284:309–314

Yamashiro H, Abe Y, Fukuda T, Kino Y, Kawaguchi I, Kuwahara Y, Fukumoto M, Takahashi S, Suzuki M, Kobayashi J, Uematsu E, Tong B, Yamada T, Yoshida S, Sato E, Shinoda H, Sekine T, Isogai E, Fukumoto M (2013) Effects of radioactive caesium on bull testes after the Fukushima nuclear plant accident. Sci Rep 3:2850

Yoshida S, Muramatsu Y, Dvornik AM, Zhuchenko TA, Linkov I (2004) Equilibrium of radiocesium with stable cesium within the biological cycle of contaminated forest ecosystems. J Environ Radioact 75:301–313

Young M, Artsatbanov V, Beller HR, Chandra G, Chater KF, Dover LG, Goh EB, Kahan T, Kaprelyants AS, Kyrpides N, Lapidus A, Lowry SR, Lykidis A, Mahillon J, Markowitz V, Mavromatis K, Mukamolova GV, Oren A, Rokem JS, Smith MCM, Young DI, Greenblatt CL (2010) Genome sequence of the Fleming strain of Micrococcus luteus, a simple free-living actinobacterium. J Bacteriol 192:841–860

Zhang P, Idota Y, Yano K, Negishi M, Kawabata H, Arakawa H, Morimoto K, Tsuji A, Ogihara T (2014) Characterization of cesium uptake mediated by a potassium transport system of bacteria in a soil conditioner. Biol Pharm Bull 37:604–607

Zhao P, Zavarin M, Leif RN, Powell BA, Singleton MJ, Lindvall RE, Kersting AB (2011) Mobilization of actinides by dissolved organic compounds at the Nevada Test Site. Appl Geochem 26:308–318

Zhu YG, Smolders E (2000) Plant uptake of radiocaesium: a review of mechanisms, regulation and application. J Exp Bot 51:1635–1645

Bioremediation and Phytoremediation: Best Approach for Rehabilitation of Soils for Future Use

Sergiy Dubchak and Olexander Bondar

1 Introduction

The radioactive contamination is one of the major issues in the biosphere pollution, which arose in the past decades. Its main sources are nuclear explosions for military and civil purposes, technological and accidental releases at the nuclear fuel cycle facilities and military nuclear installations, mining and processing of uranium, radiation accidents and disposal of radioactive waste. As a result, more than 90% of the released radionuclides were distributed in the upper 30 cm of the soil layer of fields and meadows, and within 5–10 cm of the surface in forests, rice paddies and orchards. Since most of radionuclides are adsorbed into the soil and bound strongly to clay, it is difficult to reduce the contamination level in the soil.

Up to date, a large number of different methods of radionuclide immobilization in the environment have been developed. Conventional methods used for decontamination of radioactively contaminated soils, namely chemical, physical and mechanical methods were found to be quite expensive. So, the environmentally friendly biological methods for decontamination of radioactively contaminated areas had become of great importance, and it led to intensive research of new quiet and prospective bioremediation technologies (Vidali 2001).

2 Bioremediation and Phytoremediation

Bioremediation is the overall process developed to clean the environment using biological organisms, especially microorganisms. It could be defined as the process when pollutants are biologically degraded or transformed to other forms under

S. Dubchak (✉) · O. Bondar
State Ecological Academy of Postgraduate Education and Management, Kyiv, Ukraine

D. K. Gupta, A. Voronina (eds.), *Remediation Measures for Radioactively Contaminated Areas*, https://doi.org/10.1007/978-3-319-73398-2_9

controlled conditions. Thus, bioremediation could be considered as the use of living organisms to degrade the environmental contaminants into less toxic forms. It involves naturally occurring bacteria and fungi or plants to degrade or detoxify substances hazardous to human health. Contaminant compounds are transformed by living organisms through reactions that take place as a part of their metabolic processes. Biodegradation of a compound is a result of the actions of multiple organisms. Bioremediation plants or microorganisms must impact on the pollutants and convert them to harmless products (Suresh and Ravishankar 2004).

The bioremediation can be effective only where environmental conditions permit microbial and plants growth and activity. It involves the manipulation of environmental parameters to allow microbial and plant growth and degradation to proceed at a faster rate. Bioremediation techniques are typically more economical than traditional methods such as incineration. Bioremediation is based on natural attenuation of various contaminants' impact, and nowadays it is more acceptable than other technologies.

Many different types of organisms such as plants can be used for bioremediation but microorganisms also show the great potential being nature's original recyclers. Thus, recently the new properties of *Geobacter metallireducens* bacteria were demonstrated. It was found that this species can remove uranium from drainage waters in mining operations and from contaminated groundwater (Sasikumar et al. 2003).

Bioremediation by use of plants constitutes phytoremediation. In other words, phytoremediation is a subcategory of bioremediation which only uses green plants to clean the environment. Phytoremediation is the principal component of bioremediation in case of dealing with radionuclide contamination. It is defined as any process that uses green plants, fungi, microorganisms or their enzymes to return the environment altered by pollutants to the original state (Vidali 2001).

Phytoremediation is a cost-effective plant-based approach of remediation that takes advantage of the ability of plants to concentrate elements and compounds from the environment and to metabolize various molecules in their tissues. It refers to the natural ability of certain plants called hyperaccumulators to bioaccumulators, degrade, or render harmless contaminants in soils, water, or air. These plants are capable of stimulating the biodegradation of pollutants in the soil adjacent to roots (rhizosphere), although phytoremediation is a cheap and environment friendly clean-up process for the biodegradation of soil pollutants, it takes several years.

Toxic heavy metals, radionuclides and organic pollutants are the major targets for phytoremediation. Knowledge of the physiological and molecular mechanisms of phytoremediation began to emerge in recent years together with biological and engineering strategies designed to optimize and improve phytoremediation. In addition, several field trials confirmed the feasibility of using plants for environmental cleanup (Salt et al. 1998).

The phytoremediation strategies have a number of promising advantages in comparison with other remediation techniques, in particular:

- The cost of the phytoremediation is lower than that of traditional processes both *in situ* and *ex situ*;
- The plants can be easily monitored;
- The possibility of the recovery and re-use of valuable metals (by companies specializing in "phyto mining");
- It is potentially the least harmful method because it uses naturally occurring organisms and preserves the environment in a more natural state.

However there are also several considerable limitations in use of phytoremediation techniques:

- Phytoremediation is limited to the surface area and depth occupied by the roots;
- Slow growth and low biomass require a long-term commitment;
- With plant-based systems of remediation, it is not possible to completely prevent the leaching of contaminants into the groundwater (without the complete removal of the contaminated ground, which in itself does not resolve the problem of contamination);
- The survival of the plants is affected by the toxicity of the contaminated land and the general condition of the soil;
- Bio-accumulation of contaminants, especially radionuclides, into the plants which then pass into the food chain, from primary level consumers upwards or requires the safe disposal of the affected plant material.

There are a number of alternative conventional remediation technologies which are applied to remediate environmental contamination with radionuclides such as solidification, soil washing and permeable barriers. But a majority of these technologies are quite expensive to implement and cause further impact to the already polluted environment. Phytoremediation is evolving as a cost-effective alternative to high-energy, high-cost conventional methods. It is considered to be a "Green revolution" in the field of innovative cleanup technologies (Suresh and Ravishankar 2004).

Over the past 20 years, this technology has become increasingly popular and has been employed at sites with soils contaminated with lead, uranium, and arsenic. While it has the advantage that environmental concerns may be treated in situ; one major disadvantage of phytoremediation is that it requires a long-term commitment, as the process is dependent on a plant's ability to grow and thrive in an environment that is not ideal for normal plant growth.

3 History of Phytoremediation

While phytotechnologies have gained attention over the last several years, the processes have been taking place naturally for over three centuries. Throughout the 1970s and the following decades, plants were heavily tested and used to treat soil infiltrated with metals and contaminants in wetlands. As a result, techniques for these uses are well established (Tsao 2003). Widespread use of phytoremediation by federal and state governments, as well as non-governmental organizations, began in the 1980s (USEPA 1999). The use of the term phytoremediation was initiated by the USEPA in 1991, and it was first used in open technical literature in 1993 by Cunningham and Berti. In the late 1990s new uses for phytoremediation were discovered, and it became known among innovative scientific technologies. Phytoremediation was derived from other fields such as agronomy, forestry, chemical and agricultural engineering, microbiology, and many others. Since its inception it has developed into an independent field of study and a widely applicable technology.

4 Applications of Phytoremediation Techniques

Phytoremediation has been applied at several sites on the National Priorities. The diversity of pollutants to which it can be applied crude oil, metals, radionuclides, explosives, pesticides, chlorinated solvents and numerous other contaminants is the prime reason the technology has developed rapidly (USEPA 1999). Phytoremediation applications can be classified based on the contaminant fate: degradation, extraction, containment, or a combination of these. Phytoremediation applications can also be classified based on the mechanisms involved. Such mechanisms include extraction of contaminants from soil or groundwater; concentration of contaminants in plant tissue; degradation of contaminants by various biotic or abiotic processes; volatilization or transpiration of volatile contaminants from plants to the air; immobilization of contaminants in the root zone; hydraulic control of contaminated groundwater (plume control); and control of runoff, erosion, and infiltration by vegetative covers (Tsao 2003).

In the last decades the special attention is given to application of different phytoremediation techniques to reduce impact of radionuclides in soil and minimize irradiation risks for the population. The high concentrations of radionuclides in soil and water often impact on ecosystem stability and pose a serious risk to human health. Thus, as the result of Chornobyl accident, in Ukraine over 260,000 km^2 have ^{137}Cs contamination levels more than 40 kBq/m^2 (Grodzinsky et al. 1997). This contamination level corresponds to an additional lifetime dose about 5 mSv. Radioactive contamination is also caused by nuclear explosion and accidental spills and emissions from all nuclear fuel cycle operations, and most notably mining and milling. Selection of appropriate technology for the remediation of soil and aqueous streams contaminated with radionuclides is based on the environmental chemistry of each element, character of deposition and the rate of radioactive decay. Removal of the top soil layer, or soil excavation and transfer to designated repositories remains the most common and often the most effective method of treating contaminated soil. A variety of physico-chemical methods for treatment of radionuclide contamination includes soil washing, ion exchange, leaching with chelating agents, flocculation, and reverse osmosis-ultrafiltration (Zhu and Shaw 2000).

Recently, there has been a lot of interest in biological methods for radionuclide removal. Microorganisms and cell cultures were used to remove radionuclides from aqueous streams. There is also growing interest in using plants to remove radionuclides from the environment (Entry et al. 1997). Accordingly, phytoremediation technology is being applied to a variety of radionuclide-contaminated sites. The data on the use of plants for treatment of ^{3}H, U, Pu, ^{137}Cs and ^{90}Sr were summarized by Negri and Hinchman (2000).

Nowadays several following types of phytoremediation of radionuclides are being developed:

(a) Phytoextraction, in which high biomass radionuclide accumulating plants and appropriate soil amendments are used to transport and concentrate radionuclides from the soil into the above-ground shoots, which are harvested with conventional agricultural methods,
(b) Rhizofiltration, in which plant roots are used to precipitate and concentrate radionuclides from polluted effluents,
(c) Phytovolatilization, in which plants extract volatile radionuclides (e.g., ^{3}H) from soil and volatilize them from the foliage and.
(d) Phytostabilization, in which plants stabilize radionuclides in soils, thus rendering them harmless.

4.1 Phytoextraction (Phytoaccumulation)

Phytoextraction uses plants or algae to remove contaminants from soils, sediments or water into harvestable plant biomass (Fig. 1a). Phytoextraction has been growing rapidly in popularity worldwide for the last 20 years or so. In general, this process has been tried more often for extracting radionuclides and heavy metals than for organics. At the time of disposal, contaminants are typically concentrated in the much smaller volume of the plant matter than in the initially contaminated soil or sediment.

Phytoextraction removes radionuclides from soil without destroying the soil structure and with a limited impact on soil fertility. The phytoextraction technology is also particularly appealing for treatment of large areas of low-level contamination. Phytoextraction exploits vascular plants' natural ability to take up a variety of chemical elements through the root system, deliver these elements to the vascular tissue, and transport and compartmentalize radioactive elements in the aboveground biomass.

Above-ground biomass loaded with radionuclides is harvested, processed for volume reduction and further radionuclide concentration, and disposed of as radioactive waste. Successful phytoextraction of radionuclides depends on the bioavailability of radionuclides in soil, on the rate of uptake by plant roots and efficiency of radionuclide transport through the vascular system.

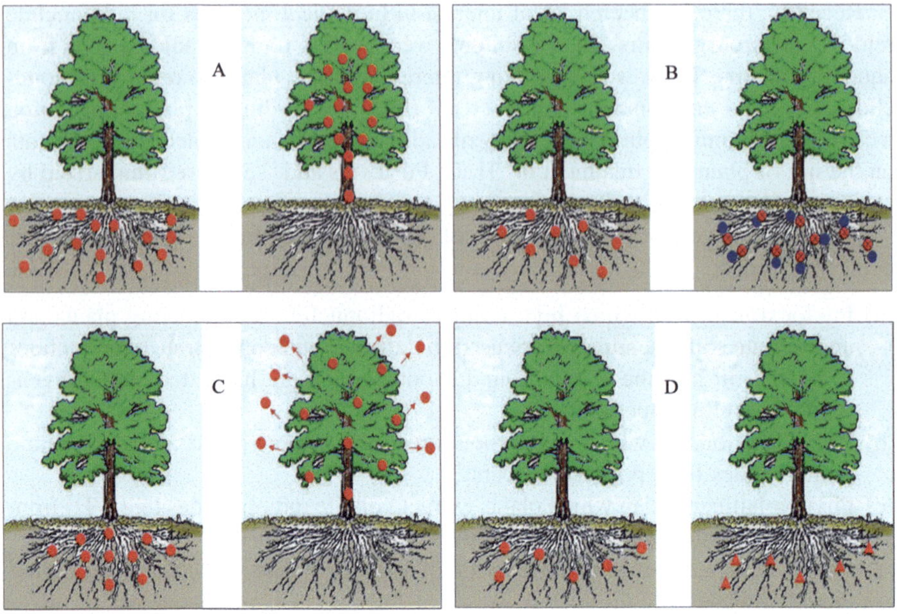

Fig. 1 Schematic representation of phytoremediation types: (**a**) – phytoextraction, (**b**) – rhizofiltration, (**c**) – phytovolatilization, (**d**) – phytostabilization

The efficiency of phytoextraction is often evaluated by the bioaccumulation coefficient (BC) (ratio of particular radionuclide concentration in plant shoots to the radionuclide concentration in soil, also known as a transfer coefficient, soil–plant transfer factor, concentration ratio) and/or total radionuclide removal per m^2 (calculated as plant biomass multiplied by radionuclide concentration). For practical purposes, total radionuclide removal per given area is the key figure, hence, the most suited phytoremediation crop must have high biomass production along with a high BC. Elements of the phytoextraction process, notably radionuclide plant uptake, were applied to ^{90}Sr, ^{95}Nb, ^{99}Tc, ^{106}Ru, ^{144}Ce, $^{226,228}Ra$, $^{239,240}Pu$, ^{241}Am, $^{228,230,232}Th$, ^{244}Cm, ^{237}Np (Nisbet and Shaw 1994; Kabata-Pendias and Pendias 1996). However, only phytoextraction of ^{137}Cs, ^{90}Sr, and $^{235,238}U$ is approaching field deployment (Dushenkov et al. 1999), being an element specific and site-specific technology. Nevertheless it is possible to formulate a general approach to developing a phytoextraction process for radionuclides, even though numerous challenges must be overcome to insure a substantial flux of radionuclide from soil to aboveground biomass.

Radiostrontium is one of a few mobile radionuclides in soil and readily available for uptake by plant roots. Usually only a tiny portion of radionuclides is in mobile form in soil. Co-precipitation with oxides and adsorption to clay minerals are believed to be the major factors reducing radionuclides bioavailability. Radionuclide speciation along with formation of complexes like in the case of U (Echevarria et al. 2001) and reduction to species with high affinity to organic matter in the case of Tc may play a crucial role in determining radionuclide mobility in soil. Bioavailability of radionuclides differs dramatically depending on soil characteristics (Fesenko et al. 1997) and properties of a specific element. It was found that ^{137}Cs transfer might correlate with soil contamination by heavy metal deposition from smelter (Bunzl et al. 1999). Type of radionuclide deposition and aging has a significant impact on radionuclide bioavailability.

It was demonstrated that ^{137}Cs accumulation by plants is determined by the content of exchangeable and mobile forms of radionuclide in the soil (Fesenko et al. 1997). In Chernobyl fallouts, for example, ^{137}Cs associated with coarse and fine dispersed fragments of irradiated fuel having a mixed uranium oxide core (fuel type of deposition) was initially less mobile in soil compared to an aerosol consisting of a core, covered by re-condensed cesium (condensed type of fallout) (Dushenkov et al. 1999). It was found that the ^{137}Cs rhizospheric mobilization was strongly correlated with the sodium tetraphenylboron-extractable ^{137}Cs, supporting that K depletion in the rhizosphere might be a principal driving force in ^{137}Cs uptake. The ^{137}Cs BC was strongly negatively correlated to the radiocesium interception potential (RIP) a common Cs binding characteristic in soil directly related to the soil vermiculite content (Delvaux et al. 2000).

A variety of factors that may help to bring radionuclides into the soil solution and make it more available for plant uptake include manipulating soil pH, adding chelators, amending soil with chemicals stimulating radionuclide desorption, interaction with microorganisms and plant exudates (Grodzinsky et al. 1997; Entry et al. 1999). For example, in hydroponic experiments it was shown that shoots to roots ratio of ^{137}Cs was significantly higher in mycorrhizal plants of *Calluna vulgaris L.* (heather) compared to non-mycorrhizal ones. In greenhouse experiments, shoots U concentration in plants grown in a U-contaminated soil increased to more than 5000 mg kg^{-1} in citric acid treated soil compared to 5 mg kg^{-1} in control pots (Huang et al. 1998).

It is generally agreed that K^+, NH_4^+, Rb^+, and Cs^+ form a homologous series of ions with considerable physico-chemical similarity. It was found that the addition of monovalent cations similar to Cs^+ physico-chemical properties resulted in the most significant levels of ^{137}Cs desorption from soil (Grodzinsky et al. 1997; Dushenkov et al. 1999). However, excessive addition of monovalent cations resulted in strong ion competition for plant uptake and did not necessarily lead to high levels of ^{137}Cs accumulation in plants. It was shown that the addition of potassium fertilizers decreased Cs uptake by vegetable crop (Lasat et al. 1998).

The majority of ^{137}Cs taken by plants tends to be tied into roots. Only about 25% of ^{137}Cs taken up by a plant is usually translocated to shoots (Clint and Dighton 1992). Ebbs et al. (1998) demonstrated in hydroponic U uptake studies that at pH 5, the uranyl (UO_2^{2+}) cation was more readily taken up and translocated by plants, than hydroxyl (pH 6) and carbonate (pH 8) U complexes. Respectively, at pH > 5 only a small fraction of U presented in plants was found in shoots. Formation of stable U-phosphate complexes in roots may prevent U translocation to aboveground plant parts. In contrast to Cs and U, almost 80% of ^{90}Sr is usually located in the shoots.

Table 1 Plant species appropriate in phytoremediation of radionuclide contaminated sites

Plant species	Radionuclide	Removal efficiency	References
Amaranthus cruentus L. *Amaranthus retroflexus L.* *Amaranthus caudatus L.*	^{137}Cs	12%	Dushenkov et al. (1999)
Helianthus annuus *Brassica juncea* *Beta vulgaris*	^{137}Cs	n/a	Willey (2001)
Helianthus annuus	^{137}Cs, ^{90}Sr	25%	Sorochinsky et al. (1998)
Kochia scoparia	^{137}Cs	2–4% yearly	Negri and Hinchman (2000)
Paspalum notatum *Sorghum halpense* *Panicum virgatum*	^{137}Cs	26–72%	Entry et al. (1999)
Helianthus annuus	^{134}Cs	8–10% yearly	Dubchak (2012)
Helianthus annuus	^{238}U	95%	Dushenkov et al. (1997)
Cladophora glomerata L. *Elodea canadensis*	^{238}U	99%	Timofeeva-Ressovskaia et al. (1962)
Brassica juncea L. *Phaseolus acutifolius*	^{90}Sr	n/a	Fuhrman et al. (2002)
Fabaceae (legumes)	^{60}Co	n/a	Reeves and Baker (2000)

Plants differ dramatically in their ability to accumulate radionuclides. For example, the ability of plant species to accumulate ^{137}Cs in the aboveground parts may differ by an order of magnitude (Grodzinsky et al. 1997). The difference in ^{137}Cs accumulation varied from 2–4 folds within cereals and reached 27 fold for all field crops (Sanzharova et al. 1997). Significant variation in radiocesium accumulation was observed in the cultivars of the same species. It was found that *Amaranthus* species *A. cruentus L., A. retroflexus L.* and *A. caudatus L.* were able to concentrate ^{137}Cs in the above-ground parts (Dushenkov et al. 1999, Table 1). This data are compared favorably to the BC of <1 for Indian mustard (*Brassica juncea L.*) and tepary bean (*Phaseolus acutifolius*) at the same plot (Lasat et al. 1998). Uranium accumulation in shoots also depends on the species (Ebbs et al. 1998). Difference in U concentrations in tested plant species was even more pronounced when soil was amended with citric acid (20 mmol kg^{-1}) (Huang et al. 1998). Thus, uranium concentration in *Brassica chinensis L.* and *Brassica juncea* was at least 100 fold higher compared with corn. Accumulation of radionuclides by plant not only heavily depends on soil and growing conditions, it is also positively correlated with radionuclide concentration in the underlining soil (Lasat et al. 1998; Dushenkov et al. 1999). This makes it difficult to normalize and compare plant species screening results obtained by many scientists in different environmental conditions.

Among other factors that may significantly influence the efficacy of phytoextraction, it is important to notice previous year's crop, type of mechanical soil amendment, irrigation, all of which may significantly influence biomass production and radionuclide uptake and translocation to aboveground biomass.

Kutlakhmedov et al. (1998) demonstrated that in the Chernobyl exclusion zone corn grown in the previous year was associated with a ten-fold increase in radiocesium removal by this plant species. Other crops preceding the phytoremediation crop in crop rotation cycle had similar but less pronounced effect. Hence, by designing a crop rotation it is possible to maximize ^{137}Cs removal from soil.

The demonstrated efficacy of ^{137}Cs phytoremediation is quite limited. It was estimated that phytoremediation of ^{137}Cs contaminated plot will return land to general use 30 years earlier than without any action (Dushenkov et al. 1999). The main advantage of phytoextraction is environmental friendliness. Traditional methods that are used for cleaning up heavy metal-contaminated soil disrupt soil structure and reduce soil productivity, whereas phytoextraction can clean up the soil without causing any kind of harm to soil quality. Another benefit of phytoextraction is that it is less expensive than any other clean-up process. The principal disadvantage of this method is that the process is controlled by plants, so it takes more time than anthropogenic soil clean-up methods.

4.2 Rhizofiltration

Rhizofiltration, use of plant roots to precipitate and concentrate radionuclides and toxic metals from polluted effluents, originated from an approach of employing hydrophytes to contain environmental contamination (Fig. 1b). In this respect it initially bordered with phytostabilization and constructed wetland technologies.

Rhizofiltration may be conducted in situ, with plants being grown directly in the contaminated water body. This allows for a relatively inexpensive procedure with low capital costs. Operation costs are also low but depend on the type of contaminant. This treatment method is also aesthetically pleasing and results in a decrease of water infiltration and leaching of contaminants (Suthersan and McDonough 1996).

First attempt to use water plants to stop radionuclide dissemination can be traced to the work initiated in early 1950s by Timofeeva-Ressovskaia et al. (1962). A study of radionuclide distribution in a freshwater pond ecosystem showed that the majority of radionuclides were adsorbed by the sludge and water plants. It was found that water plants accumulate significant amounts of radionuclides for *Cladophora glomerata L.* and *Elodea canadensis*. These results led to an idea of using a sequence of ponds with a slow water flow to clean water contaminated with radionuclides. This was the first pilot scale system for cleaning radionuclide-contaminated water. This system was tested for removal of depleted uranium. Water supplemented with depleted U at a concentration approximately 10^{-5} Ci l^{-1} was poured into the first pond at the rate of 1000 l day^{-1}. A total of 1.04 Ci of radioactivity was released into the system. Samples of water, sludge and plants were taken regularly from all parts of the system to calculate complete mass balance of radioactivity. Out of all radiation administered to the system, 97.2% were retained by the first pond. Out of that 94.9% were absorbed by sludge and water plants and 2.3% remained in water. The percent of total radioactivity that escaped the second and third pond were 1.8 and 0.7%, respectively. Thus, the three-pond system with a flow rate of 1000 l day^{-1}, removed 99.3% of the radioactivity (Table 1). A similar approach was used to study removal of ^{90}Sr and ^{137}Cs from water. Radioactive water was streamed through the sequence of 10 reservoirs with the imitation of pond sludge and vegetation. Radiocesium was not found in exiting water and only 1.5–2.0% of ^{90}Sr exited the system.

Since those pioneering experiments, excessive data has been collected on the distribution of radionuclides in ponds, lakes and wetlands (Gambrell 1994). This data may be used for modeling radionuclide distribution and constructing wetlands for retaining radionuclides. An alternative engineering approach was developed for removing contaminants from aqueous streams by terrestrial plant roots (Dushenkov et al. 2002) or seedlings (Salt et al. 1997). In controlled greenhouse experiments, Cs, Sr, and U were successfully removed from contaminated water. A different removal kinetic was observed for these elements. The plants were most effective in removing U from water. In greenhouse experiments sunflower (*Helianthus annuus L.*) proved to be more effective than beans or Indian mustard by removing greater than 95% of U from contaminated water in 24 h. In pilot rhizofiltration system sunflower plants accumulated more than 1% of U in roots (Dushenkov et al. 1997).

In the set of experiments in conducted in the Chernobyl Exclusion zone, the ability of terrestrial plants to treat surface water contaminated with ^{137}Cs and ^{90}Sr was evaluated. Radionuclide concentrations in a natural pond located 1 km west of the Chernobyl NPP were close to 80 Bq l^{-1} for ^{137}Cs and 1200 Bq l^{-1} for ^{90}Sr at the beginning of the experiment. Plants were transplanted or grown directly on 1 $\times$ 1 m rafts constructed with Styrofoam™. Plant nutrients were supplied as Hydrosol™ solution applied to a layer of Perlite on the surface of the raft. A 0.5 m^2 cavity in the center of the raft permitted plant roots to extend into the pond water. The highest BC calculated as a ratio of radionuclide content in plant tissue to the concentration in water for ^{137}Cs and ^{90}Sr were measured in sunflower roots (Dushenkov et al. 1997; Sorochinsky et al. 1998). Sunflower roots accumulated up to eight times more ^{137}Cs than the roots of timothy (*Phleum pratense L.*) or foxtail (*Alopecurus pratensis L.*). In addition, the sunflower shoots contained $2.5{\cdot}10^6$ Bq of ^{90}Sr per kg^{-1} dry weight. The BC measured for sunflower plants in these experiments demonstrated that all of radioactivity in this 75 m^2 surface area pond, estimated at $9.2{\cdot}10^6$ Bq of ^{137}Cs and $1.4{\cdot}10^8$ Bq of ^{90}Sr, could be concentrated into 55 kg dry weight of sunflower plants. The plants could then be stabilized, ashed, or vitrified and the resulting radioactive waste stored in a disposal facility.

Rhizofiltration was successfully used to remove excessive U from water at a former U-processing facility in Ashtabula, OH (Dushenkov et al. 1997). For several months a pilot rhizofiltration system effectively removed U from contaminated water. The sunflower roots used in the system proved to be effective in removing up to 95% U from water in 24 h. The U concentration exceeded 1% in dried sunflower roots. In greenhouse experiments, BC based on the ratios of U concentrations in the roots versus U concentrations in the aqueous phase reached 30,000. Thus, rhizofiltration proved to be a feasible approach in removing radionuclides from aqueous streams. Nevertheless, it requires optimization and economic evaluation against conventional technologies. Also this treatment method has its limits. Any contaminant that is below the rooting depth will not be extracted. The plants used may not be able to grow in highly contaminated areas. Most importantly, it can take years to reach regulatory levels. This results in long-term maintenance. Plants grown on polluted water and soils become a potential radiation exposure source to human and animal health, and therefore, careful attention must be paid to the harvesting process and only non-fodder crop should be chosen for the rhizofiltration remediation method.

4.3 Phytovolatilization

Phytovolatilization of radionuclides exploits a plant's ability to transpire enormous amounts of water (Fig. 1c) and is currently used for ^{3}H remediation. Tritium (^{3}H), a radioactive isotope of hydrogen, decays to a stable helium with a half-life of about 12 years. It shares physical and chemical properties of hydrogen. Tritium is found in the environment typically as tritiated water that integrated into the biogeochemical water cycle. As a weak β emitter, ^{3}H is easily shielded by air and skin and produces almost no external radiation exposure; however, its incorporation in water and organic compounds presents a health hazard when absorbed into the body.

Tritium releases by the Brookhaven National Laboratory (BNL), Savannah River Site (SRS) and other nuclear facilities has led to a raise in interest in remediation techniques for removal of titrated water from surface and ground aqueous streams. The analysis of conventional technologies for titrated water decontamination or isolation suggests that the cost is generally greater than considered practical for treatment (Fulbright et al. 1996). It was suggested that the most cost-effective approach for the minimization of ^{3}H risk is neither removal of ^{3}H from the effluent nor a typical isolation of the water, but rather a change in the path of ^{3}H exposure to the public. Calculations based on many years of experience at the SRS indicated that a 40% reduction in dose can be achieved by releasing titrated water to the atmosphere, as water vapor, as opposed to allowing it to flow off site in surface water streams (Fulbright et al. 1996).

Titrated water applied as irrigation to the floor of forests enters the hydrological cycle as it reaches the soil beneath the forest. This water evaporates from the soil surface, is absorbed into the transpiration stream by plant roots or drains through the soil to the water Table. A small portion of ^{3}H that was absorbed by plant roots remains in the plant tissues in the form of largely easily exchangeable hydroxyl ions or is incorporated into organic molecules through photosynthesis. A simple model of ^{3}H uptake in forests irrigated by titrated water was developed and analyzed (Murphy 2001). The conditions of the simulation were set to those expected for an actual irrigation scenario at the SRS with the initial ^{3}H concentration of the irrigation water at 11000 pCi ml^{-1}. In simulations, the ^{3}H concentration of the water source had been observed to decrease at a rate that approximated the decay rate of ^{3}H. The irrigation rate was set at 23 cm year^{-1}. This compares to the annual rainfall of about

120 cm year^{-1}. For this scenario the maximum ^{3}H concentration in the green wood was 1228 pCi g^{-1} occurring during the fifth year after the beginning of the irrigation. Five years after the termination of irrigation, the ^{3}H concentration had dropped to 34 pCi g^{-1}. The maximum and 5-year, post-irrigation ^{3}H inventories for the forest were 696 and 22 mCi km^{2}, respectively (Murphy 2001). The inventory of ^{3}H in the forest never exceeded 1% of the total ^{3}H supplied through irrigation. At the end of the exposure period, the inventory was less than 0.1% of the ^{3}H applied. The levels of ^{3}H calculated suggest that there is minimum risk to workers harvesting trees 5 years after the irrigation has stopped. Dose estimates for onsite workers and the public from handling the harvested trees should be considerably less than 0.01 mSv.

Phytoremediation of ^{3}H trough irrigation of forest area has been investigated at SRS for consideration as part of a system to reduce the discharge of ^{3}H from the Burial Ground Complex southwest plume, otherwise known as the Mixed Waste Management Facility (MWMF), into Four Mile Branch. The system is a combination of hydraulic control and enhanced evapo-transpiration. Tritium contaminated water is collected, moved to a location up gradient of the discharge point, and used to irrigate plants.

A phytoremediation system of high-transpiring, deep-rooted phretophytes was deployed at the 317/319 area at Argonne with the objectives to minimize water infiltration into the source soil, stabilize the tread soil surface to prevent erosion, runoff, and downstream sedimentation; hydraulically contain ^{3}H and VOCs migration to the groundwater, and continued remediation of the residual VOCs in the plume (Negri et al. 2000). The phytoremediation system was installed with savings of 33% compared to expected cost of the conventional pump-and-treat approach combined with asphalt capping. The plant-based system has a low operating and maintenance cost, providing an additional 40% savings over the lifetime of the deployment.

4.4 *Phytostabilization*

Phytostabilization focuses on the long term stabilization and containment of radionuclides. For example, the plant's presence can reduce wind erosion; or the plant's roots can prevent water erosion, immobilize the pollutants by adsorption or accumulation, and provide a zone around the roots where the pollutant can precipitate and stabilize. Unlike phytoextraction, phytostabilization focuses mainly on sequestering radionuclides in soil near the roots but not in plant tissues (Fig. 1d). Pollutants become less bioavailable, and livestock, wildlife, and human exposure is reduced. An example application of this sort is using a vegetative cap to stabilize and contain uranium mine tailings (Negri and Hinchman 2000).

Phytostabilization is particularly appropriate for radioactively contaminated sites, where one of the alternatives is to hold contaminants in place to prevent secondary contamination and exposure. Capturing radionuclides in situ is often the best alternative at sites with low contamination levels or vast contaminated areas where a large-scale removal action or other in situ remediation is not feasible. This can result in a considerable risk reduction, especially if radionuclides with relatively short half-lives are involved. Establishment of rooted vegetation prevents wind-blown dust, a major pathway for human exposure at radioactively contaminated sites. A dense root system stabilizes the soil and prevents erosion (Cunningham and Berti 2000). Plant roots also help to minimize water percolation through soil, hence reducing radionuclides leaching. Deployment of plant species capable of further renders radionuclides immobile doubles the benefits of the sites revegetation. Phytostabilization may be especially useful in controlling tailings from strip and open pit uranium mines. Overcoming U toxicity to plants and increasing fertility of tailing material are often the major challenge.

Ability of plant to transpire high water volumes is widely used for hydraulic control, which prevents migration of leachate towards groundwater or receiving waters. Fast growing deep-rooted green plants have gained popularity in their use for hydraulic control and radionuclides stabilization in soil. This approach may be effective in confining areas leaking radioactive materials. It is important to keep in mind that while phytostabilization of radionuclides may considerably reduce the environmental and human health risk it does not remove the source of radioactivity from the site. Thus, the potential risk of radiation exposure to wildlife and humans remains. It must be taken into account when decision on the best remediation approach is made.

5 Phytoremediation of Soil Contaminated with Specific Radionuclides

5.1 Phytoremediation of Cs Contaminated Soils

Pot experiments established that plant taxa differed in their Cs uptake, that ammonium salts could mobilize Cs from the soil (Lasat et al. 1997), that *Brassica oleracea* var. *capitata* (cabbage) removed the greatest amount of Cs and that additions of ammonium nitrate could increase Cs uptake by plants. Experiments in the field showed that *Amaranthus cruentus* had higher uptake than cabbage and was the taxon that eventually removed the greatest amounts of Cs from radiocaesium contaminated soil. The additions of ammonium salts to soils in the field were not as effective as in pot experiments (Lasat et al. 1998). Amongst the reasons for this might be leaching out of the root zone, the microbial transformation of ammonium salts in the field and the dislike of plants such as *Brassica oleracea* and *Amaranthus cruentus* for ammonium as a nitrogen source. The relationship between ^{137}Cs in soil and ^{137}Cs in shoot tissues of *A. cruentus* was linear, with a calculated concentration factor of 0.5. Fuhrman et al. (2002) used the equation $C_t = C_0 e^{-kt}$ (where k is plant mass/y/m^2 /soil mass × TF) to show that plant uptake would take about 18 years to reduce the ^{137}Cs in soil to 50% of its initial value at Brookhaven, assuming that TF did not change as the soil concentration decreased. The timescales suggested by the studies at Brookhaven were the first estimates of phytoremediation times for radiocaesium contaminated soils under field conditions. As the authors note, their equation is unlikely to be the most refined for estimating phytoremediation times but it, and the values from studies at Brookhaven, provided a guide for further research.

The potential of phytoremediation for Cs has also been investigated in the Chernobyl Exclusion zone by Dushenkov et al. (1999) who investigated a variety of *Amaranthus* species together with soil amendments. Amaranth species were chosen because of previous reports of high uptake by plants in this genus. Ammonium salts were shown to have the greatest desorbing power in the soils used. In field trials *A. caudatus* removed the greatest amount of Cs, but there was no increased Cs uptake into the plants with soil amendments. The highest TF recorded was 7.5 and it was estimated that, with invariant TF, "minimum possible" decontamination time would be about 7 years, considerably shorter than those reported for the soils at Brookhaven. This was related to the higher uptake of the plant taxa used and to the somewhat higher availability of Cs from the soils. British Nuclear Fuels investigated the potential of phytoremediation for Cs contaminated soils at the Bradwell Nuclear Power Station, Essex, UK, and a site which has great variation in Cs concentration. *Helianthus annuus*, *Brassica juncea* and two varieties of *Beta vulgaris* were sown and TFs measured for individual plants. A first series of experiments showed that *B. vulgaris* had the greatest phytoremediation potential and that TFs and biomass production suggested a minimum decontamination time of about 20 years was possible (Willey 2001). Interestingly, for each species the 50 plants grown on a variety of Cs concentrations took up about the same concentration of Cs irrespective of the Cs concentrations on which they were growing. The TF was significantly higher at the low Cs concentrations (Willey 2001). The TFs reported in the first experiments at this site are not as high as those reported by Dushenkov et al. (1999) and the phytoremediation times are proportionately longer. They do, however, provide the first evidence that, given sufficient desorption from soil; TFs might not change, and might even increase, as soil concentration decreases during phytoremediation. Investigations of phytoremediation at the Bradwell site are on-going. Bench scale studies at the Argonne National Laboratory-West (Idaho, USA) suggested yearly removal rates of 2–4% of radiocaesium and led to phytoremediation field trials currently using *Kochia scoparia* on ^{137}Cs contaminated soils (Negri and Hinchman 2000, Table 1).

A study done by Entry et al. (1999) included three grass species: bahia grass (*Paspalum notatum*), johnson grass (*Sorghum halpense*), and switch grass (*Panicum virgatum*). These three species were selected for this experiment because they produce a high amount of biomass in relatively short periods of time. These grasses were inoculated with two species of arbuscular mycorrhizae: *Glomus mosseae* and *Glomus intraradices*. Plants with their fungal associations were grown in the greenhouse for 6 months and above-ground biomass harvested every 2 months. At the end of the study, the above-ground portion of the three grasses that had been inoculated with fungi accumulated 41.7 to 71.7% of the total ^{137}Cs that had been added to the soil while non-inoculated plants accumulated 26.3–45.5% (Table 1).

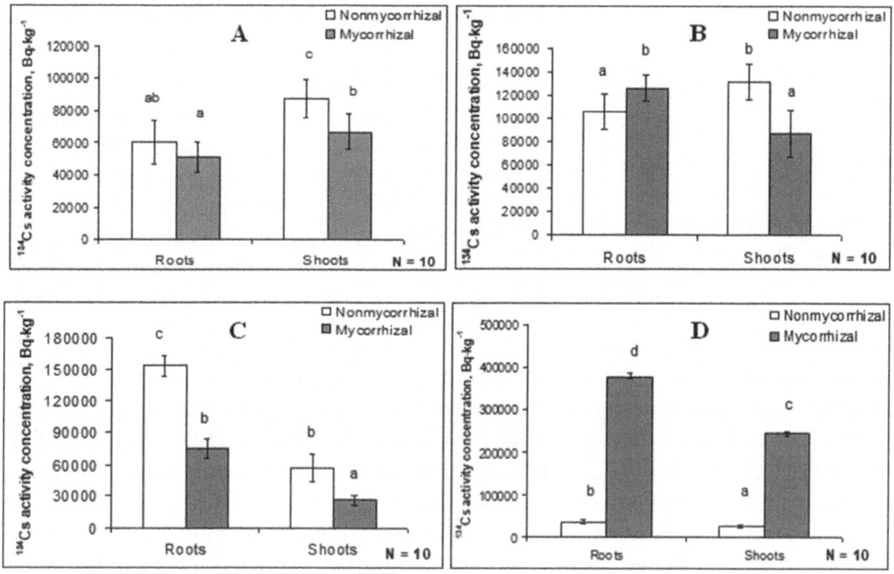

Fig. 2 ^{134}Cs activity concentration (Bq·kg^{-1}) in roots and shoots of *Plantago lanceolata* (**a**), *Medicago truncatula* (**b**), *Lolium perenne* (**c**) and *Helianthus annuus* (**d**) mycorrhizal or not with *Glomus intraradices* and grown on soil spiked with ^{134}Cs (77,000 Bq·kg^{-1})

Plants inoculated with *Glomus mosseae* resulted in the highest above ground biomass and percent accumulation of radiocaesium. Entry et al. (1999) suggested several reasons for the observed high accumulations, including optimal growing conditions, high root density, low concentrations of K and Ca in the soil, and high amounts of mycorrhizal fungi.

Dubchak (2012) in a laboratory study noticed that 30–80% decrease of ^{134}Cs activity concentration in shoots of *Plantago lanceolata*, *Medicago truncatula* and *Lolium perenne* mycorrhizal with *Glomus intraradices* in comparison with non-mycorrhizal plants. The highest (about twofold) reduction of radiocaesium concentration was found in shoots of mycorrhizal *L. perenne*. However mycorrhizal plants of *Helianthus annuus* demonstrated nearly tenfold increase of ^{134}Cs activity concentration both in their roots and shoots (Fig. 2). Thus, the sunflower plants confirmed their hyperaccumulation ability in the presence of the mycorrhiza.

These trials of phytoremediation for radiocaesium contaminated land provide a number of clear research directions. They have not suggested that phytoremediation of Cs will inevitably take so long as to always be un-worthwhile, but that TFs and/or biomass production need to be increased and sustained year on year. Root to shoot translocation of Cs is rapid so availability and uptake are the keys to achieving this. It is quite likely that, as is the case with K, manipulations at agronomic, physiological and molecular scales can be used to increase Cs uptake rates significantly.

5.2 Phytoremediation of U Contaminated Soils

The potential for phytoremediation of U from contaminated substrates has also been investigated. This is primarily because of the existence of U contaminated effluents and mine tailings but environmental contamination with depleted U from munitions is providing further impetus. Although U is not chemically similar to any ion which is metabolically essential to plants, there is some capacity for uptake of U by plants (Ebbs et al. 1998) and elevated concentrations of U have been reported in plants growing over uraniferous rocks. Investigations using plants to filter U from effluents have shown that plant roots have a significant capacity to adsorb it. For example, sunflowers have been used to remove U from contaminated effluents (Dushenkov et al. 1997). However, almost all the U remains on the external surfaces of plant roots and transfer to above ground parts is slow. Furthermore, in contaminated soils the form of U is frequently not that which plants can absorb. Huang et al. (1998) overcame this problem by the addition of citric acid to the soil, which increased soil availability and U uptake rates 1000 fold in a few days. Citric acid can be toxic to plants but further work has suggested that enhanced phytoaccumulation of U can be effected using it. Phytoextraction of U has, therefore, emerged as a possible alternative to the more widely used *ex-situ* leaching techniques, but recent reviews have concluded that further research is necessary to maximize the absorption and, in particular, translocation of U by plants (Shahandeh and Hossner 2002). The required improvement in absorption and translocation is of a greater magnitude than that for ^{137}Cs but there remains significant potential to increase U availability in soils.

5.3 Phytoremediation of Sr, Tc and Cl Contaminated Soils

The partition coefficients *K*ds are lower and TFs higher for Sr, Tc and Cl than for Cs and U but they have been less extensively investigated for phytoremediation. It was noted that there was some potential to remove ^{90}Sr from some soils using plants. There have been few subsequent estimates of phytoremediation times for soil contaminated with ^{90}Sr but comparisons of TFs (Nisbet and Woodman 2000) consistently indicate that they are likely to be less than half those for ^{137}Cs. Fuhrman et al. (2002), using field data, calculated that Sr could be phytoextracted from soils in less than half the time of ^{137}Cs at Brookhaven, NY. In some comparisons of K_ds, values for Sr can be a tenth or less than those of Cs, a difference sometimes reflected in plant uptake in the field. Other studies have confirmed these patterns and shown that taxon selection and agronomic practice can increase soil to plant transfer of ^{90}Sr (Entry et al. 1999). Given agricultural experience with Ca, there is probably room for improvement in soil to plant transfer of the radiologically significant ^{90}Sr and hence a real possibility that phytoremediation times of a few years are attainable.

^{99}Tc is generally thought to be of little radio ecological significance in terrestrial ecosystems. However, reprocessing of nuclear fuel and restrictions on discharge to aquatic systems means that much Tc will have to be put into terrestrial waste repositories. Methods for dealing with any terrestrial contamination arising from such activities are desirable. The availability of Tc is great in most aerobic soils, plants have a great uptake capacity for Tc and there is thus great potential for removing any Tc contamination from soils using phytoremediation (Bennett and Willey 2002). A number of studies have shown that plants can remove Tc almost entirely from soils in a short time.

There is similarly great potential for remediating soils contaminated with ^{36}Cl using phytoremediation. As with ^{99}Tc, much ^{36}Cl will have to be stored in terrestrial waste repositories and there are well demonstrated pathways for ^{36}Cl movement to soil surface in evapotranspiration flows. This means that Cl can move freely from depth to the soil surface (Shaw et al. 2004). Plants have a large capacity for Cl uptake, and there is therefore clearly the potential for removing ^{36}Cl contamination from soils using plants. In fact, ^{99}Tc and ^{36}Cl probably have as much potential for phytoremediation as any other environmental contaminants. They are currently of less radio ecological concern than other radionuclides but provide an opportunity for radio ecologists to demonstrate that the technology can work.

5.4 Phytoremediation of Other Radionuclides

There have been few investigations of the potential of plants to decontaminate soils of other radionuclides. Hydroponic experiments with plutonium (Pu) have shown that both plants and fungi can absorb this element (Lee et al. 2002) and there is a great deal known about the behaviour of Pu in soil and how to manipulate it. Soil availability of Pu, with the exception of sandy soils, is low (Shahandeh et al. 2001; Dulanska et al. 2012), as is plant uptake. However, it is a good example of a radionuclide for which knowledge of its behavior in soil is extensive enough to envisage the design of mobilizing agents or the engineering of uptake. Large increases in plant and fungal uptake will, however, be necessary if phytoremediation is to be a useful decontamination option for Pu. Thus, the study done by Galanda et al. (2014) considered mycoremediation properties of oyster mushrooms cultivated on soil contaminated with ^{239}Pu and ^{241}Am. These fungi demonstrated high capacity to extract radionuclides from background and tie them in their fruiting bodies with adequate big activity.

Some plants take up cobalt to satisfy metabolic demands, in particular plants in the *Fabaceae* (legumes). Thus, about 30 species of plant that hyperaccumulate Co from soils in which it occurs naturally (Reeves and Baker 2000). Cobalt hyperaccumulators can take up Co to in excess of 1000 mg kg^{-1} and suffer no detrimental effects. ^{60}Co, with its relatively short physical half-life (5.26 years), is infrequently a serious radioecological problem but it does occur in a number of effluents and contaminates a number of sites worldwide. If removal of ^{60}Co from a soil is necessary it is possible that phytoremediation might help in the future. Hyperaccumulator traits are currently the subject of a number of detailed investigations (Clemens et al. 2002) with a view to removing significant concentrations of stable Co from soils. Any advances in phytoremediation of stable Co are likely to have a great impact on the ability of plants to remove the relatively low concentrations of ^{60}Co that cause contamination problems. Ruthenium (^{106}Ru), cerium (^{144}Ce) and cadmium (^{109}Cd) are often reported to be quite available in soils whereas zinc (^{65}Zn), iron (^{59}Fe) and manganese (^{54}Mn) are less so (Sheppard and Thibault 1990) but advances in manipulating heavy metal uptake by plants could make them all suitable targets for phytoremediation.

6 Conclusions

A first phase of phytoremediation trials for radionuclides is considered to be completed now. The trials that have been carried out have mostly been focused on Cs and U but lessons have been learned that can be applied to a range of radionuclides. On the basis of the first trials, phytoremediation should not be dismissed as a potential decontamination technology, although it is still currently a technology with potential rather than proven ability. It seems likely that, even after significant refinement, decontamination times will be measured in years. If this seems slow it is important to remember that many other decontamination methods do not provide instant fixes and can also take substantial periods of time. Phytoremediation also has the advantage that it should be cheaper and less damaging to the environment than engineering based methods. Field trials of phytoremediation for radionuclides have identified the obstacles that need to be overcome for it to become a viable decontamination method. Depending on the individual radionuclide being considered these obstacles are phytoavailability and/or uptake rate.

Phytoremediation is a fast developing field. For the last 5 years over 20 projects that include at least some of the field applications of phytoremediation of radionuclides were initiated in USA, Belarus, Ukraine, United Kingdom, Yugoslavia, China and the Czech Republic. Even though there is still a lot of development of the phytoremediation technology to be done, it is clear that in the near future phytoremediation of radionuclides will become an integral part of the environment management and risk reduction process.

There is a variety of geographical and climatic conditions that need to be remediated and a variety of plant species needed for phytoremediation of radioactively contaminated areas. Plant species native to the area being restored are most desirable for remediation since they will require the least amount of external inputs and will not be a threat of becoming invasive. It is also desirable to use plants that are less susceptible to being grazed upon by animals so that radionuclides do not enter the food chain (Entry et al. 1997). Areas being remediated may need to be fenced off if it is probable that grazing will occur (Entry et al. 1999).

Once the plants are used for phytoremediation they need to be harvested or they will decompose back into the soil, redispersing radionuclides. The most common post-harvest treatment on phytoremediation projects is to incinerate the aboveground portions of plants at high temperatures, leaving only the ashes for disposal (Entry et al. 1999). Once soils do not contain significant amounts of radionuclides, further restoration can be done on the site.

Phytoremediation has many benefits over other soil remediation techniques. Since phytoremediation is in situ, it is easier and cheaper than most other remediation techniques. More information is currently needed regarding radionuclide phytoremediation. Results from greenhouse studies need to be tested in outdoor environments to determine their effectiveness.

Outdoor experiments should include various soil amendments to see which amendments favor increased radionuclide uptake in the outdoors. Such experiments may need to include long- term monitoring since radionuclide uptake will probably be slower outdoors than indoors. However, the potential for mycrorrhizal plants to be an effective remediation strategy for radionuclides appears great and will likely be enormously beneficial to cleaning the environments contaminated by recent nuclear accidents.

References

Bennett R, Willey N (2002) Soil availability, plant uptake mechanisms and soil to plant transfer of 99Tc - a review. J Environ Radioact 65:215–231

Bunzl K, Trautmannsheimer M, Schramel P (1999) Partitioning of heavy metals in a soil contaminated by slag: a redistribution study. J Environ Qual 28:1168–1173

Clemens S, Palmgren MG, Kramer U (2002) A long way ahead: engineering plant metal accumulation. Trend Plant Sci 7:309–314

Clint GM, Dighton J (1992) Uptake and accumulation of radiocaesium by mycorrhizal and non-mycorrhizal heather plants. New Phytol 121:555–561

Cunningham SD, Berti WR (2000) Phytostabilization of metals. In: Raskin I, Ensley BD (eds) Phytoremediation of toxic metals: using plants to clean up the environment. Wiley, New York, pp 71–88

Delvaux B, Kruyts N, Cremers A (2000) Rhizospheric mobilization of radiocesium in soils. Environ Sci Technol 34:1489–1493

Dubchak S (2012) Influence of arbuscular mycorrhizal fungus *Glomus intraradices* on accumulation of radiocaesium by different plant species. Probl Nucl Power Plants Chernobyl Saf 19:115–122

Dulanska S, Remenec B, Matel Ľ, Durkot E (2012) Rapid determination of ^{239,240}Pu, ^{238}Pu, ^{241}Am and ^{90}Sr in radioactive waste using combined SPE sorbents AnaLig® Pu02, AnaLig® Sr01 and TRU® resin. J Radioanal Nucl Chem 293:81–85

Dushenkov S, Vasudev D, Kapulnik Y, Gleba D, Fleisher D, Ting KC, Ensley B (1997) Removal of uranium from water using terrestrial plants. Environ Sci Technol 31:468–3474

Dushenkov S, Mikheev A, Prokhnevsky A, Ruchko M, Sorochinsky B (1999) Phytoremediation of radiocesium-contaminated soil in the vicinity of Chernobyl, Ukraine. Environ Sci Technol 33:469–475

Dushenkov V, Nanda Kumar HP, Motto H, Raskin I (2002) Rhizofiltration: The use of plants to remove heavy metals from aqueous streams. Environ Sci Technol 29:1239–1245

Ebbs SD, Brady DJ, Kochian LV (1998) Role of uranium speciation in the uptake and translocation of uranium by plants. J Exp Bot 49:1183–1190

Echevarria G, Sheppard NI, Morel J (2001) Effect of pH on sorption of uranium in soils. J Environ Radioact 53:257–264

Entry JA, Watrud LS, Manasse RS, Vance NC (1997) Phytoremediation and reclamation of soils contaminated with radionuclides. In: Kruger EL, Anderson TA, Coats JR (eds) Phytoremediation of soil and water contaminants. Washington, DC, American Chemical Society, pp 299–306

Entry JA, Watrud LS, Reeves M (1999) Accumulation of cesium-137 and strontium-90 from contaminated soil by three grass species inoculated with mycorrhizal fungi. Environ Pollut 104:449–457

Fesenko SV, Spiridonov SI, Sanzharova NI, Alexakhin RM (1997) Dynamics of ^{137}Cs bioavailability in soil-plant system in areas of the Chernobyl nuclear power plant accident zone with a different physico-chemical composition of radioactive fallout. J Environ Radioact 34:287–313

Fuhrman M, Lasat MM, Ebbs SD, Kochian LV, Cornish J (2002) Uptake of cesium-137 and strontium-90 from contaminated soil by three plant species: application to phytoremediation. J Environ Qual 31:904–909

Fulbright HH, Schwirian-Spann AL, Jerome KM, Looney BB, Brunt VV (1996) Status and practicality of detritiation and tritium reduction strategies for environmental remediation. Westinghouse Savannah River Company, Aiken, pp 1–10

Galanda D, Matel L, Strisovska J, Dulanska S (2014) Mycoremediation: the study of transfer factor for plutonium and americium uptake from the ground. J Radioanal Nucl Chem 299:1411–1416

Gambrell RP (1994) Trace and toxic metals in wetlands-a review. J Environ Qual 23:883–891

Grodzinsky D, Dushenkov S, Mikheev A, Prokhnevsky A, Ruchko M, Sorochinsky B (1997) Modification of the soil cesium-137 availability to plants. Dokl Nac Akad Nauk Ukrain 10:179–182

Huang JW, Blaylock MJ, Kapulnik Y, Ensley BD (1998) Phytoremediation of uranium-contaminated soils: role of organic acids in triggering uranium hyperaccumulation in plants. Environ Sci Technol 32:2004–2008

Kabata-Pendias A, Pendias H (1996) Trace elements in soils and plants. CRC Press, Boca Raton

Kutlakhmedov YA, Polikarpov GG, Zotov VP, Zezina NV, Mikheev AN (1998) Medico-biological consequences of the Chernobyl accident. Medecol, Kyiv

Lasat MM, Norvell WA, Kochian LV (1997) Potential for phytoextraction of ^{137}Cs from a contaminated soil. Plant Soil 195:99–106

Lasat MM, Fuhrmann M, Ebbs SD, Cornish JE, Kochian LV (1998) Phytoremediation of a radiocesium-contaminated soil: evaluation of cesium-137 bioaccumulation in the shoots of three plant species. J Environ Qual 27:165–169

Lee JH, Hossner LR, Attrep M, Kung KS (2002) Uptake and translocation of plutonium in two plant species using hydroponics. Environ Pollut 117:61–68

Murphy CE Jr (2001) An estimate of the history of tritium inventory in wood following irrigation with tritiated water. Westinghouse Savannah River Company, Aiken, pp 1–10

Negri CM, Hinchman RR (2000) The use of plants for the treatment of radionuclides. In: Raskin I, Ensley BD (eds) Phytoremediation of toxic metals: using plants to clean up the environment. Wiley, New York, pp 107–132

Negri MC, Hinchman RR, Wozniak JB (2000) Capturing a mixed contaminant plume: tritium phytoevaporation at Argonne National Laboratory's area 319. Argonne National Laboratory, Argonne, pp 1–7

Nisbet AF, Shaw S (1994) Summary of a five-year lysimeter study on the time dependent transfer of ^{137}Cs, ^{90}Sr, $^{239–240}$Pu, and ^{241}Am to crops from three contrasting soil types: distribution between different plant parts. J Environ Radioact 23:171–187

Nisbet AF, Woodman RFM (2000) Soil-to-plant transfer factors for radiocaesium and radiostrontium in agricultural systems. Health Phys 78:279–288

Reeves RD, Baker AJM (2000) Metal-accumulating plants. In: Raskin I, Ensley BD (eds) Phytoremediation of toxic metals-using plants to clean up the environment. Wiley, New York, pp 193–230

Salt DE, Pickering IJ, Prince RC, Gleba D, Dushenkov S, Smith RD, Raskin I (1997) Metal accumulation by aquacultured seedlings of Indian mustard. Environ Sci Technol 31:636–1644

Salt DE, Smith RD, Raskin I (1998) Phytoremediation. Annl Rev Plant Physiol Plant Mol Biol 49:643–668

Sanzharova NI, Fesenko SF, Lisyanskii KB, Kuznetsov VK, Abramova TN, Kotik VA (1997) Forms and accumulation dynamics of ^{137}Cs in crops after the accident at the Chernobyl nuclear power plant. Pochvovedenie 2:159–164

Sasikumar A, Sheela N, Papinazath T (2003) Environmental management: bioremediation of polluted environment. In: Bunch MJ, Suresh VM, Kumaran TV (eds) Proceedings of the third international conference on environment and health, University of York, UK, pp 465–469

Shahandeh H, Hossner LR (2002) Enhancement of uranium phytoaccumulation from contaminated soils. Soil Sci 167:269–280

Shahandeh H, Lee JH, Hossner LR, Loeppert RH (2001) Bioavailability of uranium and plutonium to plants in soil-water systems and the potential of phytoremediation. In: Gobran GR, Wenzel WW, Lombi E (eds) Trace elements in the rhizosphere. CRC Press, Boca Raton, pp 93–126

Shaw G, Wadey P, Bell JNB (2004) Radionuclide transport above a near-surface water table: IV. Soil migration and crop uptake of Chlorine-36 and Technetium-99, 1990 to 1993. J Environ Qual 33:2272–2280

Sheppard MI, Thibault DH (1990) Default soil solid/liquid partition coefficients, Kds, for four major soil types: a compendium. Health Phys 59:471–482

Sorochinsky BV, Mikheev AN, Kuchko MV, Prokhrevsky AT (1998) Decontamination of small water reservoirs of the 10-km zone of Chernobyl NPP by rhizofiltration. Problems of Chernobyl Exclusion Zone, pp 97–102

Suresh P, Ravishankar GA (2004) Phytoremediation-A novel and promising approach for environmental clean-up. Crit Rev Biotechnol 24:97–124

Suthersan SS, McDonough J (1996) Remediation engineering: design concepts. CRC Press, Boca Raton

Timofeeva-Ressovskaia EA, Agafonov BM, Timofeev-Ressovsky NV (1962) On radioisotopes fate in water bodies. Proc Inst Biol 22:49–67

Tsao DT (2003) Overview of phytotechnologies. In: Tsao DT (ed) Advances in biochemical engineering/biotechnology book series, phytoremediation. Springer 78:1–50

USEPA (1999) Phytoremediation resource guide. EPA/542/B99/003 (1999). Available online at http://www.epa.gov/tio

Vidali M (2001) Bioremediation an overview. Pure Appl Chem 73:1163–1172

Willey NJ (2001) Phytoremediation-Solar powered soil decontamination. Biochemist pp 18–20

Zhu YG, Shaw G (2000) Soil contamination with radionuclides and potential remediation. Chemosphere 41:121–128

Water Decontamination at Radioactively Contaminated Lands

Anna V. Voronina, Nickolay D. Betenekov, and Vladimir S. Semenishchev

1 Introduction

Radioactive contamination of the environment occurs as a result of nuclear weapon tests, uranium and thorium ores mining, daily work of nuclear fuel cycle enterprises, radiation accidents and deposition of radioactive waste. Types of radioactive fallouts may vary depending on cause and conditions of radionuclides release. Radionuclides may be released into surface waters immediately or through the atmosphere; they can be washed out from radioactively contaminated objects (buildings, roads, forests, soils, etc.). This results in radionuclides migration in soil and lithosphere with surface and underground waters.

Catastrophic radiation accidents with radionuclides release resulted in the most dramatic radioactive contamination of lands and natural waters by anthropogenic radionuclides including transuranium elements and fission products. A higher level of radioactive contamination is typical for water in rivers and lakes as compared with coastal seawater. At the radioactively contaminated lands, contamination of natural water bodies including sources of drinking water occurs. A certain time after the accident, radionuclides migrate into underground water.

The level of radioactive contamination of natural water depends on characteristics of radioactive release. Before the explosion in 1957 at the Mayak PA, the tank with high-level radioactive waste (HLW) contained approximately 74 PBq of beta emitting fission products (Batorshin and Mokrov 2013). Relatively short-lived radionuclides dominated in this tank over long-lived ones: ^{144}Ce (66%), ^{95}Zr (25%) and ^{106}Ru (3.7%) have given the main contribution to activity of the HLW. By contrast, the contribution of activity of long-lived radionuclides was as following: ^{90}Sr - 5.4%,

A. V. Voronina (✉) · N. D. Betenekov · V. S. Semenishchev
Radiochemistry and Applied Ecology Department, Physical Technology Institute, Ural Federal University, Ekaterinburg, Russia
e-mail: ndbetenekov@urfu.ru

D. K. Gupta, A. Voronina (eds.), *Remediation Measures for Radioactively Contaminated Areas*, https://doi.org/10.1007/978-3-319-73398-2_10

^{137}Cs – 0.35%, plutonium – 0.0043%. After the explosion, the main activity have fallen into the lakes Uruskul, Berdenish, Kazhakul and Alabuga locating in the head part of the East-Urals Radioactive Trace (Karavaeva et al. 1994; Lagunov and Smagin 2007). Smagin (2013) reported that 10 of October 1957 beta activity of water was $5.3 \cdot 10^4$ Bq L^{-1} in Berdenish lake and $1.5{\cdot}10^5$ Bq L^{-1} in Uruskul lake.

Nearly 7.4 PBq of radionuclides transferred into natural waters as a result of the Chernobyl disaster (Israel and Petrov 1988). Primary contamination of rivers and the Kiev water pond was conditioned by the release of volatile radionuclides and radioactive aerosols; a certain time after the accident, radionuclides released to rivers from contaminated soils through rainwater. The level of radioactive contamination of the Kiev water pond by ^{137}Cs was 20 Bq L^{-1} in July 1986; in May 1987, this activity decreased to 1 Bq L^{-1} (Lattrell 1993).

Natural water bodies and underground waters may be located in zones of influence of nuclear fuel cycle enterprises. For example, the zone of influence of the Mayak PA includes Techa River where low-level liquid radioactive waste was discharged in 1948–1952; Lake Kyzyltash (V-2) was exploited as a cooling pond since 1947. Approximately 88 PBq of radionuclides were released into surface waters because of normal work of Mayak PA (Sadovnikov et al. 2002). In 1980–1988 activity of certain radionuclides in water of Lake Kyzyltash was as following: ^{90}Sr $= 200 \pm 75$ Bq L^{-1}, ^{137}Cs $= 150 \pm 35$ Bq L^{-1}, ^{106}Ru $= 200 \pm 60$ Bq L^{-1}, ^{51}Cr $= 250 \pm 75$ Bq L^{-1} (Smagin 2013). Medium-level liquid radioactive waste from Mayak PA was discharged into Lake Karachay (V-9). Totally, 4.6 PBq of ^{90}Sr, 450 tons of uranium and 1.2 PBq of ^{60}Co migrated from this lake to underground waters during the period of its exploitation (Aleksakhin et al. 2007). The average concentrations of radionuclides in water migrating from Lake Karachay to subsurface in 1970–1980 were as following: ^{90}Sr $= 1{\cdot}10^6$ Bq L^{-1}, ^{137}Cs $= 8{\cdot}10^5$ Bq L^{-1}, ^{106}Ru $= 4{\cdot}10^6$ Bq L^{-1}, ^{60}Co $= 1{\cdot}10^5$ Bq L^{-1}, ^{3}H $= 6{\cdot}10^5$ Bq L^{-1}, uranium = 62 mg L^{-1} (Aleksakhin et al. 2007). Since 2015, the lake is completely filled in by concrete blocks and rocks, with no visible water on the surface.

As a result of the Fukushima Daiichi accident, the release of the main dose contributing radionuclides ^{137}Cs and ^{131}I into the environment was 100–500 and 6–20 PBq respectively (Aleksakhin and Sychov 2013). Hamada and Ogino (2012) reported that the maximal activity of ^{131}I (965 Bq L^{-1}) was detected on 20 of March 2011 in samples of tap water from Fukushima prefecture; in Tokio this activity was 210 Bq L^{-1}. Rivers Abukuma, Mano, Nitta, Ota, Ukeda, Kodaka, Osami, Same and Kido flow through contaminated territories of Fukushima prefecture. These river systems are now long-term sources of secondary contamination of natural waters including the Pacific Ocean (Konoplev 2016). The most contaminated Ukeda River flows at lands with ^{137}Cs activity in soils over 3000 kBq m^{-2}. Concentrations of dissolved radiocesium in coastal regions and the Abukuma River system were 2.0–330 mBq L^{-1} (Yoshimura et al. 2015).

During nuclear weapon tests in 1949–1989 at the Semipalatinsk Test Site (STS), $2.8{\cdot}10^{15}$ Bq of ^{137}Cs, $1.7{\cdot}10^{15}$ Bq of ^{90}Sr, $2.6{\cdot}10^{13}$ Bq of $^{239,\,240}$Pu were released into the environment (Nazarbayev et al. 2016). There are a number of zones with elevated concentrations of radionuclides at the territory of STS. Particularly, elevated concentrations of anthropogenic radionuclides in water of waterways were found at the territory near Degelen Site; the average activities are ^{90}Sr – n·10^{3}÷ n·10^{5} Bq L^{-1}, ^{137}Cs – n·10^{4}÷n · 10^{5} Bq L^{-1}, $^{239+240}$Pu – up to n·10^{4} Bq L^{-1}, ^{3}H – up to n · 10^{5} Bq L^{-1} (Lukashenko 2016). There are 12 waterways at the Degelen Site; all of them are free-access objects of water consumption, which can be potentially used by population. Thus, permanent radiation control is necessary in zones of waterways being connected with nuclear test sites.

Contamination of natural water with natural radionuclides is another one problem unsolved. These natural radioactive waters may be used as sources of drinking water. Natural radionuclides may appear in natural waters due to either natural geochemical processes or anthropogenic dissipation of naturally occurred radioactive materials (NORMs). The presence of uranium is usually conditioned by dissolution of uranium-containing minerals because normally uranium presents in crystal lattice of a mineral. After dissolution, uranium migrates very well in natural waters with various chemical compositions. Radioactive daughter decay products of uranium are usually located outside the crystal lattice of a mineral; therefore, transport of these daughters (e.g. Ra-226) occurs due to processes of leaching. Migration of radon occurs due to physical processes conditioning its emanation from rocks and transfer through capillaries and cracks. Water contamination by natural radionuclides is typical for many countries in the world. In water of tributaries of Syrdarya River (Kazakhstan) activity of ^{226}Ra varies from 0.02 to 53.2 Bq L^{-1} depending on location and season (Zhanbekov and Mukataeva 2010; Zhanbekov et al. 2012). Elevated activities of natural radionuclides was found in Khyargas Lake in Western Mongolia because this lake feeds by the main rivers of Western Mongolia (Zavkhan and Khovd) that flow through uranium containing regions (Isupov et al. 2011). Uranium concentration in water of Lake Sasykkol (Kazakhstan) is 30 mg L^{-1} that corresponds to approximately 370 Bq L^{-1} (Razykov et al. 2002). In water of salty lakes in Qinghai and Xinjiang regions (China), uranium concentration reaches up to 60 mg L^{-1} (nearly 740 Bq L^{-1}) (Mingkuan Qin 2009). Isupov et al. (2014) reported that in water of Lake Bezymyannoye (Chuiskaya valley, Russia) uranium content is near to 2.2 mg L^{-1}. Presence of elevated amounts of uranium in rocks may also cause transfer of radon into ground air and underground waters. For example, in Sverdlovsk region (Ural Mountains, Russia), more than 350 water sources with elevated concentrations of natural radionuclides were found. Most of these sources are artesian wells and springs. Activity of ^{222}Rn in some of these sources exceeds 1000 Bq L^{-1} (Semenishchev et al. 2017).

Uranium or thorium ores mining and processing usually results in anthropogenic dissipation of natural radionuclides. Former or active uranium factories (mines, pits, industrial sites of ore processing and hydrometallurgical plants, tailing dumps, etc.) are large-scale sources of natural radionuclides contaminating natural waters. As an example, Lauria and Godoy (2002) reported that ^{228}Ra activity in a seawater from a lagoon in Brazil near a monazite enrichment plant reached up to 2000 mBq L^{-1}.

2 Decontamination of Natural Water with Various Salt Content from Radionuclides

Partitioning and speciation of radionuclides should be taken into account in solving the problem of radioactive natural waters decontamination.

Sorption is a widely used method for decontamination of aqueous media. It can be realized either in batch conditions, when a sorbent is added to water body for a long-term period, or in dynamic conditions using filters or columns filled by a sorption material. A great variety of sorption materials complicates the choice of a sorbent for natural water decontamination. Large volume of radioactive contaminated waters requires using relatively low-cost, effective and environmentally friendly sorbents. These sorbents should also possess good chemical stability and mechanical strength. Use of these sorbents should not result in a secondary contamination of water by products of sorbents dissolving and leaching. Natural sorbents or sorbents based on natural materials are preferable for deactivation of natural waters. These sorbents are compatible with the environment and more eco-friendly as compared with synthetic ones.

Natural aluminosilicates such as clinoptilolite, vermiculite, zeolite, etc. possessing a high sorption affinity to cesium and strontium radionuclides are successfully used in deactivation of natural waters for a long time (Vdovina et al. 1976; Tsitsishvili et al. 1985; Tarasevich 1996; Dyer et al. 2006). However, low mechanical strength, relatively low selectivity and reversibility of sorption result in low effectivity of water decontamination as well as complicate their use in dynamic conditions.

Sorbents based on some natural and synthetic supports are interesting for deactivation of natural waters with various salt content. Particularly, thin-layer inorganic sorbents (TLIS) and surface-modified sorbents may be successfully used for this purpose.

In order to synthesize TLIS, a thin film of a sparingly soluble inorganic compound is coated onto an inert or sorption-active support; the thin film should possess a high selectivity for a radionuclide (Betenekov et al. 1977, 1982). TLIS prospective for decontamination of natural waters were obtained using nonwoven filtering materials such as polyethylene terephthalate (NTFM) and polypropylene membranes (MP) produced by EMD Millipore company, USA (Voronina 1996) as well as aluminosilicates (Voronina 1995; Betenekov et al. 2004). Only sorption-active materials can be used for synthesis of the surface-modified sorbents; among these there are hydrated titanium and zirconium oxides (Voronina et al. 2012), technical cellulose (Voronina et al. 2013a), aluminosilicates (Voronina et al. 2013b, 2015b). Voronina et al. (2017) described the effect of synthesis method on differences in structure and sorption characteristics of TLIS and surface-modified sorbents based on clinoptilolite. Sorbents based on these supports possess better kinetic characteristics and available sorption capacity as compared with granulated analogues; furthermore, these sorbents show high chemical, mechanical and radiation stability and good selectivity.

Sorbents based on nonwoven materials are especially interesting for developing filters for decontamination of natural and drinking water. These sorbents combine sorption characteristics of thin films and sanitary-hygienic quality of the support material. Filters based on membranes can retain colloidal particles and suspended matter due to Van der Waals, electrostatic, hydrophobic and hydrogen interactions. These membranes are ideal supports for TLIS because they are chemically and thermally stable, possess good mechanical characteristics, do not swell and clod. Furthermore, these membranes are commercially available as cartridge filters; they do not contain toxic components (in contrast to many organic and inorganic ion exchangers) and do not release any sediment into solution. All these advantages conditioned a wide use of membranes in water treatment systems (Brock 1987). Thin films of Ni, Cu, Zn, Fe, Ti and Mn hydroxides and basic salts as well as ferrocyanides were coated onto the surface of nonwoven materials and membranes from aqueous solutions using the thin-layer modification method. Hydroxides and basic salts were synthesized using carbamide as a precipitating agent. Films of manganese hydroxide were synthesized through reduction of potassium permanganate in acidic solutions. Mixed ferrocyanides and phosphates were obtained by chemical modification of hydroxides films in potassium ferrocyanide or phosphate solutions.

Use of natural aluminosilicates (such as clinoptilolite and glauconite) as a support for sorbents is conditioned by their selectivity for cesium and strontium as well as their successful chemical and mechanical stability, low cost and ability to improve such indicators of water quality as turbidity, color, content of iron and manganese ions (Kuznetsov et al. 1974).

2.1 Decontamination of Fresh Water from Radionuclides

Table 1 shows distribution coefficients of ^{137}Cs, ^{90}Sr and ^{144}Ce radionuclides from tap water (K_d) for some TLIS based on nonwoven filtering materials being prospective for production of individual water filters (IWF). Tap water may be considered as an analogue of natural fresh water.

Table 2 presents equilibrium distribution coefficients of ^{137}Cs and ^{90}Sr radionuclides sorption from tap water by sorbents based on natural aluminosilicates. The data are given for thin-layer sorbents TH-Cl, TP-Cl and MD-Cl as well as for surface-modified sorbents NPF-Cl, IPF-Cl and NPF-GL.

Table 3 shows equilibrium distribution coefficients of natural radionuclides sorption from tap water by various sorbents based on supports. All data on radium and lead sorption are published for the first time. References are given for distribution coefficients of uranium and thorium.

The results of study of sorption properties of certain materials with respect to natural radionuclides have shown that the NPF-Cl, TH-Cl and TP-Cl are the most effective inorganic sorbents for uranium and thorium separation from natural waters. Chemical modification of the TH-Cl sorbent into the TP-Cl sorbent allows for the increase of distribution coefficient and static exchange capacity for uranium; for

Table 1 Distribution coefficients of ^{137}Cs, ^{90}Sr and ^{144}Ce radionuclides from tap water for some TLIS based on nonwoven filtering materials; pH 7.2 ± 0.2, time of phases contact was 60 min

Sorbent	K_d, mL g^{-1} of a film-forming metal		
	^{137}Cs	^{90}Sr	^{144}Ce
Copper-potassium ferrocyanide based on NTFM (CPF-NTFM)	$2.1 \cdot 10^5$	$1.6 \cdot 10^4$	$1.0 \cdot 10^5$
Zinc-potassium ferrocyanide based on NTFM (ZPF-NTFM)	$1.9 \cdot 10^5$	$1.5 \cdot 10^4$	$7.7 \cdot 10^4$
Nickel-potassium ferrocyanide based on NTFM (NPF-NTFM)	$1.5 \cdot 10^5$	$3.5 \cdot 10^4$	$4.6 \cdot 10^4$
Iron hydroxide based on NTFM (IH-NTFM)	$7.8 \cdot 10^2$	$3.2 \cdot 10^3$	$7.2 \cdot 10^3$
Iron-potassium ferrocyanide based on NTFM (IPF-NTFM)	$5.6 \cdot 10^4$	$4.2 \cdot 10^3$	–
Iron phosphate based on NTFM (IP-NTFM)	$3.4 \cdot 10^4$	$1.2 \cdot 10^5$	$1.6 \cdot 10^5$
Titanium hydroxide based on NTFM (TH-NTFM)	$1.7 \cdot 10^4$	$3.0 \cdot 10^4$	$9.2 \cdot 10^4$
Titanium phosphate based on NTFM (TP-NTFM)	$2.9 \cdot 10^4$	$1.2 \cdot 10^5$	$4.6 \cdot 10^5$
Manganese dioxide based on NTFM (MD-NTFM)	$1.8 \cdot 10^4$	$2.3 \cdot 10^4$	$2.1 \cdot 10^4$
Manganese dioxide based on Millipore membrane (MD-MP)	–	$1.0 \cdot 10^5$	–
Manganese phosphate based on Millipore membrane (MP-MP)	–	$2.1 \cdot 10^3$	–
Iron phosphate based on Millipore membrane (IP-MP)	–	$7.9 \cdot 10^3$	–
Iron-potassium ferrocyanide based on Millipore membrane (IPF-MP)	$4.0 \cdot 10^6$		

Adopted from Voronina (1996)

Table 2 Equilibrium distribution coefficients of ^{137}Cs and ^{90}Sr radionuclides sorption by sorbents based on natural aluminosilicates; tap water, time of phases contact = 1 week

Sorbent	K_d, mL g^{-1} of a sorbent		References
	^{137}Cs	^{90}Sr	
Titanium hydroxide based on clinoptilolite (TH-Cl)	$1.3 \cdot 10^4$	$3.5 \cdot 10^3$	Voronina (1995, 1996)
Titanium phosphate based on clinoptilolite (TP-Cl)	–	$4.7 \cdot 10^3$	Voronina (1996)
Manganese dioxide based on clinoptilolite (MD-Cl)	–	$7.0 \cdot 10^3$	Voronina (1996)
Nickel-potassium ferrocyanide based on clinoptilolite (NPF-Cl)	$1.0 \cdot 10^7$	$7.9 \cdot 10^3$	Voronina et al. (2015a, b)
Iron-potassium ferrocyanide based on clinoptilolite (IPF-Cl)	$7.9 \cdot 10^4$	–	Voronina et al. (2015b)
Nickel-potassium ferrocyanide based on glauconite (NPF-Gl)	$1.0 \cdot 10^5$	$1.0 \cdot 10^2$	Voronina et al. (2015a, b)

these sorbents, exchange capacities are 28.9 and 41 mg of U g^{-1} of sorbent respectively. The MD-MP and TH-Cl sorbents showed the highest selectivity for radium and lead radionuclides.

It should be mentioned that modified sorbents based on clinoptilolite are multipurpose sorbents possessing a high selectivity and capacity for cesium, strontium, uranium, thorium, radium and lead radionuclides.

Table 3 Equilibrium distribution coefficients of natural radionuclides sorption from tap water; time of phases contact = 1 week

	K_d, mL g^{-1} of a sorbent				
Sorbent	U	Ra	Pb	Th	References
MD-MP	–	$2.7 \cdot 10^{4*}$	$3.1 \cdot 10^{4*}$	–	Unpublished data
MD-NTFM		$6.7 \cdot 10^{3*}$	$3 \cdot 10^{3*}$		Unpublished data
TH-Cl	$3.5 \cdot 10^3$	$7.3 \cdot 10^2$ $1.8 \cdot 10^{4*}$	$9 \cdot 10^2$ $2.2 \cdot 10^{4*}$	$3.7 \cdot 10^3$	Betenekov et al. (2001).
TP-Cl	$4.5 \cdot 10^3$	–	–	–	Betenekov et al. (2001)
NPF-Cl	$9.6 \cdot 10^5$	–	–	–	Malikova et al. (2017)
NPF-Gl	$3.5 \cdot 10^3$	–	–	–	Malikova et al. (2017)

*mL g^{-1} of a film-forming metal.

2.2 Decontamination of Seawater from Radionuclides

Deactivation of highly mineralized waters, e.g. seawater is a more difficult task than decontamination of fresh waters. Elevated concentrations of Na^+, K^+ and Ca^{2+} suppress sorption of cesium and strontium radionuclides due to the competition effect. High concentration and variety of anions (such as Cl^-, HCO_3^-, SO_4^{2-}, SiO_3^{2-}, etc.) in highly mineralized waters conditions a great variety of radionuclides speciation. Modified aluminosilicates NPF-Cl and NPF-Gl showed high decontamination of seawater from cesium and strontium radionuclides. These sorbents possess a high exchange capacity for cesium and strontium radionuclides (capacities for cesium are as high as 402 ± 28 and 211 ± 15 and for strontium – 560 ± 39 and 37 ± 4 mg g^{-1} for NPF-Cl and NPF-Gl sorbents respectively) and high distribution coefficients for these radionuclides in presence of analogue ions (Kulyaeva and Voronina 2015). Cesium distribution coefficients in simulated seawater is $5.0 \cdot 10^4$ and $3.0 \cdot 10^4$ L kg^{-1} for the NPF-Cl and NPF-Gl sorbents respectively.

3 Decontamination of Drinking Water at Radioactively Contaminated Lands

The problem of decontamination of drinking water from radionuclides may occur not only after serious radiation accidents but also at radon hazardous areas. In the first case, drinking water will be contaminated by fissionable materials (uranium and plutonium), fission products (^{137}Cs, ^{90}Sr, ^{144}Ce etc.) and activation products (^3H, ^{60}Co etc.); whereas, in the second case, it will contain natural radionuclides such as ^{238}U, ^{226}Ra and their short-lived daughter decay products.

Long-lived fission products ^{137}Cs and ^{90}Sr are the main radionuclides conditioning a long-term contamination of extended areas in Russia after Chernobyl and Kyshtym disasters. In case of Fukushima-1 disaster, contamination of territories in Japan is conditioned mainly by ^{137}Cs. Every country has its own permissible levels of radionuclides activity in drinking water. Temporary permissible levels may be accepted during the first time after an accident. Table 4 gives an overview of the maximal permissible activities of radionuclides in drinking water for several countries being contaminated as a result of radionuclides release. It can be seen that permissible activities are significantly increased after an accident and then decreased over years after the accident.

Until 1990, regulations of the USSR being passed by the Ministry of Public Health in 1986–1987 were valid in territories of Ukraine and Belarus (Kenigsberg 2008). A significant toughening of limits was done in 1990. Increasing the quality of drinking water to compliance with regulatory requirements is one of the main tasks for reducing population doses at radioactively contaminated lands.

3.1 Drinking Water Deactivation Using Local Installations

After a radiation accident, the neighbor population receives elevated doses of internal irradiation. Availability of individual water filters (IWF) for population and collective water filters (CWF) for water treatment plants would allow eliminating radionuclides from drinking water at contaminated lands and therefore minimizing internal doses due to prevention of radionuclides consumption with drinking water. Creation of local systems for deactivation of drinking water would also help to decrease internal dose rates at radon hazardous areas where drinking water usually contains elevated activities of natural radionuclides.

IWF and CWF for deactivation of drinking water can be developed using sorption materials. Development of these filters requires an intensive study of radionuclides behavior in sorption systems taking into account their concentrations, speciation as well as composition and parameters of water. The sorbents used in the filters should effectively concentrate radionuclides, have an extended resource and should not decrease the quality of drinking water. Radionuclides may present in drinking water as hydrated cations, complex ions and radioactive colloids. The variety of radionuclides speciation will affect the choice of a sorbent.

Filters for deactivation of drinking water should separate a wide spectrum of radionuclides from large volumes of water. Creation of a complex system allowing both deactivation and improve of other indicators of water quality is another one interesting research topic. Sanitary and hygienic attestation of IWF is the necessary final stage of the development process.

3.1.1 Individual and Collective Filters for Radionuclides Elimination from Drinking Waters

Individual water filters were produced using some of sorbents described above. Comparative life tests of these filters were performed.

Table 4 The maximal permissible activities of radionuclides in drinking water

Country	Radionuclide	Permissible activity of a radionuclide in water, Bq L^{-1}	Type of regulation	References
Russia (USSR)	^{137}Cs	370	Temporary, 'VDU-86', (30.05.86)[a]	Onischenko (2009)
		18.5	Temporary, 'VDU-91' (22.01.91)	Onischenko (2009)
		11	Permanent	NRB-99/2009
	^{90}Sr	0.37	Temporary, 'VDU-91' (22.01.91)[b]	Onischenko (2009)
		4.9	Permanent	NRB-99/2009
	^{238}U	3.0	Permanent	NRB-99/2009
	^{226}Ra	0.49	Permanent	NRB-99/2009
	^{222}Rn	60	Permanent	NRB-99/2009
Japan	Radioiodine	300	Temporary (17.03.2011)	Hamada and Ogino (2012)
	Plutonium and other transuranium elements	1		
	Cesium radionuclides	200		Akahane et al. (2012), MHLW (2011), and Aleksakhin and Sychev (2013)
	Uranium isotopes	20		
Belarus	^{137}Cs	18.5	Permanent, 'RDU-96'	Kenigsberg (2008)
		10	Permanent, 'RDU-99'	RDU-99 (in redaction of 2006)
	^{90}Sr	0.37	Permanent, 'RDU-96'	Kenigsberg (2008)
		0.37	Permanent, 'RDU-99'	RDU-99 (in redaction of 2006)
Ukraine	^{137}Cs	2	Permanent	Kenigsberg (2008)
	^{90}Sr	2	Permanent	
Kazakhstan	^{137}Cs	11	Permanent	GN-2015
	^{90}Sr	4.9		
	^{238}U	3		
	^{226}Ra	0.49		
	^{239}Pu	0.55		

[a]Since 30 of May 1986 in the USSR, only gross beta activity was controlled; therefore, the regulation VDU-86 limited only gross beta activity, whereas later VDU limited cesium activity in drinking water

[b]^{90}Sr was not controlled until 1991 (Onischenko 2009)

IWF with thin-layer inorganic sorbents based on nonwoven filtering materials and modified clinoptilolite were developed by the Radiochemistry and Applied Ecology Chair of Ural Federal University (UrFU, previously Radiochemistry Chair of Ural State Technical University). The results of IWF development and life tests are described by Voronina (1996).

Filters developed by the research group of Perm State Technical University (PSTU) as well as commercially available filters produced by Millipore (USA) and Compomet Cantec (USSR/Canada) companies were tested at the same conditions for comparison. Overall, ten IWFs were tested; their content and main characteristics are described below.

IWF-1. A mixture of an ISM-S sorbent (developed by Chemistry department of PSTU) and a BAU-A activated carbon is used as an active material in this filter. The ISM-S sorbent is a derivative of manganese oxide with the following chemical composition: (Na, K)·x·MnOy, where $x = 0.25 \div 0.3$ and $y = 1.9 \div 2.1$ with the humidity of up to 20%. The ISM-S sorbent is selective for strontium radionuclides; the filter contains 800 cm^3 of the ISM-S sorbent and 400 cm^3 of the BAU-A activated carbon.

IWF-2. The filter contains 1 l of a mixture of equal weights of two sorbents, a BAU-MF activated carbon and a FCU sorbent developed by PSTU. Weight of the FCU sorbent is 150 g. The FCU sorbent is the BAU-MF activated carbon impregnated by the mixed copper-potassium ferrocyanide (4.1%) and iron ferrocyanide (2.0%). Content of the active component is ~ 11–12%wt. The FCU sorbent is selective for cesium radionuclides.

IWF-3. The filter contains a mixture of ISM-S sorbent (800 cm^3), FCU sorbent (400 cm^3) and BAU-A activated carbon (a little addition). This combined filter is intended for cesium and strontium elimination from drinking water.

IWF-4. Filters Rosa-1 was produced by Soviet-Canadian company Compomet Cantec. According to the product sheet, the filter contains activated carbon impregnated by silver with addition of a ferrocyanide sorbent. The filter is intended for domestic decontamination of water from various pollutants including radioisotopes of cesium, strontium and other heavy metals. Resource of the filter is mentioned to be 3100 liters.

IWF-5. This system for water decontamination consisted of two consequently attached filters. The first filter contained activated carbon; whereas the second one contained TH-Cl sorbent developed by Radiochemistry Chair of USTU. The TH-Cl sorbent is a thin-layer sorbent consisting of natural clinoptilolite with a thin layer of titanium hydroxide. This sorbent is selective for radionuclides of strontium, cesium, rare earth elements, uranium and plutonium. Thus, the filtering system provides two stages of water treatment: biological decontamination occurs at the first stage and deactivation occurs at the second stage. The system contained 1000 cm^3 of the TH-Cl sorbent and 240 cm^3 of activated carbon.

IWF-6. A cartridge filter produced by Millipore Company (USA). The active component of the filter is a polypropylene membrane with the average pore size of 0.5 μm. The filter provides deep decontamination of water from microorganisms, colloidal particles and suspended matter.

IWF-7. The cartridge filter produced by Millipore Company (USA) was modified by Radiochemistry Chair of USTU. A thin layer of iron-potassium ferrocyanide was coated onto the polypropylene membrane with the average pore size of 0.5 μm in order to make it selective for cesium radionuclides. Combination of mechanical filter and sorption-active film provides the possibility of separation of radionuclides presenting as both dissolved and colloidal forms.

IWF-8. This is the combined filtering system consisting of polypropylene membrane (Millipore, USA) with coated iron-potassium ferrocyanide film on it (the same as IWF-7) and the TH-Cl sorbent. The IWF-8 was developed by Radiochemistry Chair of USTU for elimination of natural (uranium) and anthropogenic (cesium, strontium, rare earth elements and plutonium) radionuclides as well as heavy metals, colloids and suspended matter from drinking water.

IWF-9. This combined filtering system consists of polypropylene membrane (Millipore, USA) with coated manganese hydroxide film on it and the MD-NTFM sorbent. The thin film of manganese hydroxide provides selectivity for strontium radionuclides. The filter was developed by Radiochemistry Chair of USTU.

IWF-10. This combined filtering system consists of polypropylene membrane (Millipore, USA) with coated manganese hydroxide film on it and the TH-Cl sorbent. The IWF-10 was developed by Radiochemistry Chair of USTU for elimination of natural (uranium, thorium, radium, lead) and anthropogenic (cesium and strontium) radionuclides as well as heavy metals, colloids and suspended matter from drinking water.

3.1.2 The Method for Filters Life Test

It was necessary to develop special methods of life test of IWF taking into account mineral and radionuclide composition of drinking water. Low levels of activity of initial drinking water before treatment (5–500 Bq L^{-1}) and significantly lower activities of filtrate after treatment do not allow direct measurement on a gamma spectrometer because of its low sensitivity. Measurement of pure beta emitting radionuclides is even more complicated. Single determination of a radionuclide sorption degree from a limited volume of water cannot be used as a reliable indicator of a filter quality.

Elimination of ^{137}Cs and ^{90}Sr radionuclides by IWF is performed under conditions of frontal chromatography. Approximate preliminary evaluation of resource of a filter may be performed using the value of retention volume. The volume of water passed until breakthrough concentration gives a reliable experimental evaluation of filter resource (Betenekov et al. 1977).

The Eq. (1) allows calculating the value of retention volume *VmR* in frontal chromatography under the conditions of equilibrium sorption dynamics:

$$V_{mR} = V_m + K_d m_c \qquad (1)$$

where Vm is free volume of the column (volume of aqueous phase in the column), L; mc is weight of the sorbent (kg); K_d is distribution coefficient, L kg^{-1}.

Experimentally found activities of radionuclides during the first period after the Chernobyl disaster in 1986 in water of Pripyat River (up to 1000 Bq L^{-1}) or Kiev water pond (up to 100 Bq L^{-1}) (Kryshev et al. 1991) may be used as probable initial activity of contaminated drinking water (A_0). The maximal permissible activities of radionuclides in drinking water according to permanent regulations such as NRB-99/2009 (11 Bq L^{-1} for ^{137}Cs (NRB-99/2009)) or temporary regulations such as VDU may be used as probable breakthrough activity in deactivated water (A_b). Thus, in emergency cases, the IWF should provide decrease of radionuclides activity in water by orders of one or two magnitudes (decontamination factors should be at least 10–100).

In our experiments, the initial activity of radionuclides in water before treatment was accepted to be 3000 Bq L^{-1}. Initially, it was suggested to load through filters only tap water spiked with ^{137}Cs or ^{90}Sr. However, calculations in accordance with the Eq. (2) showed that in case of $VmR = 1.5{\cdot}10^4$ L and weight of a sorbent in a filter $m = 0.5$ kg activity of the filter in the end of experiment should be as high as $9{\cdot}10^7$ Bq kg^{-1}.

$$A_m = m^{-1} \cdot A_{V(0)} \cdot V_{mR} \tag{2}$$

Very high activity of the filter would complicate the work of personnel and require a protection shield for decrease of external dose rate. Since real activities were used in calculations above, these calculations show that in emergence cases the spent IWF will become an additional source of external irradiation of consumers (population). Therefore, protection shield as well as official collection and treatment (or regeneration) of spent IWF will be required.

An alternative method for filters life test was used in order to eliminate the problems described above (Voronina 1996). This method is based on alternation of loading spiked and non-spiked water. Use of this method allows decreasing activity of spent filter by an order of magnitudes (Betenekov et al. 2006). Radionuclides pre-concentration from both initial water and filtrate before measurement allows performing the experiments with lower activity of water.

In the experiments on determination of filters resource, 10 L of spiked tap water were loaded through a filter with the flowrate of 10 L hour-1. The spiked water contained ^{137}Cs and 0.1 mg L^{-1} of stable cesium and/or ^{90}Sr and 0.1 mg L^{-1} of stable strontium. Then 200 L of non-spiked tap water containing only stable cesium and/or strontium was loaded through the IWF followed by a new portion of 10 L of the spiked tap water. These cycles were repeated until achieving the predefined breakthrough activity of a radionuclide in filtrate. In accordance with the principles of the radioactive tracer's method (Lukyanov et al. 1985) chemical behavior of a chemical element including its degree of sorption is identical for stable and radioactive isotopes of this element; therefore, sorption of cesium or strontium should be the same for both spiked and non-spiked water. According to Kulsky et al. (1980), natural drinking water contains 10^{-6} – 10^{-4} and 10^{-5} – 1 mg L^{-1} of stable cesium and strontium respectively. The results of experiments performed on small columns have shown that the resource of a filter experimentally determined using this method totally coincides with this value determined by loading only spiked solutions (Voronina 1996).

The filtrates after loading spiked water were collected into a 10 L container in order to analyze activity of ^{137}Cs and/or ^{90}Sr using a method of radiochemical analysis based on radionuclides pre-concentration on selective sorbents. Methods of rapid and quantitative separation of radionuclides from aqueous solutions (Betenekov et al. 1979; Betenekov and Egorov 1997) were used for analysis of samples of drinking water. Initial water or filtrate was loaded through two sequential 40 mL sorption columns with sorbents NPF-C and KU-2 in order to pre-concentrate radionuclides. The NPF-C sorbent is selective to ^{137}Cs, whereas the KU-2 sorbent is a strong acid cation exchange resin providing separation of ^{90}Sr. After pre-concentration, the sorbents were extracted from columns, dried and measured. The NPF-C was measured on a NaI(Tl)63x63 scintillation gamma spectrometer AMA-03F; a low background beta radiometer was used for measurement of KU-2 sorbent. Standard samples (40 mL of sorbents with known activity) and pure samples of the sorbents were used for determination of counting coefficients and background count rate respectively. Volumes of standard ^{137}Cs and ^{90}Sr solutions were equal in both preparation of standard sample of radioactive sorbent and spiking water sample. Weight of a sorbent in the standard sample was also the same as that used in the column.

Count rate (I, counts per second) of the standard sample of a sorbent with known activity (A_{st}, Bq) were measured. Own count rate of the standard sample (I_{st}, counts per second) were calculated according to the Eq. (3):

$$I_{st} = I - I_b \tag{3}$$

where I_b is the background count rate of pure sorbent (this value takes into account the radioactivity that may present in pure sorbent), counts per second.

Counting coefficients K_{count} were calculated according to the Eq. (4):

$$K_{count} = I_{st}/A_{st}, \text{count per decay} \tag{4}$$

The experimentally determined values of counting coefficients were: $K_{count}{}^{NPF\text{-}C} = (0.041 \pm 0.002)$ count per decay, $K_{count}{}^{KU\text{-}2} = (0.104 \pm 0.005)$ count per decay.

Also, chemical yields of ^{137}Cs and ^{90}Sr radionuclides (Y) in the process of sorption on the NPF-C and KU-2 sorbents were determined in another series of experiments.

Using the values of counting coefficients and chemical yields of radionuclides in sorption process, activities of filtrates (A_f), sorption degrees of ^{137}Cs and ^{90}Sr (S) as well as decontamination factors (D) were calculated for each sampling point.

Activities of filtrate (Bq) were calculated taking into account the value of a radionuclide yield in accordance with the Eq. (5):

$$A_f = I_c/(K_{\text{count}} \cdot S) \tag{5}$$

where $I_c = I - I_b$ is the count rate of the concentrate (NPF-C or KU-2 sorbents after radionuclide preconcentration) after deduction of background count rate I_b.

Degrees of a radionuclide sorption by IWF (S) were calculated according to the Eq. (6):

$$S = (A_0 - A_f)/A_0, \tag{6}$$

where $A_0 = A_{V(0)} \cdot V$; $A_f = A_{V(f)} \cdot V$; V is the volume of water sample.

Decontamination factors (D) for the process of drinking water deactivation using IWF were calculated in accordance with the Eq. (7):

$$D = A_0/A_f \tag{7}$$

According the described method, life tests of individual water filters produced by various organizations were performed.

3.1.3 The Results of Comparative tests of Various Filters

3.1.3.1 The Results of Life Tests of IWF for Decontamination of Drinking Water from Anthropogenic Radionuclides

The results of life tests of some individual filters are given in Table 5. The values of sorption degrees and decontamination factors are given for the first and the last portions of treated water.

Decontamination factor of at least 10 was the criterion of successful work of the IWF. IWF-8 and IWF-5 provided the best elimination of strontium from the first volumes of drinking water. The resource of IWF-8 was found to be 4000 L; whereas, for the IWF-5 the resource of work with decontamination factor not less than 10 was only 2000 L. IWF-1 and IWF-3 also showed high decontamination factors for water containing strontium; their resource was approximately twice of the resource of IWF-4. However, use of these filters resulted in the increase of pH value of the first 100 L of filtrate up to 9 and even more that makes this water to be unsuitable for drinking.

The IWF-2, IWF-3, IWF-5, IWF-7, IWF-8 and IWF-10 filters showed high decontamination factors with respect to ^{137}Cs and a significant resource. These IWFs have not exhausted their capacity even after loading mentioned in Table 5; therefore, only minimal resources were determined.

The IWF-5 was a filtering system consisting of two sequentially assembled filters. The method used in the experiments did not allow assessment of sorption characteristics of each filter; therefore, only result of their joint work was obtained. Therefore, after experiments, the spent filter was disassembled and samples of activated carbon were measured on beta radiometer and gamma spectrometer.

Table 5 The results of life tests of individual filters for decontamination of drinking water from radionuclides (Adopted from Voronina 1996; Betenekov et al. 2004)

	Type of IWF (producer)	Water loaded, L	Sorption degree		Decontamination factor	
			Sr	Cs	Sr	Cs
1	ISM-S (PSTU)	4000	>0.99 ÷ 0.94	–	162 ÷ 17	–
2	FCU (PSTU)	4000	–	0.99 ÷ 0.98	–	237 ÷ 44
3	ISM-S+FCY (PSTU)	3800	>0.99 ÷ 0.93	0.99 ÷ 0.97	142 ÷ 15	114 ÷ 39
4	"Rosa" ("Kompomet Cantec")	800	0.23 ÷ 0.04	0.77 ÷ 0.64	1.3 ÷ 1.04	4.3÷2.8
5	TH-Cl (USTU*)	2060	>>0.99 ÷ 0.90	0.99 ÷ 0.94	1000 ÷ 10	100 ÷ 17
6	Polypropylene filter («Millipore»)	1.0	0.27	0.1	1.4	1.1
7	IPF-MP (USTU)	4150	–	0.97 ÷ 0.91	–	33 ÷ 11
8	IPF-MP + TH-Cl (USTU)	4000	>>0.99 ÷ 0.90	0.99 ÷ 0.94	1000 ÷ 10	1000 ÷ 17
9	MD-MP + MD-NTFM (USTU)	260	0.96 ÷ 0.02	–	27 ÷ 1.0	–
10	MD-MP + TH-Cl (USTU)	>2100	0.98 ÷ 0.93	>>0.99 ÷ 0.91	48 ÷ 5.0	1000 ÷ 11

The results have shown that the activated carbon did not adsorb ^{90}Sr at all. Activity of ^{137}Cs in the activated carbon was 0.1% of total activity of water loaded through the filter. Thus, the high values of decontamination factor being observed for the IWF-5 were conditioned only by the presence of the TH-Cl sorbent. Activated carbon did not adsorb cesium and strontium; it just functioned as a filter for biological decontamination.

The characteristics of the IWF-4 were significantly lower than those being indicated in technical sheet for this filter. As it might be seen in Table 5, the IWF-4 provided the maximal decontamination factors as low as 4.3 and 1.3 for cesium and strontium respectively.

As expected, decontamination factors and sorption degrees were significantly different in cases of the untreated Millipore filter (IWF-6) and the Millipore filter modified by the thin film of iron-potassium ferrocyanide (IWF-7). The IWF-6 provides only mechanical decontamination resulting in elimination of colloidal particles and suspended matter; the values of decontamination factor and separation degree were low: $D = 1.1$ ($S = 0.1$) in case of cesium and $D = 1.4$ ($S = 0.27$) in case of strontium. The presence of the sorption-active film on the surface of the filter allowed retaining 96–97% of activity due to additional adsorption of dissolved forms of radionuclides (see Table 5). Thus, membrane filters (such as IWF-6) are suitable as a preliminary mechanical filter for elimination of colloidal particles and suspended matter; however, their use as a separate filter for water deactivation is not effective since it does not provide achieving the chosen criterion ($D \geq 10$).

The IWF-8 and IWF-10 are multipurpose filters for decontamination of drinking water from both natural and anthropogenic radionuclides. These filters provided decontamination of 4000 L of drinking water from cesium or 2000 L of drinking water from cesium and strontium. The determined resource of the IWF-10 (2100 L) is an underestimated value because we did not achieve the complete exhaustion of exchange capacity of this filter. The IWF-8 and IWF-10 are able to separate both dissolved radionuclides and radioactive colloids.

The results of life tests of the filters confirmed effectiveness of the developed method for deactivation of drinking water as well as showed the real possibility of resource assessment and use the filters developed by PSTU and USTU (UrFU) for elimination of cesium and strontium radionuclides from drinking water.

Furthermore, change of hydraulic resistance of the filters was assessed in the experiments according to change of flowrate. It was determined that hydraulic resistance remained the same for the IWF-7, IWF-8, IWF-9 and IWF-10 filters containing TLIS based on nonwoven materials. In case of the IWF-1, IWF-2 and IWF-5 filters, hydraulic resistance increased with the increase of water loading; this may be conditioned by either mechanical abrasion of grains of the TH-Cl sorbent resulting in the appearance of fine fraction or conglomeration of ISM-S and FCU sorbents.

3.1.3.2 The Assessment of Use of Sorbents in Filters for Decontamination of Drinking Water Containing Natural Radionuclides

Advanced technologies of synthesis of TLIS allowed modification of various filtering membranes in order to make them selective for radionuclides. Life tests of filters for separation of natural radionuclides were not performed. Sorption of cesium, strontium, bismuth and lead was studied using untreated and modified filters Crystallic FB-2 produced by JSC Tenzor (Dubna, Russia). The filter is intended for decontamination of water from colloids, suspended matter and iron hydroxide as well as for decrease of turbidity, color and hardness of water. The average pores size of the filter is 5 μm. The Crystallic FB-2 filter was chemically modified by thin films of manganese dioxide (MD), iron-potassium ferrocyanide (IPF), iron hydroxide (IH) and iron phosphate (IP). For comparison, the Millipore filter with the average pore size of 0.5 μm was assembled in the same container. 700 mL of water containing ^{137}Cs, ^{90}Sr or ^{222}Rn (with daughter radionuclides ^{214}Pb and ^{214}Bi) was loaded through the filters. Activities of radionuclides were as following: ^{137}Cs and ^{90}Sr $= 37{,}000$ Bq L^{-1}, ^{214}Pb $= 417$ Bq L^{-1}, ^{214}Bi $= 486$ Bq L^{-1}. Water samples containing cesium, bismuth and lead were measured on a semiconductor gamma spectrometer; strontium was measured on a low background semiconductor beta radiometer UMF-1500. Using the results of measurements, sorption degrees of the radionuclides were calculated takin into account decay of ^{214}Pb as well as decay and ingrowth of ^{214}Bi. The results are given in Table 6.

Table 6 shows that decontamination of water by MP and Crystallic FB-2 filters occurs due to mechanical separation of colloids and suspended matter. The

Table 6 The results of tests of individual water filters for decontamination of radioactive drinking water

Type of IWF	Sorption degree S, %			
	Cesium	Strontium	Bismuth	Lead
MP	–	–	7	8.7
Crystallic FB-2	13	20	–	–
IPF-K	85	4	52.8	81.5
IH-K	–	–	83.9	73.2
IP-K	12	9	–	–
MD-K	61	50	69.6	67.3

Crystallic FB-2 filter has pores ten times larger than Millipore membrane; therefore, it does not provide retention of radioactive colloids containing lead and bismuth radionuclides with the size less than 1 μm. Chemical modification of these filters by TLIS results in increase of their selectivity for cesium, strontium, lead and bismuth.

It is obvious that sorbents separating natural radionuclides from fresh waters with high distribution coefficients (see Table 3) can be successfully used in individual water filters for decontamination of drinking water. In this case, radionuclides presenting as radioactive colloids should be taken into account; thus, the filtering system should combine membrane and sorption steps of decontamination. Either membrane modified by TLIS or combination of membranes (or modified membranes) with modified aluminosilicates may be used in these filtering systems.

3.1.4 Filters Sanitary Qualifying and Treatment of Spent Filters for Water Deactivation

Filters containing sorbents may become a secondary source of contamination of drinking water by chemical agents because of destruction of sorbents. Therefore, sanitary qualifying of filters (or sorbents) for decontamination of drinking water is required.

It is necessary to draw a special attention to the question of official collection of spent IWF and the possibility of their regeneration and reuse. The spent IWF can be considered as low-level or even medium-level solid radioactive waste and thus it will be an additional source of external irradiation of a user. Therefore, it is necessary to create an official system for collection of spent IWF with further deposition or regeneration. In the last case, radioactive solutions after regeneration of IWF should be also treated, solidified and deposited. Regenerated IWF may be sold again for reuse.

In case of exploitation of IWF-1, IWF-2, IWF-3, IWF-4 water filters, change or regeneration of sorption material is not possible; therefore, consumers may receive an increased external irradiation dose after saturation of the filter by radionuclides.

Voronina (1996) studied the process of regeneration of some spent IWF containing ^{137}Cs and ^{90}Sr. The filter loaded by TH-Cl sorbent (IWF-5) was regenerated by ammonium nitrate solutions. The residual activity of the TH-Cl sorbent in the IWF-5 filter after regeneration was $6.7 \cdot 10^5$ Bq kg^{-1}. After regeneration, the same water containing ^{137}Cs and ^{90}Sr was loaded through the IWF-5 filter again. Degrees of cesium and strontium sorption from water in the beginning of sorption were 98% and 97% respectively. This fact indicates the possibility of multiple reuse of the IWF-5 filter after regeneration.

Sodium hydroxide solution was used for regeneration of spent IWF-7. Radionuclides desorption into 7 bed volumes (B.V.) war near to 90%. The residual activity of the filter was 0.3% of activity of the spent filter. Regeneration of IWF-10 was performed by hydrochloric acid solution; degree of desorption into 4 B.V. was 90%.

3.2 Large-Scale Deactivation of Drinking Water at Water Treatment Plants

A system of protective measures should be provided for prevention of radionuclides transfer to sources of drinking water. After the Chernobyl disaster, in winter 1986–1987, a system of protective and filtering dams was built in order to prevent contamination of Kiev water pond in snowmelt period (Lattrell 1999). Creation of filtering dams allowed decreasing contamination of water used at water treatment plants.

Traditionally, natural aluminosilicates loaded into sorption columns are used for decontamination of drinking water at water treatment plants in some cities. After nuclear accidents, when radionuclides appeared in sources of drinking water, these columns worked as deactivation columns. Modified sorbents based on natural aluminosilicates that are described above can also be used for large-scale deactivation of drinking water at water treatment plants. The sorbents can be loaded into columns with a higher size than individual and collective filters. However, in case of using these columns, the problems of radiation safety of personnel and further treatment of spent columns should be solved.

4 Conclusions

Modified sorbents based on nonwoven materials and aluminosilicates can be used for decontamination of radioactive water under various salt content. They are able to eliminate natural and anthropogenic radionuclides from seawater and fresh water including drinking water. Typical distribution coefficient of various radionuclides for these sorbents vary from $n \cdot 10^3$ to $n \cdot 10^5$ L kg^{-1}. The sorbents possess high exchange capacity, chemical stability, mechanical strength and environmental safety. Filters for decontamination of radioactive drinking water were developed using the sorbents; life tests of these filters were performed. It was experimentally shown that resource of the filters exceeds 4000 liters; whereas decontamination factors remained to be more than 10. The modified sorbents based on natural aluminosilicates (clinoptilolite and glauconite) can be used also for decontamination of natural water bodies, creation of protection dams and geochemical barriers as well as for water decontamination at water treatment plants.

The experience of serious radiation accidents has shown that provision of population with quality drinking water is necessary at radioactively contaminated lands. Therefore, it is necessary to create a buffer stock of IWF for deactivation of drinking water in emergency cases. Among the possible consumers of IWF, there are civil defense service, Ministry of Emergency, enterprises of nuclear fuel cycle, water treatment plants and other organizations that are interested in local systems for decontamination and deactivation of drinking water. The question of official collection and treatment (or recycling) of spent IWF containing radionuclides should be solved by national regulation authorities.

Furthermore, use of IWF for elimination of natural radionuclides from drinking water at radon hazardous territories will decrease doses of internal irradiation due to consuming contaminated water.

References

Akahane K, Yonai S, Fukuda S, Miyahara N, Yasuda H, Iwaoka K, Matsumoto M, Fukumura A, Akashi M (2012) The Fukushima nuclear power plant accident and exposures in the environment. Environmentalist 32:136–143

Aleksakhin RM, Sychev VG (2013) Radioecological aspects of agricultural remediation following the accident at the Fukushima Daichi NPP. Fertility 4:2–6

Aleksakhin RM, Glagolev AV, Drozhko EG, Zinin AI, Zinina GA, Ivanov IA, Mokrov YG, Orlova EI, Samsonov BG, Samsonova LM, Stukalov PM (2007) Water pond-9 – a deposit of liquid radioactive waste and its impact on the geological environment, Moscow, Ozersk, Russia

Batorshin GS, Mokrov YG (2013) The experience of liquidation of results of the accident at the Mayak Production Association in 1957. Rad Saf Quest 1:13–20

Betenekov ND, Egorov YV (1997) Thin-layer inorganic adsorbents in radiochemical analysis. J Anal Chem 52:1020–1025

Betenekov ND, Egorov YV, Medvedev VP (1977) Static criteria of sorption on microconstituents from aqueous solution. In: Volkhin VV (ed) Inorganic ion exchangers (synthesis, structure and characteristics). Zvezda, Perm

Betenekov ND, Egorov Yu V, Puzako VD, Remez VP, Stepanets OV, Khitrov LM, Cheremukhin Yu G (1979) A method for concentration of cesium radionuclides from seawater. Soviet Patent No 828464

Betenekov ND, Vasilevsky VA, Egorov YV, Nedobukh TA (1982) Radiochemical study of hydroxide films. Part I Sedimentation of titanium hydroxide films onto the surface of various supports and physicochemical properties of sediments and films. Radiochemistry 24:419–429

Betenekov ND, Voronina AV, Kutergin AS, Nedobukh TA, Gileva YE (2001) The development of technologies for synthesis of sorbents and sorption filters for radiochemical analysis as well as for decontamination of drinking and waste water from radionuclides. Trans Saint-Petersburg Mining Inst 149:16–18

Betenekov ND, Voronina AV, Kutergin AS (2004) Comparative studies of filters for decontamination of drinking water from radionuclides. Bull Ural State Tech Univ 17:65–70

Betenekov ND, Voronina AV, Nogovitsyna EV, Chopko NN (2006) Peculiarities of dynamic tests of highly specific sorbents. Sorpt Chromato Proces 6:1115–1118

Brock TD (1987) Membrane filtration: a user's guide and reference manual. Science Tech Inc, Madison

Dyer A, Chimedtsogzol A, Campbell L, Williams C (2006) Uptake of caesium and strontium radioisotopes by natural zeolites from Mongolia. Micropor Mesopor Mat 95:172–175

Hamada N, Ogino H (2012) Food safety regulations: what we learned from the Fukushima nuclear accident. J Environ Radioact 111:83–99

Hygienic regulations 155 (2015) Sanitary-hygienic requirements for radiation safety. Kazakhstan Republic, 27.02.2015

Israel YA, Petrov VN (1988) Procedures on assessment and measures of safety for the population after the Chernobyl accident. In: Harley JH, Schmidt GD, Silini G (eds) Radionuclides in the food chain. Springer, Berlin, pp 285–290

Isupov VP, Lyakhov NZ, Shatskaya SS, Chupakhina LE, Vladimirov AG, Moroz EN, Shvartsev SL, Kolpakova MN, Ariunbileg S, Kuibida LV (2011) Uranium in saline lakes of Northwestern Mongolia. Doklady Earth Sci 437:352–356

Isupov VP, Shatskaya SS, Borodulina IA (2014) Uranium in mineralized lakes of Western Mongolia and adjacent territory of Russia: resources, sources of accumulation, routes of innovative development. Chem Sust Develop 22:429–436

Karavaeva EN, Kulikov NV, Molchanova IV, Pozolotina VN, Pushkov (1994) Accumulation and distribution of long-living radionuclides in the forest ecosystems of the Kyshtym-accident zone. Sci Total Environ 157:147–151

Kenigsberg JE (2008) Hygienic standards for Cs-137 and Sr-90 in the foodstuffs: experience of Belarus after the Chernobyl accident. Rad Hygien 1:32–51

Konoplev AV (2016) Comparative analysis of radioactive cesium wash-off from contaminated catchment areas after accidents at the Fukushima Dai-ichi and Chernobyl nuclear power plants. Geochem Int 54:522–528

Kryshev II, Aleksakhin RM, Ryabov IN, Mahon'ko KP, Taskaev AI, Chumak VK, Borovoi AA, Gagarinsky AY, Demin VF, Sazykina TG, Zarubin OL, Blinova LD, Nikitin AI, Smirnov VV, Priester BS, Sanzharova NI, Perepelyatnikova LV, Astasheva NP, Grebenshchikova NN (1991) Radioecological consequences of the Chernobyl accident. Kurchatov Institute of Atomic Energy, Moscow

Kulsky LA, Goronovsky IT, Koganovsky AM, Shevchenko MA (1980) Handbook on properties, analytical methods and decontamination of water. Naukova Dumka, Kiev, USSR

Kulyaeva IO, Voronina AV (2015) Study of sorption characteristics of natural and modified clinoptilolite and glauconite. In: Proceedings of the 2nd international youth conference: physics, technology, innovation. Ural Federal University, Ekaterinburg, Russia, pp 175–176

Kuznetsov YV, Shchebetkovsky VN, Trusov AG (1974) Basis of water decontamination from radioactive pollutants. Atomizdat, Moscow

Lagunov AV, Smagin AI (2007) The role of the East Ural Nature Reserve in the system of protected areas of Urals. Rad Saf Quest (Special issue):63–76

Lattrell SP (1993) Description of accidents with a significant radionuclides release. In: Warner F, Harrison RM (eds) Radioecology after Chernobyl. Biogeochemical pathways of artificial radionuclides. Wiley, New York

Lauria DC, Godoy JM (2002) Abnormal high natural radium concentration in surface waters. J Environ Radioactiv 61:159–168

Lukashenko SN (2016) Radioecological justification for the return of the Semipalatinsk test site to economic use: fundamental and applied aspects. PhD thesis, Obninsk, Russia

Lukyanov VB, Berdonosov SS, Bogatyrev IO, Zaborenko KB, Iofa BZ (1985) Radioactive tracers in chemistry. High School, Moscow, USSR

Malikova EM, Lushnikova MV, Voronina AV (2017) The study of chemical and radionuclide composition of liquid radioactive waste from a special drainage tank at the JSC INM. In: Proceedings of the 4th international youth conference: physics, technology, innovation. Ural Federal University, Ekaterinburg, Russia, pp 230–231

MHLW (2011) Foodsafety. http://www/mhlw.go.jp/english/foodsafety/at/110318.pdf.2011/

Mingkuan Qin (2009) Current progresses and rrospects on unconventional uranium resources (UUR) of China. Technical Meeting on Uranium from Unconventional Resources. Vienna, IAEA Headquarters

Nazarbayev NA, Shkolnik VS, Batyrbekov EG, Berezin SA, Lukashenko SN, Skakov MK (2016) Conducting a complex of scientific, technical and engineering works to bring the former Semipalatinsk test site into a safe state. Part 1. Kazakhstan National Nuclear Center, Kurchatov, Russia

NRB-99/2009 (2009) Russian regulations on radiation safety–99/2009. Energoatomizdat, Moscow

Onischenko GG (2009) Radiation hygienic consequences of the accident at the Chernobyl NPP and the tasks of their minimization. Rad Hygien 2:5–13

Razykov ZA, Bakulina GK, Niyazmatova DT, Belyaev AP, Yunusov MM (2002) Separation of uranium from natural uranium-containing waters with a complex salt composition. Bull Tajikistan Acad Sci 45:60–65

RDU-99 (2006) Belarussian republican permissible levels of cesium-137 radionuclides and strontium-90 content in food and drinking water (RDU-99) (in redaction of 2006), Minsk, Belarus

Sadovnikov VI, Glagolenko Yu V, Drozhko EG, Mokrov Yu G (2002) Modern state and ways of solving the problem of the Techa Tandem Reservoir System. Rad Saf Quest 1:3–14

Semenishchev VS, Betenekov ND, Nikiforov AF, Glazyrin SV, Tomashova LA (2017) Radon in groundwater drinking water of Sverdlovsk oblast: determination of activities and methods of disposal. Water Indust Russia 2:108–119

Smagin AI (2013) Ecology of water bodies at the zone of anthropogeinc radioactive geochemical anomaly at the South Urals. SUSU Publisher, Chelyabinsk

Tarasevich YI (1996) Use of natural sorbents as deactivating agents in liquidation of consequences of the Chernobyl disaster. J Water Chem Techno 18:127–131

Tsitsishvili GV, Andronikashvili TG, Kirov GN (1985) Natural zeolites, chemistry. USSR, Moscow

Vdovina ED, Radyuk RI, Sultanov AS (1976) Use of natural zeolites from Uzbekistan for decontamination of low-level radioactive wastewater. Part I, sorption of radioactive cesium. Radiochemistry 3:422–423

Voronina AV (1995) Sorption of cesium and strontium from low-activity fresh waters. Radiochemistry 37:538

Voronina AV (1996) Synthesis, study of properties and application fields of thin-layer inorganic sorbents based on nonwoven materials. PhD thesis, UGTU, Ekaterinburg, Russia

Voronina AV, Semenishchev VS, Nogovitsyna EV, Betenekov ND (2012) A study of ferrocyanide sorbents on hydrated titanium dioxide support using physicochemical methods. Radiochemistry 54:69–74

Voronina AV, Chaikina TI, Nikiforov AF, Driker BN, Vurasko AV, Frolova EI (2013a) Sorbents based on technical cellulose for decontamination of radioactively contaminated waters and rehabilitation of natural water bodies. Water Indust Russia 5:45–53

Voronina AV, Savchenko MO, Semenishchev VS, Nikiforov AF (2013b) The assessment of effectiveness of use of clinoptilolite and ferrocyanide sorbents based on it for decontamination of aqueous media and radionuclides immobilization. Water Indust Russia 6:88–94

Voronina AV, Blinova MO, Kulyaeva IO, Sanin PY, Semenishchev VS, Afonin YD (2015a) Sorption of cesium radionuclides from aqueous solutions onto natural and modified aluminosilicates. Radiochemistry 57:446–452

Voronina AV, Blinova MO, Semenishchev VS, Gupta DK (2015b) Returning land contaminated as a result of radiation accidents to farming use. J Environ Radioact 144:103–112

Voronina AV, Gorbunova TV, Semenishchev VS (2017) The effect of the synthesis method on the parameters of pore structure and selectivity of ferrocyanide sorbents based on natural minerals. J Radioanal Nucl Chem 312:241–254

Yoshimura K, Onda Y, Sakaguchib A, Yamamoto M, Matsuura Y (2015) An extensive study of the concentrations of particulate/dissolved radiocaesium derived from the Fukushima Dai-ichi

nuclear power plant accident in various river systems and their relationship with catchment inventory. J Environ Radioact 139:370–378
Zhanbekov HN, Mukataeva ZS (2010) Radiochemical condition monitoring data for Syrdaria water basin. Water Chem Ecol 5:2–9
Zhanbekov HN, Mukataeva ZS, Lakhbaeva ZA (2012) Radioactive pollution of river Syrdarya by anthropogenic factors. Water Chem Ecol 9:14–17

Treatment of Radioactive Waste After Rehabilitation of Contaminated Areas

Anna V. Voronina, Vladimir S. Semenishchev, and Dharmendra K. Gupta

1 Rehabilitation Activities at Radioactively Contaminated Lands

Radioactive contamination means presence of radionuclides on the surface or inside of a material, in water, air, human body or any other object if activity exceeds maximal allowable level (Sanitary regulations SP.2.6.1.2612-10 2013). Lands being contaminated by radionuclides because of work of nuclear or non-nuclear enterprises, radiation accidents, deposition of radioactive waste and nuclear weapon tests should be rehabilitated.

The aim of rehabilitation is to make radioactively contaminated territories suitable for living or agricultural use from the point of view of radiation safety. Rehabilitation activities should result in a decrease of dose exposure for population down to allowable limits as well as the maximally achievable decrease of contamination of the environment. The complete elimination of radioactive contamination is not necessary (IAEA 2008).

Rehabilitation activities may include the following works: conservation of radioactively contaminated objects, deactivation of buildings, equipment, transport vehicles, soil, ground, etc., rehabilitation of aqueous ecosystems, decontamination of drinking water, rehabilitation of forest and meadow ecosystems as well as rehabilitation of agricultural landscapes and returning soils to farming use. The list of rehabilitation activities depends on the type of radioactive contamination

A. V. Voronina (✉) · V. S. Semenishchev
Radiochemistry and Applied Ecology Department, Physical Technology Institute, Ural Federal University, Ekaterinburg, Russia

D. K. Gupta
Institut für Radioökologie und Strahlenschutz (IRS), Gottfried Wilhelm Leibniz Universität Hannover, Hannover, Germany

D. K. Gupta, A. Voronina (eds.), *Remediation Measures for Radioactively Contaminated Areas*, https://doi.org/10.1007/978-3-319-73398-2_11

and the list of radionuclides as well as on the characteristics of the territory and objects to be rehabilitated.

Complete or partial rehabilitation of lands may be performed. As a result of complete rehabilitation, the territory is completely returned to farming use. If the complete rehabilitation is impossible or not economically sounds, then partial rehabilitation is recommended. In this case, a part of the territory is returned to farming use; whereas, another part is conserved and transformed to a relatively safe state for a population and the environment (Romanovich et al. 2016).

2 Characterization of Radioactive Waste Being Formed as a Result of Rehabilitation of Radioactively Contaminated Lands

Formation of radioactive waste (RW) is one of the main problems appearing as a result of rehabilitation of radioactively contaminated lands. According to the strategy of International Atomic Energy Agency (IAEA), the methods of RW treatment should include collection, sorting, conditioning, storage and final deposition of the waste. Large volumes of RW of various activity levels and chemical composition are formed after rehabilitation of radioactively contaminated lands. These RWs may include deactivation solutions, radioactive ground and soil removed from industrial territories, civil objects or agricultural fields as well as radioactive spent sorption materials, filters, etc. Organizing temporary and long-term storage, treatment and disposal is required for these types of waste.

The choice of a rehabilitation method should aim to minimize the amount of radioactive waste formed as a result of rehabilitation. In certain cases, elimination of removable radioactive contamination from buildings, equipment and materials by deactivation allows decreasing radioactive contamination down to allowable limits and returning these materials to industrial use. Treatment of RW formed after rehabilitation will result in a decrease of deposited RW. However, a significant part of RW is difficult for treatment.

Deactivation is elimination or decrease of radioactive contamination from a surface or any medium. Deactivation allows eliminating only removable (non-fixed) contamination. In this case, radioactive matter is able to be transferred onto other objects after contact; furthermore, this contamination is eliminated after deactivation. In the case of irremovable (fixed) radioactive contamination, radioactivity does not transfer onto contacting objects and remains after deactivation (Sanitary regulations SP.2.6.1.2612-10 2013). If the activity level of fixed radioactive contamination in an object exceeds allowable limits and deactivation failed, this object is supposed to be solid radioactive waste being destined for deposition or burial. Durability of fixation of radionuclides on a surface and effectiveness of decontamination depend on radionuclides concentration, state of their aggregation, chemical properties, durability of contact, material of the surface and some other conditions.

Liquid and dry methods of deactivation are used. Technical Reports Series No. 395 (IAEA 1999) gives a detailed overview of methods of deactivation. Liquid deactivation is performed using a deactivating solution, pure water steam (water emulsion and steam decontamination) or foamed deactivating solution (foam deactivation); this type of deactivation results in formation of liquid radioactive waste (LRW). Some pilot tests of deactivation methods based on tributil phosphate, phosphorus- and fluorine-containing acids (Shadrin et al. 2007) and hexafluoroacetylacetone (Murzin et al. 2007) solutions in liquid (supercritical) carbon dioxide are described. However, in practice, all methods of liquid deactivation use various aqueous solutions. Dry deactivation uses mechanical (blowing out with compressed air, sucking off with vacuums, removing by sweeping, shaking out, knocking out, rubbing with rags, cutting off, scraping off), thermal (calcination of contaminated objects), sorption (deposition on contaminated surfaces of dry sorption-active materials) and other methods. Dry decontamination results in formation of solid radioactive waste.

2.1 *Liquid Radioactive Waste After Deactivation*

In cases of deactivation works at populated areas and industrial site territories, LRW containing surfactants and complexing agents, inorganic acids (nitric, sulfuric and hydrochloric acids), oxidants (hydrogen peroxide, potassium permanganate), strong bases (sodium hydroxide, etc.), organic solvents (dichloroethane, trichloroethane, acetone, kerosene, etc.).

There are two types of techniques of deactivation of soils and ground – in-situ and ex-situ techniques. In-situ deactivation means performing deactivation works immediately at the contaminated area. In case of ex-situ deactivation, radioactive soil or ground is removed, transferred to storage and then decontaminated at a plant.

In-situ deactivation does not require large capital investments, since measures to remove radionuclides are performed directly in the upper layer of contaminated soil. Among other advantages of this method, there are absence of expenses for removal and transport of the upper layer of the soil. The main disadvantages of in-situ methods are low effectiveness and large volume of secondary RW being formed after deactivation.

In case of eluent deactivation, the upper layer of a contaminated soil is washed by chemical agents dissolving and desorbing radionuclides form soil aggregates. Both organic and inorganic compounds including biological matter (chelates siderophores) may be used as eluents. Treatment of a contaminated area is performed either by sprinkling or by addition of the eluate through gutters. Solution drainage is performed by pumping from a specially organized system of gutters, pits or boreholes. Presence of clay aquifuges in soils or ground results in a decrease of eluent loss. Usual loss of eluent does not exceed 5%. The main disadvantage of this method is formation of elevated quantity of low-level LRW that should be treated later. Activities of soils decrease only by 10–20% after eluent deactivation. Washing soil and ground by a large volume of eluents for a long time (for example, sprinkling for several days) results in activity decrease by up to 40%. The deactivating solution may be decontaminated and recycled.

Dmitriev et al. (2001) and Barinov et al. (2003) suggested the electrokinetic method of deactivation. In this method, migration of ions of radionuclides and fine dispersed particles due to electric potential results in deactivation of soils and ground. The processes of electro migration and electro osmosis occur in a contaminated ground under electric field of direct current between electrodes being located at the contaminated zone. This method is suitable for deactivation of waterproof clayey soils. Solutions of desorbing electrolytes may be added to the soil for intensification of this process due to increase of migration of radionuclides. Overall, this method allows decreasing activity of a soil by up to 20%. Addition of desorbing agent to the contaminated ground prior to electrokinetic deactivation results in increase of deactivation by up to 40% and the average rate of decontamination from 0.23 to 0.5% per day (Barinov et al. 2003).

In case of ex-situ deactivation, the upper layer of a soil is removed and treated under industrial or lab conditions. Reagents being used for deactivation of soils and ground are described in 3.1.

2.2 *Solid Radioactive Waste After Deactivation*

2.2.1 Radioactive Contaminated Soil and Ground

Removal of the upper layer of a soil is the most effective method of rehabilitation of soils and ground being contaminated as a result of radiation accidents and nuclear weapon tests. In case of removal of up to 4 cm of the upper layer of a soil, the volume of radioactive soil is up to 400 tons per hectare (Aleksakhin and Sychev 2013). For example, the volume of radioactive ground to be removed at the 80 km zone of the Fukushima-1 nuclear power plant is estimated to be approximately $3.4 \cdot 10^7$ m^3 (Nakano and Yong 2013). Activity of this type of waste can reach up to 10^5–10^6 Bq kg^{-1} (Endo et al. 2012). Moskalchuk (2015) estimated the cost of rehabilitation of 1 m^2 of a soil via removal of its upper layer to be approximately 1000 USD. This cost includes removal of the upper layer of contaminated soil, transport and deposition; it does not include expenses for restoration of the natural fertility of the soil. The removed soil is considered to be a solid radioactive waste; therefore, it should be either deposited or treated.

2.2.2 Organic Waste

Organic RW are formed as a result of phytoremediation of soils as well as rehabilitation of forests, meadows and agricultural lands.

Plants that are able to extract radionuclides by roots from the soil solution are used for phytoremediation of radioactively contaminated soils. Vozzhenikov et al. (1997) reported that sunflower, rapeseed, calendula, alfalfa and clover accumulate ^{137}Cs, whereas alfalfa, clover, peas and fenugreek accumulate ^{90}Sr. The worldwide experience of phytoremediation of radioactively contaminated soils showed low effectiveness of this method; however, study of this method is still in progress. Aleksakhin (2009) showed that the average transfer of ^{137}Cs to agricultural plants at the Chernobyl Exclusion Zone was 0.07%; for a comparison, the natural radioactive decay of ^{137}Cs results in the decrease of its activity by 2.28% per year. Amaranth and sunflower were sown at the territories contaminated after the Fukushima-1 radiation accident. Transfer of ^{137}Cs and ^{134}Cs radionuclides to the raw biomass was 0.02% of activity of these radionuclides in the soil (Aleksakhin and Sychev 2013). In this method, plants biomass with elevated activity is considered as solid RW; further treatment of this biomass is required.

Removal of forest floor, leaves from trees (defoliation), plants and turf from meadows are used as methods of rehabilitation of radioactively contaminated forests and meadows. Defoliation may be used as a rehabilitation measure in urban zones (Warner and Harrison 1993). Radioactive biomaterials are the main RW after this type of rehabilitation measures.

2.2.3 Spent Sorbents

Natural water bodies were significantly contaminated as a result of radiation accidents. Voronina et al. (Unpublished) characterized the level of radioactive contamination of natural water bodies at the influence zones of Mayak PA, Chernobyl NPP and Fukushima-1 NPP as well as Semipalatinsk Test Site. The sorption method is the most effective for deactivation of radioactive natural waters (including contaminated drinking water) and treatment of LRW after deactivation works (Myasoedova and Nikashina 2006; Leont'eva et al. 2016; Voronina et al. 2017). Spent radioactive sorption materials and filters are accumulated after sorption decontamination of aqueous media. All these materials require further treatment and utilization.

Thus, liquid RW containing surfactants, complexing agents, inorganic acids, bases, salts and organic solvents as well as solid RW such as ground, soil, forest floor, plants biomass, spent sorbents, etc. are accumulated as a result of deactivation measures at radioactively contaminated areas. The sub-chapter 3 characterizes methods of treatment of these types of RW. After treatment and conditioning, all RW should be disposed.

Raw materials, products and equipment with irremovable (fixed) radioactive contamination are also considered as solid radioactive waste that should be disposed. The approaches to disposal and burial of this type of waste are described in the subchapter 4.

3 Methods for Treatment of Radioactive Waste After Rehabilitation

3.1 Decontamination of Radioactively Contaminated Soils and Grounds

Physical and chemical methods are used for elimination of radionuclides from contaminated soils and ground. The physical method of removing the most contaminated fine fraction from a ground via wet gravity separation was realized as an experimental-industrial installation for ground deactivation in the National Research Center "Kurchatov Institute" (Russia). The method results in the decrease of activity of a ground by 4–5 times, whereas the circulating water remains almost clear during several working cycles of the installation (Volkov et al. 2007). Overall, the method allows reducing the volume of RW to be long-term deposited approximately by five times. More than 2000 m^3 of radioactive ground was treated by the wet gravity separation method at the (Federal State Unitary Enterprise) FSUE "Radon" (Moscow, Russia); approximately 400 m^3 of the most contaminated fraction of the ground was sent to long-term deposition (Volkov et al. 2007). The method of wet gravity separation yields a low decontamination factor and it is suitable for deactivation of grounds with a low activity level.

Technologies of reagent deactivation such as treatment of soils and grounds by acids, bases and salts solutions are used for a deeper decontamination of radioactive soil and ground. Among reagents being used for cesium radionuclides elimination from a ground, there are aqueous solutions of the mixture of oxalic acid with ammonium, potassium or sodium fluoride or silicofluoride (Sobolev et al. 2000); mixture of nitric or hydrochloric acid with ammonium, potassium or sodium fluoride or silicofluoride; mixture of sulfuric and phosphoric acids at the molar ratio of 1:3 and the temperature of 50–100 °C (Dmitriev et al. 2006); 1 M ammonium acetate solution (Szabo et al. 1997; Kim et al. 2007); nitric acid with addition of ammonium fluoride (Prozorov et al. 2002). The rate and degree of deactivation depend on the reagents used. Some methods result in decontamination degree as low as 10–20% or insignificantly higher. For example, Chirkst et al. (2001) described a pilot technology for ^{90}Sr removal from contaminated soils using nitric acid and iron (III) chloride solutions; testing this technology under field conditions resulted in decontamination factors of about 3. However, detailed studies and correct choice of the type of reagent and composition of the deactivating solution provided decontamination factors for a soil contaminated by cesium, radium, plutonium and americium

Table 1 Characteristics of methods of soils and grounds deactivation

Method of deactivation	Decontamination factor	Volume reduction factor
Wet gravity separation	5	8
Reagent treatment	40	50
Combined scheme	200	20

Adopted from Nikolaevskii et al. (2011b)

radionuclides as high as 90–99% for one deactivation step (Sklifasovskaya et al. 2009; Nikolaevsky et al. 2011a). The combined scheme of grounds deactivation using wet gravity separation and reagent deactivation allows decontaminating grounds with a higher activity level. Table 1 lists characteristics of methods of soils and grounds deactivation.

Creation of mobile installations allows decreasing expenses for transport a radioactively contaminated soil and ground to the location of its treatment and deposition. The FSUE "Radon" developed and tested a mobile installation for decontamination of 250 kg of ground per hour via wet gravity separation method (Nikolaevskii et al. 2011b).

3.2 Decontamination of Solutions After Deactivation

Sorption and precipitation methods are widely used for radionuclides separation from solutions after deactivation. Acids, bases, salts, surfactants and organic complexing agents presenting in spent deactivation solutions negatively affect the process of decontamination. LRW containing surfactants and organic complexing agents (ethylene diamine tetraacetate (EDTA), oxalic acid, etc.) are the most difficult for treatment. In presence of surfactants and organic complexing agents, effectiveness of decontamination is decreased dramatically, time of settling and volume of precipitates are increased; irreversible sorption of surfactants by sorbents results in a decrease of sorption kinetics (Milyutin et al. 2015). Therefore, the choice of conditions of precipitation or sorbents with high distribution coefficients is necessary for decontamination of these difficult types of LRW. Alternatively, destructive methods may be used for preliminary elimination of surfactants, complexing agents and other organic compounds from waste before treatment.

Decontamination of LRW by the methods of chemical precipitation and coprecipitation was intensively studied for last decades (Novikov and Ruzankin 1972; Plotnikov and Safonov 1983; Plotnikov et al. 1989; Milyutin et al. 2004; Flouret et al. 2012). In spite of simplicity of realization, precipitation methods are quite effective for separation of hydrolyzable radionuclides such as REEs, Zr, Nb, Pu, Am, U; however, their effectiveness is low for Cs and Sr radionuclides. Furthermore, the effectiveness of precipitation strongly depends on anionic composition of LRW. Varying the quantity of precipitating agent and pH of solution, optimal conditions of precipitation may be created. Formation of a relatively large volume of secondary RW as precipitates and suspensions is the main disadvantage of this method.

Nikolaevskii et al. (2006) describes a practical example of decontamination of solutions after deactivation using the precipitation method. The technological solution after reagent deactivation of a ground contaminated mostly by ^{137}Cs was treated by ferrocyanide decontamination in a thickener via addition of $K_4[Fe(CN)_6]$ to the solution. After decontamination, the deactivated solution was returned to radionuclides leaching from the contaminated ground. Approximately 97–99% of ^{137}Cs was transferred from the technological solution to the ferrocyanide precipitate. Activity of the solution decreased from 100 kBq L^{-1} to 1 kBq L^{-1} and even less. The ferrocyanide precipitate with the activity of 16–28 MBq kg^{-1} (more than 100 times higher than initial activity of the contaminated ground) was transported to the radioactive waste storage (Nikolaevskii et al. 2006).

Sorption is more practically feasible method of deactivation providing higher decontamination factors than precipitation. Effectiveness and cost of sorbents determine their practical use. Ferrocyanide sorbents based on natural aluminosilicates and hydrated titanium and zirconium dioxides can be used for cesium radionuclides separation from solutions after deactivation with a complex chemical composition.

Table 2 presents distribution coefficients ^{137}Cs cesium from deactivation solutions containing 1 M ammonium acetate and EDTA. The T-3A and T-35 sorbents are produced by Termoxid Company (Zarechny, Russia); all other sorbents excluding natural aluminosilicates are developed by Radiochemistry and Applied Ecology Department of Ural Federal University.

Both synthetic sorbents (such as T-35, NPF– hydrated titanium dioxide) and natural aluminosilicates (clinoptilolite and glauconite) are suitable for deactivation of solutions containing EDTA. In case of treatment of solution after reagent deactivation of soils and grounds by 1 M ammonium acetate solution, distribution coefficients of cesium for all studied sorbents were lower than in low mineralized

Table 2 Distribution coefficients of cesium from deactivation solutions

Sorbent	1 M ammonium acetate	EDTA 0.01M	EDTA 0.1M
Termoxid T-3A	$2.2 \cdot 10^2$	$2.6 \cdot 10^3$	$8.8 \cdot 10^2$
Termoxid T-35	$1.0 \cdot 10^3$	$1.4 \cdot 10^3$	–
NPF – hydrated titanium dioxide	$2.0 \cdot 10^3$	$2.4 \cdot 10^3$	$4.9 \cdot 10^2$
Clinoptilolite	$5.4 \cdot 10^2$	$6.6 \cdot 10^3$	$7.1 \cdot 10^2$
NPF-clinoptilolite	$5.3 \cdot 10^2$	$7.4 \cdot 10^2$	–
IPF-clinoptilolite	$3.0 \cdot 10^2$	$1.6 \cdot 10^3$	–
Glauconite	$2.6 \cdot 10^2$	$1.2 \cdot 10^3$	$2.0 \cdot 10^3$
NPF-glauconite	$4.9 \cdot 10^2$	$4.8 \cdot 10^2$	$6.9 \cdot 10^2$

Adopted from Nogovitsyna et al. (2016)

solutions. This can be explained by the competitive sorption of NH_4^+ ions due to the fact that its ionic radius is very similar to radius of Cs^+ ion. The NPF – hydrated titanium dioxide showed the highest distribution coefficient for cesium among other studied sorbents.

It should be mentioned that the ferrocyanide sorbents (NPF– hydrated titanium dioxide, NPF-clinoptilolite, NPF-glauconite) showed high distribution coefficients for ^{60}Co, ^{90}Sr and alpha emitters, besides ^{137}Cs separation from solutions, when these radionuclides jointly present in LRW (Malikova et al. 2017).

The NPF– hydrated titanium dioxide sorbent can be successfully used for cesium separation from strong acidic solutions (up to 7 M nitric acid), alkalescent solutions (up to pH 11) and solutions containing up to 1 g L^{-1} of surfactants and 0.05 mol L^{-1} of EDTA (Semenischev et al. 2013). Increase of HNO_3 concentration to 3 mol L^{-1} results in the decrease of distribution coefficient (K_d) of cesium from $3.8 \cdot 10^4$ L kg^{-1} for neutral solution to $3.7 \cdot 10^3$ L kg^{-1}; further increase of acidity does not affect the K_d value. Concentration of surfactants up to 1 g L^{-1} does not affect sorption of ^{137}Cs; distribution coefficient remains to be as high as $3.1 \cdot 10^4$ L kg^{-1}. In case of solutions containing up to 0.05 mol L^{-1} EDTA, K_d values are $1 \cdot 10^4$–$6 \cdot 10^4$ L kg^{-1} depending on initial concentration of cesium in the solution.

Among the methods used for destruction of surfactants and other organic compounds before deactivation, there are reagent oxidation, ozonization, ultraviolet irradiation, radiolysis and thermal decomposition (Bobrov et al. 2014; Milyutin et al. 2015).

3.3 Treatment of Spent Sorption Materials and Filters

Spent sorbent and filters containing radionuclides are special types of radioactive waste requiring special methods of treatment. Usually, spent sorbents contain high radioactivity levels. The IAEA Technical Report No. 408 (IAEA 2002) recommends the concept of postponed decision with respect to this type of RW. The spent sorbents were transferred to the long-term storage without preconditioning or transported from sorption columns to underground tanks and stored under a layer of water considering further possible treatment, conditioning and deposition of these sorbents.

Storage of spent sorbents for decay of radionuclides is suitable only as an intermediate operation if relatively short-lived radionuclides present in the sorbents. If spent sorbents contain long-lived radionuclides (^{137}Cs, ^{90}Sr, actinides, etc.), IAEA recommends immobilization and further long-term storage or final deposition. Immobilization is the process of incorporation of radionuclides into a matrix for consolidation or a container for temporal storage or final deposition. The following preliminary treatment of spent sorbents may be performed before immobilization: dehydration, drying, calcination, grinding and chemical or thermal destruction. In certain cases, immobilization can be performed immediately after dehydration without any other additional treatment.

Cement, bitumen and some polymers such as epoxy resins, polyester, polyethylene, polyurethane and others are used as matrices for immobilization of spent ion exchange resins. The methods of treatment of spent ion exchange resins including their immobilization have found application on an industrial scale and are described in a series of IAEA Technical Reports (IAEA 1983, 1988, 1993a, b, c) and Los Alamos publication (LA Report 1997). As an example, according to the technology developed in Great Britain, the spent resin is dried by hot air at the temperature of 100 °C, compressed into briquettes and then cemented in a container.

Spent inorganic sorbents containing long-lived radionuclides can be transferred to the long-term storage or final deposition in a leak-proof container after immobilization into cement matrix or without this operation. Immediate deposition of an inorganic sorbent is suitable in case of a high chemical, thermal and radiation stability of the sorbent and low leaching rate of radionuclides. In this case, inorganic sorbents based on titanium and zirconium oxides and modified aluminosilicates possessing a high stability and low leaching rates of radionuclides and being suitable for final deposition are very interesting for liquid RW treatment. Radionuclides are strongly retained in the sorbents; no operations for their immobilization result in a significant reduction of their volume/weight and decrease of operational expenses. Inorganic sorbents being suitable as matrices for separation and immobilization of cesium and strontium radionuclides are described in a series of publications (Voronina et al. 2013, 2015 Voronina and Semenishchev 2015). Integral rate of cesium leaching from sorbents saturated up to the level of high active RW is $3.3 \cdot 10^{-10}$ g cm^{-2} day^{-1} for the NPF-clinoptilolite and $2.6 \cdot 10^{-10}$ g cm^{-2} day^{-1} for the IPF-clinoptilolite (Voronina et al. 2013); for the T-55 sorbent, leaching rate of cesium and strontium radionuclides are $(3.7 \pm 0.2) \cdot 10^{-10}$ g cm^{-2} day^{-1} and $(8.2 \pm 0.1) \cdot 10^{-12}$ g cm^{-2} day^{-1} respectively (Voronina and Semenishchev 2015).

3.4 Treatment of Radioactive Biological Materials

Radioactive biological waste being formed as a result of rehabilitation measures (plants biomass, contaminated agricultural products) can be burned in an open fire or treated by plasma combustion together with other solid RW. Combustion reduces the volume of biological waste by a factor of 20–100; more than 90% of radioactivity remains in the ash residues. If volatile radionuclides present in waste, the necessity of these radionuclides trapping should be taken into account in the project of the gas cleaning system. The ash residues after combustion are immobilized into a cement or glass with further long-term storage or final deposition.

The FSUE "Radon" developed and exploited a technology of plasma treatment of RW based on a "Pluto" installation. This installation has successfully undergone the Russian State Ecological examination and is licensed for RW treatment. Plasma treatment does not require careful waste sorting, provides reduction of gas volume by 1.5–2 times. A high degree of radionuclides incorporation into the ash residues is provided because of using plasma heat sources instead of fuel heaters. The ash residue after plasma treatment has a structure similar to the structure of volcanic glass; it has enough chemical stability against radionuclides leaching, therefore it can be immediately disposed in metal containers. Polkanov (2012) reported that leaching rates of radionuclides were $(0.3–5)\cdot10^{-6}$ g cm^{-2} day^{-1} for ^{137}Cs and $(0.8–2)\cdot10^{-7}$ g cm^{-2} day^{-1} for ^{239}Pu. The percentage of vegetation and berries in the RW for plasma treatment may reach up to 5%. The average humidity of waste is approximately 40%; however humidity of some packages may reach 90%. According to the license conditions, activity of solid RW suitable for plasma treatment is limited by $2.2\cdot10^{5}$ Bq kg^{-1} of alpha emitting radionuclides and $3.7\cdot10^{6}$ Bq kg^{-1} of beta emitting radionuclides (Polkanov 2012). Taking into account disposal of ash residues in metal containers, the factor of volume reduction varies from 40 to 50.

4 Disposal of Radioactive Waste After Rehabilitation

Removal of the upper layer of radioactively contaminated soil and ground results in formation of colossal volume of solid RW that should be treated or disposed. In-situ deposition of untreated soil at the most contaminated zone is practiced in order to decrease financial expenses for waste transport and treatment as well as for organizing radioactive waste storage. Batorshin and Mokrov (2013) reported that 329,000 m^3 of the upper soil layer was removed and disposed at abandoned pits and ditches. At the territory of the East Urals Radioactive Trace, contaminated soil layer of up to 5–10 cm was removed into the underploughinng layer (80–100 cm) in order to decrease transfer of radionuclides into farming products (Teplyakov et al. 1997). This method does not prevent radionuclides migration outside contaminated zones through ground and underground waters. Nakano and Yong (2013) suggested disposing low active soils immediately at the contaminated zone underground after isolation from ground water. Authors advised to mix contaminated soil with sorbents for strong retention of radionuclides and cover the contaminated soil by a layer of pure soil with a depth of at least two times more than roots length. Deposition of high active soil can be performed only after capsulation at special landfills with multibarrier system of environment protection.

5 Conclusions

The methods of treatment of radioactive waste being formed as a result of rehabilitation of radioactively contaminated lands should follow the concept of IAEA providing collection, sorting, treatment, storage and final disposal of RW. Precipitation and sorption are used as the main methods for treatment of liquid RW after liquid deactivation. The choice of conditions of radionuclides co-precipitation depends on reagents presenting in LRW. Surfactants and organic complexing agents may suppress effectiveness of decontamination. Sorbents being used for decontamination of spent deactivating solutions should possess high distribution coefficients of radionuclides in presence of reagents used for deactivation. It is preferable to use multipurpose sorbents having affinity to several radionuclides and possessing chemical stability against radionuclides leaching. These characteristics allow considering these sorbents as sorption-active matrices for radionuclides immobilization; therefore, these sorbents can be deposited in metal containers without additional pretreatment. Destructive methods may be used for preliminary elimination of surfactants, complexing agents and organic compounds from liquid radioactive waste before chemical treatment.

Soil and ground removed from radioactively contaminated lands can be deactivated using either reagent method or combination of wet gravity separation and reagent treatment. Radioactive biological materials are incinerated with further immobilization of ash residues into cement. Alternatively, plasma treatment resulting in formation of chemically stable ash can be used for utilization of radioactive biological materials.

Secondary radioactive waste being formed after various liquid and solid RW treatment as well as materials, products and equipment with fixed (irremovable) contamination should be disposed. In situ disposal can be used to decrease expenses for treatment. It should be mentioned that in situ disposal is suitable only for low-level RW and it does not prevent radionuclides migration in the environment. The multibarrier principle of protection should be followed in disposal of high-level radioactive waste.

References

Aleksakhin RM (2009) Radioactive contamination of soils as a type of their degradation. Pedology 12:1487–1498

Aleksakhin RM, Sychev VG (2013) Radioecological aspects of agricultural remediation following the accident at the Fukushima Daichi NPP. Fertility 4:2–6

Barinov AS, Prozorov LB, Nikolayevskii VB, Scheglov MY, Shevtsova EV (2003) The method of electrokinetic decontamination of ground from radioactive and toxic matter. Patent of Russian Federation No 2211493

Batorshin GS, Mokrov YG (2013) The experience of liquidation of results of the accident at the Mayak Production Association in 1957. Rad Saf Quest 1:13–20

Bobrov PA, Kozlov PV, Slyunchev OM, Gorbunova OA (2014) Methods of treatment of liquid radioactive waste containing surfactants OTI, Ozersk, Russia

Chirkst DE, Litvinova TE, Cheremisina OV, Streletskaya MI (2001) Pilot technology for soil deactivation to remove ^{90}Sr. Radiochemistry 43:540–544

Dmitriev SA, Prozorov LB, Shcheglov MY (2001) Electrokinetic method for decontamination of ground from radionuclides. Rad Saf Quest 1:42–49

Dmitriev SA, Nikolaevsky VB, Prozorov LB, Shcheglov MY (2006) Method for cleaning soils from cesium radionuclides. Patent of the Russian Federation No 227491512

Endo S, Kimura S, Takatsuji T, Nanasawa K, Imanaka T, Shizuma K (2012) Measurement of soil contamination by radionuclides due to the Fukushima Dai-ichi nuclear power plant accident and associated estimated cumulative external dose estimation. J Environ Radioact 111:18–27

Flouret J, Barré Y, Muhr H, Plasari E (2012) Design of an intensified co-precipitation reactor for the treatment of liquid radioactive wastes. Chem Eng Sci 77:176–183

IAEA (1983) International Atomic Energy Agency. Conditioning of Low- and Intermediate-level Radioactive Wastes: Technical Report Series No. 222. IAEA, Vienna

IAEA (1988) International Atomic Energy Agency. Immobilization of low and intermediate level radioactive wastes with polymers: technical report series no. 289. IAEA, Vienna

IAEA (1993a) Improved cement solidification of low and intermediate level radioactive wastes: Technical report series no. 350. IAEA, Vienna

IAEA (1993b) Bituminization processes to condition radioactive wastes: technical report series no. 352. IAEA, Vienna

IAEA (1993c) International Atomic Energy Agency. Containers for packaging of solid low and intermediate level radioactive wastes: technical report series no. 355. IAEA, Vienna

IAEA (1999) State of the art technology for decontamination and dismantling of nuclear facilities. Technical reports series no. 395, IAEA, Vienna

IAEA (2002) Application of ion exchange processes for the treatment of radioactive waste and management of spent ion exchangers: technical report series no. 408. IAEA, Vienna

IAEA (2008) Safety standards series "Release of sites from regulatory control on termination of practices". Safety Guide ⍰.WS.G.5.1. Vienna

Kim GN, Choi WK, Jung CH, Moon JK (2007) Development of a washing system for soil contaminated with radionuclides around TRIGA reactors. J Indust Engg Chem 13:406–413

LA Report (1997) Cement waste-form development for ion-exchange resins at the rocky flats plant: rep. LA-13226, Los Alamos National Laboratory, Los Alamos

Leont'eva TG, Baklai AA, Moskalchuk LN (2016) Prospective of using clay-salt slurries of JSC "BELARUSKALIY" for cleaning water environments and ecosystems from radiocesium. Trans Belarus State Tech Univ 3:74–80

Malikova EM, Lushnikova MV, Voronina AV (2017) The study of chemical and radionuclide composition of liquid radioactive waste from a special drainage tank at the JSC INM. In: Proceedings of the 4th international youth conference: physics. Technology. Innovation. PTI-2017. Ural Federal University, Ekaterinburg, pp 230–231

Milyutin VV, Gelis VM, Klindukhov VG, Obruchikov AV (2004) Co-precipitation of micro amounts of Cs with ferrocyanides of various metals. Radiochemistry 46:479–480

Milyutin VV, Ryabchikov BE, Kozlov PV (2015) Modern methods of liquid radioactive waste treatment. OTI, Ozersk

Moskalchuk LN (2015) Scientific substantiation of the use of solid wastes of mining enterprises by developing the technology for obtaining and using organomineral sorbents for the rehabilitation of soils contaminated with radionuclides. PhD thesis technical sciences. Minsk, Belarus. http://www.ipkonran.ru/docs/moskalchuk/auto.pdf

Murzin AA, Babain VA, Shadrin AY, Kamachev VA, Romanovskii VN, Starchenko VA, Podoinitsyn SV, Revenko YA, Logunov MV, Smart NG (2007) Deactivation in sub- and supercritical carbon dioxide. Radiochemistry 44:410–415

Myasoedova GV, Nikashina VA (2006) Sorption materials for separation of radionuclides from water media. Russ Chem J 5:55–63

Nakano M, Yong RN (2013) Overview of rehabilitation schemes for farmlands contaminated with radioactive cesium released from Fukushima power plant. Engg Geol 155:87–93

Nikolaevskii VB, Rodionov VV, Shcheglov VY (2006) Reagent soil deactivation technology development. Environ Saf 1:40–42

Nikolaevskii VB, Sklifasovskaya YG, Sabodina MN, Klochkova NV (2011a) Reagent-based remediation of soil contaminated with transuranium elements. Atom Ener 111:29–36

Nikolaevskii VB, Poluektov PP, Arustamov AE (2011b) Prospective soils decontamination technology. Nucl Environ Saf 4:114–117

Nogovitsyna EV, Voronina AV, Gritskevich EY, Nikiforov AF (2016) Use of inorganic sorbents for treatment of liquid radioactive waste with various composition. Water Indust Russ 5:102–110

Novikov AI, Ruzankin VI (1972) Coprecipitation of fission products with magnesium hydroxide. Radiochemistry 4:505–511

Plotnikov VI, Safonov II (1983) Radiochemical study of co-precipitation of trace amounts of certain hydrolysable elements with metals hydroxides and oxides. Radiochemistry 2:161–170

Plotnikov VI, Kamaeva K, Myasishchev AV (1989) Co-precipitation of strontium with individual and mixed hydroxides of certain methods. Radiochemistry 3:85–89

Polkanov MA (2012) Plasma-pyrolytic treatment of solid rad waste. Nucl Environ Saf 1:89–93

Prozorov LB, Komarova NI, Molchanova TV (2002) Acid deactivation of soils with ammonium fluoride to remove ^{137}Cs. Radiochemistry 44:518–522

Romanovich IK, Stamat IP, Sanzharova NI, Panov AV (2016) Criteria for the rehabilitation of objects and territories contaminated with radionuclides as a result of past activities: Part 1. Selection of indicators to justify the criteria for rehabilitation. Radiat Hygen 9:6–15

Sanitary regulations SP.2.6.1.2612-10 (2013) The main sanitary regulations for providing radiation safety (OSPORB-99/2010). Public Health Ministry of Russia, Moscow

Semenischev VS, Voronina AV, Bykov AA (2013) The study of sorption of caesium radionuclides by "T-55" ferrocyanide sorbent from various types of liquid radioactive wastes. J Radioanal Nucl Chem 295:1753–1757

Shadrin AY, Murzin AA, Shafikov DN, Kamachev VA, Bondin VV, Bychkov SI, Efremov IG (2007) Deactivation of metals, fabrics, and soils in a liquid CO_2 medium. Theoret Found Chem Engg 41:786–792

Sklifasovskaya YG, Nikolaevskii VB, Prozorov LB (2009) Reagent treatment of ^{226}Ra-contaminated soils in the dynamic mode. Radiochemistry 51:418–421

Sobolev IA, Prozorov LB, Komarova NI, Molchanova TV (2000) The method of deactivation of soils from cesium radionuclides. Patent of the Russian Federation No. 2152650

Szabo G, Guczi J, Nisbet A (1997) Investigation of the soil phase speciation of Sr-90, Cs-137, Pu-239 and Am-241 in soils determined by extraction and ultra-filtration methods. J Radioanalyt Nucl Chem 226:255–259

Teplyakov IG, Romanov GN, Spirin DA (1997) Returning of lands in East-Ural radioactive trace to farming use. Rad Saf Quest 3:33–41

Volkov VG, Zverkov YA, Ivanov OP, Koltyshev SM, Semenov SG, Stepanov VE, Shisha AD (2007) Decontamination of radioactively contaminated soil at the Russian Science Center Kurchatov Institute. Atom Ener 103:979–985

Voronina AV, Semenishchev VS (2015) Sorption-active matrix based on titanium hydroxide for concentration and joint immobilization of cesium and strontium radionuclides. J Radioanal Nucl Chem 303:229–236

Voronina AV, Savchenko MO, Semenishchev VS, Nikiforov AF (2013) The assessment of effectiveness of use of clinoptilolite and ferrocyanide sorbents based on it for decontamination of aqueous media and radionuclides immobilization. Water Indust Russia 6:88–94

Voronina AV, Blinova MO, Semenishchev VS, Gupta DK (2015) Returning land contaminated as a result of radiation accidents to farming use. J Environ Radioact 144:103–112

Voronina AV, Kutergin AS, Semenishchev VS, Nogovitsyna EV, Nikiforov AF (2017) Sorbents for radiocaesium removal from natural water and soil. In: Gupta DK, Walther C (eds) Impact of cesium on plants and the environment. Springer, Cham, pp 231–252

Vozzhenikov GS, Aleksandrova ZN, Vozzhenikov EG (1997) A method of soils deactivation. Patent of the Russian Federation No. 2077749

Warner F, Harrison RM (1993) Radioecology after Chernobyl. Biogeochemical pathways of artificial radionuclides. Wiley, New York

Modelling the Effect of Mechanical Remediation on Dose Rates Above Radiocesium Contaminated Land

Alex Malins, Hiroshi Kurikami, Akihiro Kitamura, and Masahiko Machida

1 Introduction

In the aftermath of the March 2011 accident at the Fukushima Daiichi Nuclear Power Plant (FDNPP) the Japanese Government legislated for the remediation of contaminated land across North East Japan (Government of Japan 2011). Remediation efforts were organized separately for the Special Decontamination Area (SDA) surrounding the FDNPP site (the evacuation zone), and the Intensive Contamination Survey Areas (ICSAs) further afield in Iwate, Miyagi, Fukushima, Tochigi, Ibaraki, Gunma, Saitama and Chiba Prefectures. Remediation of the SDA was managed directly by the Government under the auspices of the Ministry of the Environment (MoE), while remediation of the ICSAs was devolved to local municipalities, backed by Government funding and MoE technical assistance (Ministry of the Environment 2013).

In 2011 and 2012 Japan Atomic Energy Agency (JAEA) conducted research into remediation options for radiocesium contaminated land in the *Decontamination Pilot Project*. Approaches were tested applicable for different land types (forests, farmland, residential areas, public infrastructure etc.) and contamination levels, and generating different wastes as a byproduct. Systematic results on clean-up

A. Malins (✉) · M. Machida
Center for Computational Science and e-Systems, Japan Atomic Energy Agency, Kashiwa, Chiba, Japan
e-mail: malins.alex@jaea.go.jp; machida.masahiko@jaea.go.jp

H. Kurikami · A. Kitamura
Center for Computational Science and e-Systems, Japan Atomic Energy Agency, Kashiwa, Chiba, Japan

Fukushima Environmental Safety Center, Japan Atomic Energy Agency, Tamura-gun, Fukushima, Japan
e-mail: kurikami.hiroshi@jaea.go.jp; kitamura.akihiro@jaea.go.jp

D. K. Gupta, A. Voronina (eds.), *Remediation Measures for Radioactively Contaminated Areas*, https://doi.org/10.1007/978-3-319-73398-2_12

effectiveness, costs and byproduct volumes from different approaches were summarized in an English language report (JAEA 2015b).

By March 2017 an estimated $24 billion (USD) had been spent remediating the SDA and ICSAs (Ministry of the Environment 2017). One of the most expensive aspects of the program was the management and storage of large volumes of contaminated topsoil stripped from farmland, residential gardens, public parks, schoolyards etc. (Yasutaka et al. 2013; Yasutaka and Naito 2016). For example the stripping of topsoil down to 5 cm from 85 km^2 of farmland in the SDA alone will have generated over 4,000,000 m^3 of waste, which is a significant proportion of the estimated total of 16,000,000 m^3 of waste generated across Japan by August 2017 (Ministry of the Environment 2017).

The debate on the costs and benefits of the remediation program in Japan should consider not only the averted radiation dose (Miyazaki and Hayano 2017), but also social and economic factors pertinent for the recovery of the contaminated areas (Oughton 2013). It is clearly beneficial however for remediation to maximize contaminant removal (or the reduction in the air dose rate), while minimizing *over-remediation*. This is where soil is stripped or mixed to a depth beyond which there is any further contaminant removal and/or reduction in the air dose rate.

This chapter outlines some previous modelling for the change in air dose rates that occur upon four different mechanical remediation strategies for radiocesium contaminated land (Malins et al. 2016a, b). A new analysis on the effectiveness of remediation is presented eliminating previous modelling assumptions about the depth distribution of radiocesium within soil. The new analysis directly employs measured depth distributions of radiocesium in soil in North East Japan to model remediation performance. The results determine whether remediation performance is affected by the downward migration of radiocesium in soil over the initial five year period following the FDNPP accident.

2 Depth Distributions of ^{134}Cs and ^{137}Cs in Un-Remediated Soils Across North East Japan

One of the pertinent factors that must be considered when designing a remediation program for a site contaminated with radiocesium is the depth distribution of the contamination within the soil prior to remediation. Since December 2011 the Japanese Ministry of Education, Culture, Sports, Science and Technology (MEXT) and the Nuclear Regulation Authority (NRA) have funded the sampling of ^{134}Cs and ^{137}Cs activity depth distributions at un-remediated sites within 80 km of FDNPP (Matsuda et al. 2015). At each site a scraper plate apparatus was used to remove soil layers from the surface downwards. The ^{134}Cs and ^{137}Cs activity concentrations in the soil layers (Bq kg^{-1} wet weight) were then measured using High Purity Germanium (HPGe) gamma spectroscopy. The results from eight sampling campaigns running until September 2015 were analyzed in this paper

Table 1 Details of the eight soil sampling campaigns for the depth distribution of ^{134}Cs and ^{137}Cs at un-remediated sites within 80 km of FDNPP

Campaign	Period	No. sites analyzed	Mean ^{137}Cs mass depth (g/cm^2)	Mean $\dot{H}$*(10) at 1 m (μSv h^{-1})
1	Dec 12–22, 2011 and Apr 17–19, 2012	83	0.99	0.60[a]
2	Aug 21 to Sep 26, 2012	81	1.09	0.51
3	Nov 26 to Dec 21, 2012	82	1.25	0.47
4	Jun 4–27, 2013	82	1.35	0.49
5	Oct 28 to Nov 29, 2013	80	1.58	0.40
6	Jul 14–24, 2014	76	1.81	0.36
7	Nov 4–13, 2014	76	1.70	0.35
8	Aug 24 to Sep 25, 2015	72	1.96	0.22

[a]Mean of measurements from 41 sites only [$\dot{H}$*(10) was not measured at all sites in campaign 1]

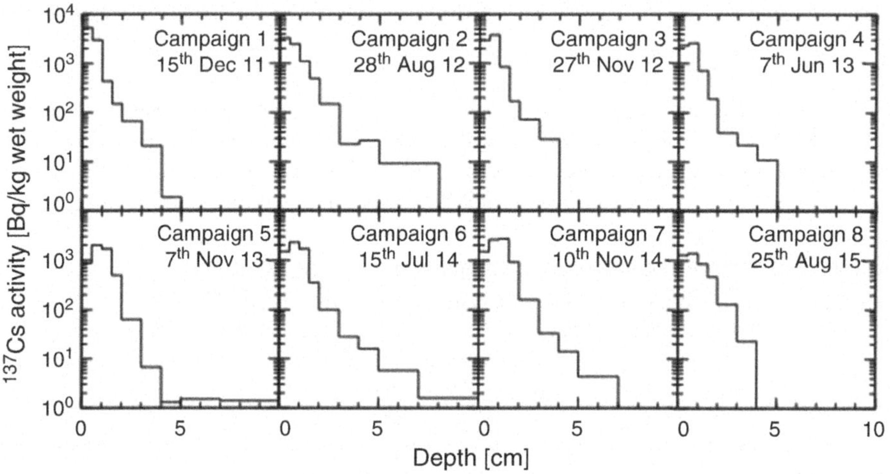

Fig. 1 Evolution of the ^{137}Cs depth distribution in the grounds of a religious assembly hall, 5 km south of Koriyama City, Fukushima Prefecture

(Table 1). Raw data from the campaigns are published online at JAEA's Database for Radioactive Substance Monitoring Data (JAEA 2015a).

The evolution of the ^{137}Cs depth distribution at a typical site is shown in Fig. 1. For the first two sampling campaigns the depth distribution shows a characteristic exponential shape (ICRU 1994). By the third campaign a peak in the ^{137}Cs activity concentration is apparent below the ground surface. The distribution continues to show a peak below the surface through to the eighth sampling campaign.

The trend of the results is for the radiocesium inventory to migrate slowly downwards in the ground over time. Figure 2a shows the mean mass depth of the ^{137}Cs inventory within the ground for both the example site in Fig. 1 and for all sites taken

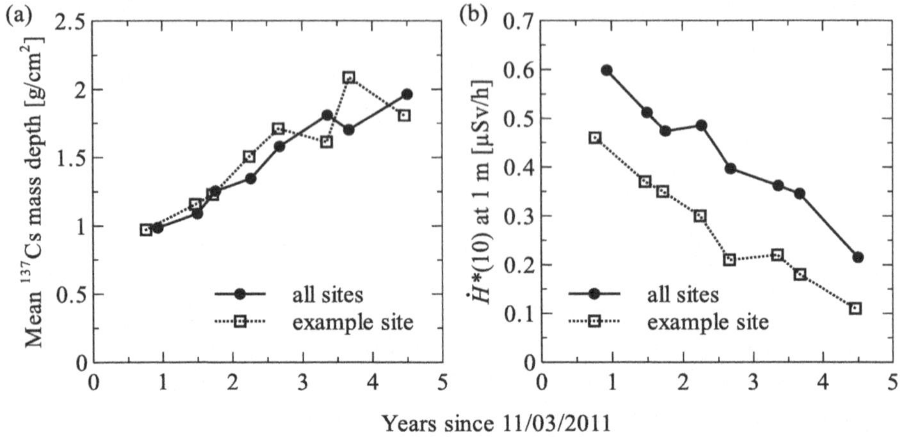

Fig. 2 (**a**) Mean mass depth of the ^{137}Cs inventory across all un-remediated sites and at the assembly hall 5 km south of Koriyama City. (**b**) Change in the air dose rate [$\dot{H}$*(10)] measured at 1 m

collectively. The air dose rate, specifically the ambient dose equivalent rate $\dot{H}$* (10) (μSv h^{-1}), at 1 m above the ground tends to decrease over time (Fig. 2b). This is predominantly because of radioactive decay and downward migration of the inventory within the soil increasing the self-shielding of the radiocesium gamma rays (Mikami et al. 2015; Malins et al. 2016b). The two characteristic types of depth distribution seen in the sampling campaigns have been fitted with exponential and hyperbolic secant functions (Matsuda et al. 2015; Malins et al. 2016b):

$$A(\zeta) = A_0 \exp(-\zeta/\beta), \tag{1}$$
$$A(\zeta) = A_0 \cosh(\zeta_0/\beta)\ \mathrm{sech}(-(\zeta - \zeta_0)/\beta). \tag{2}$$

Here $A(\zeta)$ is the ^{137}Cs activity concentration (Bq kg^{-1}), as a function of the soil mass depth ζ (g/cm^2), A_0 is the activity concentration (Bq kg^{-1}) at the surface, and β and ζ_0 (g/cm^2) are fitting parameters characterizing the penetration of the ^{137}Cs inventory into the ground. In particular ζ_0 is the mass depth of the peak in the activity distribution within the hyperbolic secant model. Note that in Eqs. (1) and (2) mass depths (i.e. the mass of soil between the surface and a given depth below the surface, per unit area) are used instead of physical depths as mass depth is more closely correlated with the amount of shielding provided by soil than physical depth (ICRU 1994). Fitting the measured depth distributions with these two empirical functions enabled the characterization of the rate of downward migration of the inventories within soil (Matsuda et al. 2015), and a systematic analysis of the effectiveness of different remediation strategies for reducing air dose rates (Malins et al. 2016a).

3 Mechanical Remediation Strategies

In the main four methods have been employed for the remediation of farms, schoolyards, gardens, parks etc. in North East Japan after the FDNPP accident (Yasutaka et al. 2013; JAEA 2015b; Ministry of the Environment 2015, 2017). The first two methods are variations on topsoil removal (Fig. 3a). Here the topsoil is stripped down to a set depth and this waste topsoil is sent for disposal. Variations of topsoil stripping include recovering the stripped ground with a clean layer of topsoil brought in from elsewhere (denoted strategy A1), or leaving the stripped ground in a bare state (A2).

Topsoil stripping is effective at lowering dose rates as it segregates the radiocesium contamination from the environment. Another option is to redistribute the contamination deeper within the ground such that the radiocesium gamma rays are attenuated by soil before they reach the surface. The soil layer interchange method is an example of the latter option (A3, Fig. 3b). Soil layer interchange involves first digging out a layer of topsoil and placing this to one side. Then a layer of subsoil is excavated from the pit that has been created. The pit is then refilled first with the soil that was initially topsoil, and finally to the surface with the soil that was initially subsoil.

The fourth mechanical remediation method is topsoil mixing (A4, Fig. 3c). In this method the topsoil is mixed down to a set depth using a rotovator or a plow pulled by a tractor. A notable disadvantage of soil layer interchange and topsoil mixing methods (A3 and A4) over topsoil stripping (A1 and A2) is that the radiocesium still remains within the environment after remediation, albeit at a greater soil mass depth below the surface. This may potentially affect the viability of agricultural products produced on farmland remediated by methods A3 or A4.

Yasutaka et al. (2013) estimated the total cost of remediating the agricultural land in the SDA by topsoil stripping (A1 and A2) as between 1022 billion and 2270 billion Japanese yen (~$10 billion to $23 billion USD). This compares with 62 billion yen (~$620 million USD) for the soil layer interchange method (A3), and 7 billion

(a) Topsoil stripping (A1 recovered & A2 bare) | (b) Soil layer interchange (A3) | (c) Topsoil mixing (A4)

Fig. 3 Mechanical remediation strategies for radiocesium contaminated soil. (Photos reproduced from JAEA 2015b) yen (~$70 million USD) for the soil mixing method (A4). Topsoil stripping is thus one to two orders of magnitude more expensive than the soil layer interchange and topsoil mixing remediation strategies.

4 Remediation Effect on Air Dose Rates (Modelling of Empirical Activity Depth Distributions)

Malins et al. (2016a) performed a systematic analysis of the change that occurs in the 1 m ambient dose equivalent rate [$\dot{H}*(10)$] upon remediation by each of the four mechanical strategies (A1 to A4) using the results of Monte Carlo radiation transport calculations. A simplified land geometry was modelled consisting of an infinite slab of soil and air (Malins et al. 2016b). The spatial distribution of the ^{134}Cs and ^{137}Cs activity concentration was assumed to be uniform in the initial state prior to remediation, i.e. constant Bq/m^2. The initial radiocesium depth distribution was modelled as either an exponential [Eq. (1)] or the hyperbolic secant [Eq. (2)] distribution. This method of calculating $\dot{H}*(10)$ using an approximated environmental geometry and radiocesium distribution was shown to give reasonably accurate results when its predictions were compared against actual measurements of $\dot{H}*(10)$ at un-remediated sites within 80 km of FDNPP (Malins et al. 2016b).

Dose rates after remediation were calculated by altering the radiocesium depth distribution within a square area of land that was modelled as having been remediated by one strategy (A1, A2, A3 or A4). Taking the topsoil mixing method as an example, the radiocesium distribution in the mixed topsoil layer was modelled as a homogeneous distribution throughout the layer (i.e. perfect mixing of the soil). Results from this strategy for modelling remediation were consistent with measured reductions in air dose rates upon remediation of test sites in the Decontamination Pilot Project (JAEA 2015b; Malins et al. 2016a).

The effect that various parameters have upon the amount that dose rates decrease was examined systematically. Variables considered included the initial distribution of radiocesium into the ground prior to remediation, by varying β and ζ_0 in Eqs. (1) and (2), the area of land that was remediated, and the mass depth to which soil was remediated.

The performance of each remediation strategy was characterized in terms of the ratio of the ^{134}Cs (^{137}Cs) component of $\dot{H}*(10)$ post-remediation to its initial value:

$$R_{134} = {}^{134}{}_{r}\dot{H}*(10) / {}^{134}{}_{i}\dot{H}*(10), \quad (3)$$

$$R_{137} = {}^{137}{}_{r}\dot{H}*(10) / {}^{137}{}_{i}\dot{H}*(10). \quad (4)$$

Here the subscripts r and i denote the remediated and initial air dose rates, respectively. R_{134} and R_{137} are termed *residual dose rate factors*. As the residual dose rate factors represent the relative change in the radiocesium components of $\dot{H}*(10)$ rather than $\dot{H}*(10)$ as a whole, they apply generally rather than being specific to any particular remediation site. Assuming the natural background component of the air dose rate [$^{nat}\dot{H}*(10)$] at a given site is unchanged by remediation, the air dose rate after remediation is:

$$_r\dot{H}*(10) = R_{134} \times {}^{134}{}_i\dot{H}*(10) + R_{137} \times {}^{137}{}_i\dot{H}*(10) + {}^{nat}\dot{H}*(10). \qquad (5)$$

Figure 4 shows residual dose rate factors for remediation of soil by methods A1 to A4. The initial radiocesium depth distribution was modelled as an exponential distribution with $\beta = 2.0$ g/cm^2. The residual dose rate factors for ^{134}Cs and ^{137}Cs coincide as the energies of the primary gamma rays emitted by these radioisotopes are comparable.

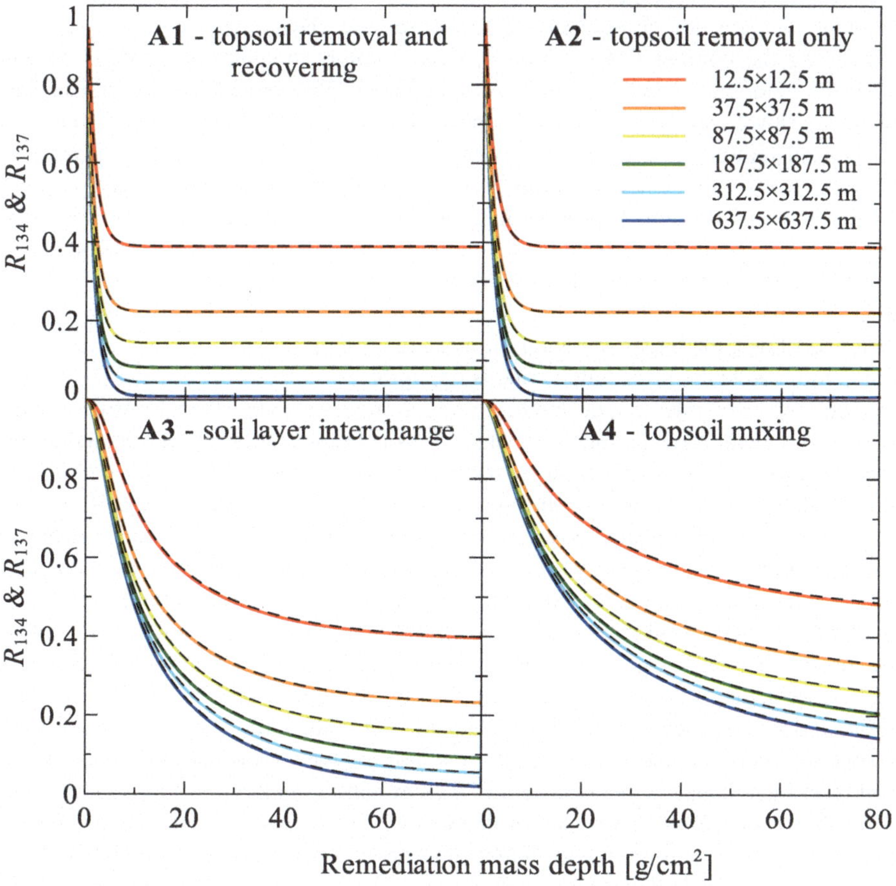

Fig. 4 Residual dose rate factors as a function of remediation mass depth and the size of the remediated area, given an initially exponential radiocesium depth distribution ($\beta = 2.0$ g/cm^2). Solid lines indicate results for ^{137}Cs, while dashed lines are those for ^{134}Cs. The legend in the upper right panel lists the various sizes of the remediated area. (Figure adapted from Malins et al. 2016a)

In general the topsoil stripping methods (A1 and A2, top panels of Fig. 4) are more effective for reducing air dose rates than the soil layer interchange (A3, lower left panel of Fig. 4) or topsoil mixing (A4, lower right panel of Fig. 4) methods. This is clear when comparing data points between the panels at equal remediation mass depth and remediated land area.

The residual dose rate factors for the topsoil removal methods plateau out for remediation mass depths greater than ~10 g/cm^2. Essentially all the radiocesium has been removed from remediation zone upon remediation to these mass depths. Further removal of topsoil thus conveys no further reduction in the air dose rate.

The reason that the residual dose rate factors plateau at values above zero is because of radiocesium that remains outside the remediated area of land. Cesium-134 and ^{137}Cs gamma rays have a mean free path in air on the order of 100 m (Malins et al. 2015), and very large areas need to be remediated to completely eliminate this residual radiocesium component of the dose rate. This effect can be seen by examining the colored lines in Fig. 4. Cooler colors indicate increasingly larger areas of land that were modelled as remediated. The remediation performance increases (lower residual dose rate factors) with the size of the remediated area.

Topsoil removal and recovering with a clean soil layer (A1) only outperforms topsoil removal alone (A2) within a narrow range of remediation mass depths (between 0 and 10 g/cm^2). In this range recovering with a clean soil layer leads to slightly lower residual dose rate factors than without replacing the stripped soil. This is because of shielding provided by the covering layer of clean soil. However the effect is small, and moreover there is no radiological benefit of recovering with a clean soil layer for larger remediation mass depths, as the amount of radiocesium remaining inside the remediated zone is negligible.

The soil layer interchange method (A3) is more effective than topsoil mixing (A4) given equal remediation parameters. This is because, given equal remediation mass depths, soil layer interchange redistributes the radiocesium deeper within the ground than topsoil mixing, resulting in a larger shielding effect by the soil.

The curves of the residual dose rate factors for methods A3 and A4 decrease more slowly with increasing remediation mass depth than for the topsoil removal methods. As a consequence larger remediation mass depths are necessary with soil layer interchange and topsoil mixing to achieve the same residual dose rate factor given by topsoil stripping. Taking the top red curve in Fig. 4 as an example (12.5 m by 12.5 m remediation area), topsoil stripping to 10 g/cm^2 (A1 or A2) results in a residual dose rate factor of 0.4 (i.e. 60% reduction in the radiocesium component of [$\dot{H}^*(10)$]. The remediation mass depth required to give a corresponding result with soil layer interchange is on the order of 80 g/cm^2, and even larger for topsoil mixing.

Malins et al. (2016a) presented further results on residual dose rate factors. When the initial depth distribution was modelled as an exponential function, residual dose rate factors were calculated for relaxation mass depths of $\beta = 1.0$ and 5.0 g/cm^2, in addition to $\beta = 2.0$ g/cm^2 (Fig. 4). When modelling the initial depth distribution with the hyperbolic secant function, both β and ζ_0 were varied between 1.0 and 5.0 g/cm^2.

The effect of migration of the radiocesium inventory downwards in soil over time on the performance of remediation was investigated by considering different possible rates of evolution of the radiocesium depth distribution. Downward migration of radiocesium was modelled by increasing the relaxation mass depth (β) of the exponential distribution in Eq. (1). Relaxation mass depths applicable at different lags post fallout were drawn from values published in ICRU (1994), based on a global literature search, and by reverse modelling of the dose rate attenuation factors derived by Likhtarev et al. (2002) for sites in Ukraine affected by Chernobyl accident fallout.

Figure 5 shows the results of this analysis, considering up to 20 years migration of radiocesium within soil following its deposition. The results are sensitive to the model chosen for the evolution of the radiocesium depth profile over time, i.e. the speed at which downward migration occurs. With the faster scenarios for the

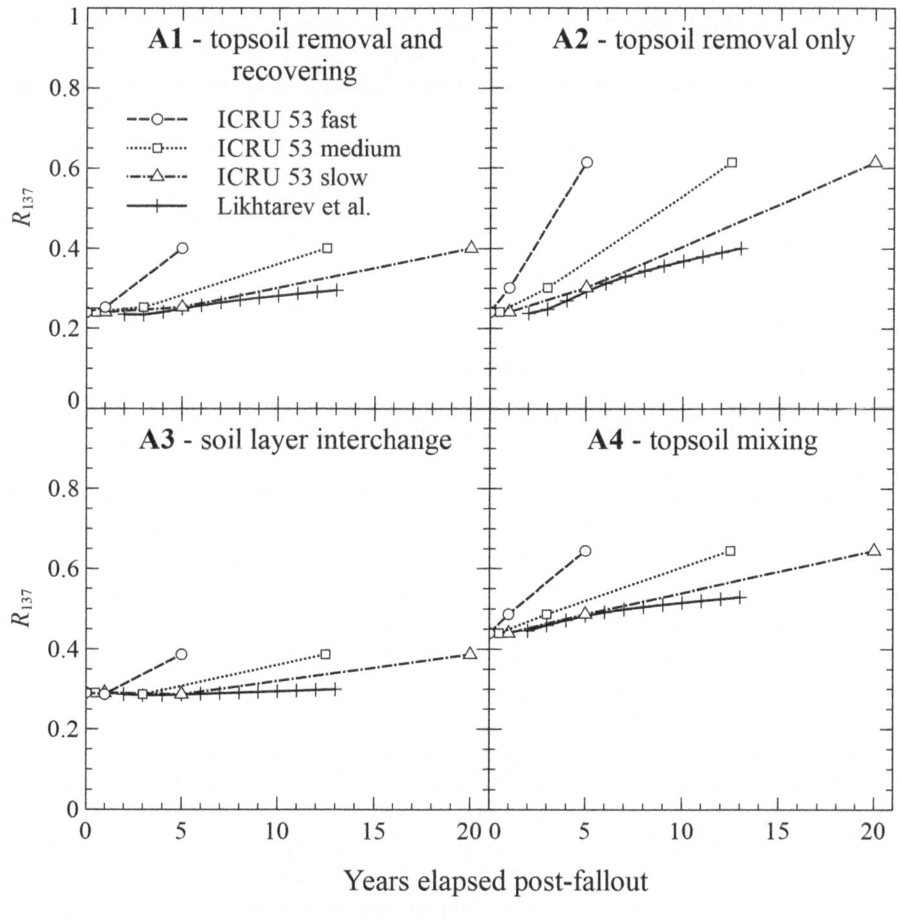

Fig. 5 The change in ^{137}Cs residual dose rate factors under four different models for the migration of radiocesium downwards in soil over time. In all cases the modelled area of remediated land was 37.5 m by 37.5 m. The remediation mass depths were fixed at 6.7 g/cm^2 for strategies A1 and A2, 40.3 g/cm^2 for strategy A3, and 33.8 g/cm^2 for A4. (Figure adapted from Malins et al. 2016a)

downward migration (ICRU 53 fast and medium models in Fig. 5), the performance of remediation decreases the later it is undertaken after the initial contamination event. This is indicated by the increase in the ^{137}Cs residual dose rate factors in Fig. 5. One option to recover the performance of remediation undertaken at late times after the initial contamination event would be to increase the (mass) depth to which soil is remediated.

5 Remediation Modelling Using Measured Activity Depth Profiles

A limitation of the results described in the previous section is the assumption that the initial depth distribution of the radiocesium takes a perfectly exponential or hyperbolic secant form. In reality radiocesium depth distributions always differ from these empirical models. This limitation may be significant as long tails have been observed in depth distributions of radiocesium fallout worldwide (Antonopoulos-Domis et al. 1995; Matsuda et al. 2015; Takahashi et al. 2015). The long tails are not well characterized by a single exponential function (Kurikami et al. 2017).

A study was therefore undertaken of the effectiveness of remediation without resorting to employing an empirical model for the radiocesium depth distribution prior to remediation. Measurement data on depth distributions at the un-remediated sites within 80 km of FDNPP (Table 1) were used to model the initial radiocesium depth distribution prior to remediation. In particular the thickness, mass, and ^{134}Cs and ^{137}Cs activity concentration of each soil layer was used to define a step-wise activity distribution. Remediation was then modelled within a 37.5 m by 37.5 m area of land using methods A1 to A4.

Topsoil stripping (A1 and A2) was modelled down to physical depths of 3, 5 and 7 cm below the surface. The corresponding remediation depths for soil layer interchange and topsoil mixing were 10, 20 and 30 cm. The post-remediation distributions for the radiocesium inventory, and other applicable modelling factors, were defined in an identical manner to Malins et al. (2016a).

The results are presented in Fig. 6, with individual panels showing data for methods A1 to A4. Remediation performance was assessed in terms of the ratio of total radiocesium component (^{134}Cs and ^{137}Cs) of $\dot{H}*(10)$ post-remediation to its initial value prior to remediation. Each data point represents the mean of all residual dose rate factors from one soil sampling campaign. The data are plotted as a function of the number of years elapsed since March 2011.

The overarching result is that the residual dose rate factors show little sensitivity to the time point at which remediation is undertaken in the initial five year period following the fallout. Although the radiocesium inventory did tend to migrate downwards within the soil over this period (Fig. 2a), the results of Fig. 6 indicate that migration is insignificant over the scale of the remediation depths.

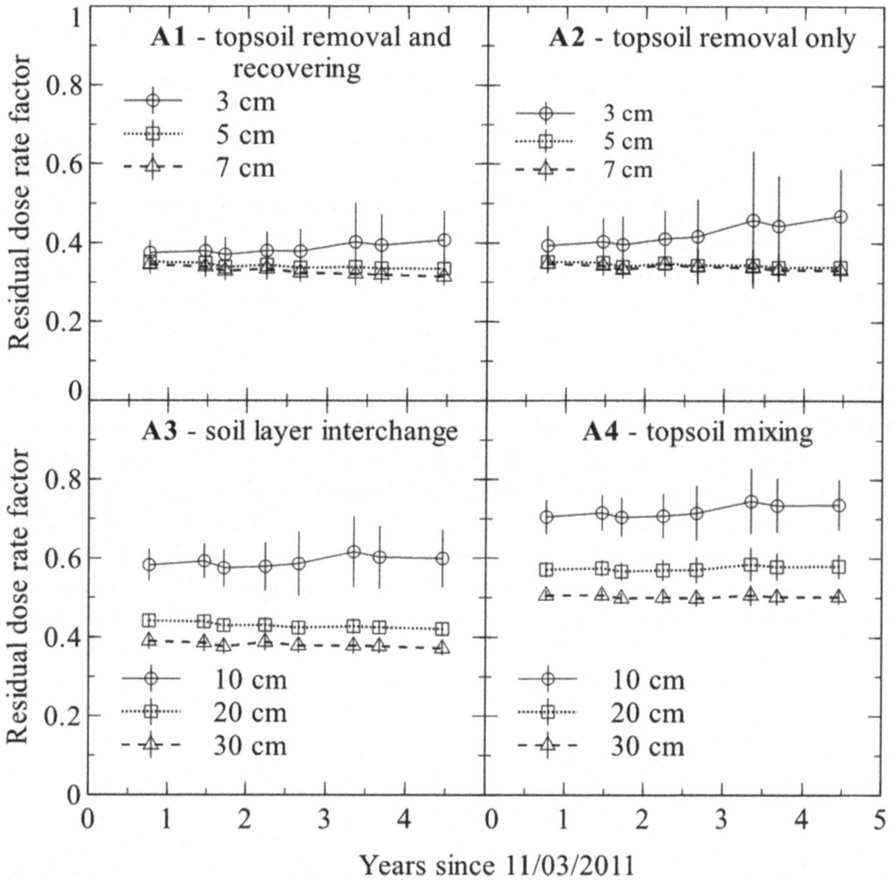

Fig. 6 Mean residual dose rate factors for the radiocesium component of $\dot{H}^{*}(10)$ upon modelled remediation of with 80 km of FDNPP. Different panels show results for A1 to A4 methods of remediation. Lines and markers represent different physical depths to which the soil was remediated (see legends). Error bars show plus/minus one standard deviation from the mean. In all cases the size of remediated area was 37.5 m by 37.5 m

The possible exceptions are the results for topsoil removal down to 3 cm (solid lines and circle markers, Fig. 6a, b). Here the residual dose rate factors increase by a small amount with increasing number of years after the fallout. This indicates that a significant proportion of the radiocesium inventory migrated downwards below 3 cm over the five year period. Remediation by topsoil removal in the latter years of the five year period therefore does not remove the entire radiocesium inventory. The residual dose rate after remediation is thus larger than if remediation had been undertaken at an earlier time.

There is only a small difference in the performance of the two topsoil removal methods when stripping 5 cm or more of topsoil (cf. Fig. 6a, b). This reinforces the previous conclusion that once the majority of the radiocesium inventory has been removed from the environment, i.e. the inventory of the top few centimeters of topsoil, there are diminishing returns from stripping further topsoil or by recovering with a clean soil layer.

By comparing Figs. 5 and 6 it is clear that the trend of the residual dose rate factors for North East Japan for the first five years after the contamination fallout are most consistent with the ICRU 53 slow and Likhtarev et al. models in Fig. 5. Therefore, excluding the case of topsoil stripping down to 3 cm alone, there is no indication that increasing the remediation depth over the first five years following the FDNPP accident was necessary to maintain remediation performance.

The performance of remediation undertaken five or more years after the FDNPP accident will depend on the development of the radiocesium depth profile within the soil. It is known that cesium binds strongly to clay minerals in soil (Okumura et al. 2013; Mukai et al. 2014; Fuller et al. 2015). This suggests that future changes to the radiocesium depth distribution will depend on natural biological and physical mixing of the soil, rather than physiochemical migration of the radiocesium ions (Kurikami et al. 2017). Past experience of Chernobyl fallout in Eastern Europe (IAEA 2006), and Mayak contamination of the Techa River, South Ural Mountains, Russia (Akleyev and Kisselyov 2002) has shown radiocesium tends to remain in the upper 20 cm of soil, even 35 years after the initial contamination event. This suggests mechanical soil remediation will remain a viable strategy for lowering air dose rates in as yet un-remediated areas in North East Japan over the coming years.

6 Conclusions

Mechanical remediation of radiocesium contaminated soil was one of the most expensive aspects of the remediation program undertaken in Japan after the FDNPP accident in 2011. This chapter summarized modelling work for the relationship between different mechanical remediation strategies for radiocesium contaminated soil and reductions in air dose rates. By using a modelling approach it was possible to quickly and cheaply quantify the effect of various factors on the performance of the remediation work, including the amount of soil that is remediation (both in terms of land area and soil depth) and the initial depth distribution of radiocesium in soil.

Topsoil removal, with or without recovering with a clean soil layer, tends to be more effective at lowering air dose rates than either soil layer interchange or soil mixing. This is because radiocesium is removed and separated from the environment with the former strategies. In order to attain the levels of dose rate reduction achieved by stripping circa 10 g/cm^2 of topsoil, it is necessary to remediate down to much larger soil mass depths with either the soil layer interchange or topsoil mixing methods.

Radiocesium activity depth distributions from North East Japan were modelled to determine the effect of downward migration of radiocesium in soil on remediation performance. The results suggest remediation performance is independent of the time at which it is undertaken within five years from March 2011. The exception to this conclusion was topsoil stripping (A1 or A2) to 3 cm. Migration of radiocesium below 3 cm depth in the initial five year period means topsoil stripping only down to 3 cm becomes less effective the later it is undertaken.

Acknowledgements We thank S. Nakama, K. Saito and K. Miyahara for their advice and assistance to this work.

References

Akleyev AV, Kisselyov MF (2002) Medical-biological and ecological impacts of radioactive contamination of the Techa River. Fregat, Chelyabinsk

Antonopoulos-Domis M, Clouvas A, Hiladakis A, Kadi S (1995) Radiocesium distribution in undisturbed soil: measurements and diffusion-advection model. Health Phys 69:949–953

Fuller AJ, Shaw S, Ward MB, Haigh SJ, Mosselmans JFW, Peacock CL, Stackhouse S, Dent AJ, Trivedi D, Burke IT (2015) Caesium incorporation and retention in illite interlayers. Appl Clay Sci 108:128–134

Government of Japan (2011) Act on special measures concerning the handling of environment pollution by radioactive materials discharged by the NPS accident associated with the Tohoku District-Off the Pacific Ocean Earthquake That Occurred on March 11, 2011

IAEA (2006) STI/PUB/1239: environmental consequences of the Chernobyl accident and their remediation: 20 years of experience. Report of the Chernobyl Forum Expert Group 'Environment'. http://www-pub.iaea.org/MTCD/publications/PDF/Pub1239_web.pdf

ICRU (1994) ICRU 53: Gamma-Ray spectrometry in the environment

JAEA (2015a) Database for radioactive substance monitoring data-depth distribution in soil. http://emdb.jaea.go.jp/emdb/en/

JAEA (2015b) JAEA-Review 2014-051: remediation of contaminated areas in the aftermath of the accident at the Fukushima Daiichi Nuclear Power Station: overview, analysis and lessons learned Part 1: a report on the 'Decontamination Pilot Project'. Technical report, March, Japan Atomic Energy Agency. https://doi.org/10.11484/jaea-review-2014-051

Kurikami H, Malins A, Takeishi M, Saito K, Iijima K (2017) Coupling the advection-dispersion equation with fully kinetic reversible/irreversible sorption terms to model radiocesium soil profiles in Fukushima prefecture. J Environ Radioact 171:99–109

Likhtarev IA, Kovgan LN, Jacob P, Anspaugh LR (2002) Chernobyl accident: retrospective and prospective estimates of external dose of the population of Ukraine. Health Phys 82:290–303

Malins A, Okumura M, Machida M, Takemiya H, Saito K (2015) Fields of view for environmental radioactivity. In: Proceedings of the international symposium on radiological issues for Fukushima's revitalized future, pp 28–34

Malins A, Kurikami H, Kitamura A, Machida M (2016a) Effect of remediation parameters on in-air ambient dose equivalent rates when remediating open sites with radiocesium-contaminated soil. Health Phys 111:357–366

Malins A, Kurikami H, Nakama S, Saito T, Okumura M, Machida M, Kitamura A (2016b) Evaluation of ambient dose equivalent rates influenced by vertical and horizontal distribution of radioactive cesium in soil in Fukushima Prefecture. J Environ Radioact 151:38–49

Matsuda N, Mikami S, Shimoura S, Takahashi J, Nakano M, Shimada K, Uno K, Hagiwara S, Saito K (2015) Depth profiles of radioactive cesium in soil using a scraper plate over a wide area

surrounding the Fukushima Dai-ichi Nuclear Power Plant, Japan. J Environ Radioact 139:427–434
Mikami S, Maeyama T, Hoshide Y, Sakamoto R, Sato S, Okuda N, Sato T, Takemiya H, Saito K (2015) The air dose rate around the Fukushima Daiichi nuclear power plant: its spatial characteristics and temporal changes until December 2012. J Environ Radioact 139:250–259
Ministry of the Environment (2013) Decontamination guidelines. Technical report. http://josen.env.go.jp/en/framework/pdf/decontaminationguidelines 2nd.pdf
Ministry of the Environment (2015) FY2014 decontamination report. Technical report. http://josen.env.go.jp/en/cooperation/pdf/decontaminationreport1503 01.pdf
Ministry of the Environment (2017) Progress on off-site cleanup and interim storage in Japan (August 2017). Technical report
Miyazaki M, Hayano R (2017) Individual external dose monitoring of all citizens of Date City by passive dosimeter 5 to 51 months after the Fukushima NPP accident (series): II. Prediction of lifetime additional effective dose and evaluating the effect of decontamination on individual dose. J Radiol Prot 37:623–634
Mukai H, Hatta T, Kitazawa H, Yamada H, Yaita T, Kogure T (2014) Speciation of radioactive soil particles in the Fukushima contaminated area by IP autoradiography and microanalyses. Environ Sci Technol 48:13053–13059
Okumura M, Nakamura H, Machida M (2013) Mechanism of strong affinity of clay minerals to radioactive cesium: first-principles calculation study for adsorption of cesium at frayed edge sites in muscovite. J Phys Soc Jpn 82:033802
Oughton DH (2013) Social and ethical issues in environmental remediation projects. J Environ Radioact 119:21–25
Takahashi J, Tamura K, Suda T, Matsumura R, Onda Y (2015) Vertical distribution and temporal changes of ^{137}Cs in soil profiles under various land uses after the Fukushima Dai-ichi Nuclear Power Plant accident. J Environ Radioact 139:351–361
Yasutaka T, Naito W (2016) Assessing cost and effectiveness of radiation decontamination in Fukushima Prefecture, Japan. J Environ Radioact 151:512–520
Yasutaka T, Naito W, Nakanishi J (2013) Cost and effectiveness of decontamination strategies in radiation contaminated areas in Fukushima in regard to external radiation dose. PLoS One 8(9): e75308

Practical Experience in Applying Remediation Methods at the Semipalatinsk Test Site (Kazakhstan)

Andrey V. Panitskiy, Sergey N. Lukashenko, Erlan G. Batyrbekov, Sergey A. Berezin, Murat A. Umarov, Yuriy G. Strilchuk, and Vitaliy V. Yakovlev

1 Introduction

Over the period of 1949 to 1989, 456 nuclear tests (616 nuclear devices) were conducted at Semipalatinsk test site (STS). Nuclear testing at STS can be divided into 2 stages: stage 1–nuclear explosions in the atmosphere; stage 2– underground nuclear explosions (in tunnels and boreholes).

On «Experimental Field» ground, during the period from 1949–1962, 116 atmospheric tests were conducted for the purpose of studying nuclear effects. Among them-30 aboveground tests and 86 air tests. As from 1961, 340 more underground nuclear tests were conducted. 209 underground nuclear tests were conducted in tunnels of «Degelen» site to test nuclear charges and ammunition, for large-scale testing of underground structures of various design solutions and their technological equipment and other sites. Underground nuclear tests were conducted in boreholes of «Balapan» (105 tests), «Sary-Uzen» (24 tests) and «Telkem» sites (2 tests of soil outburst) to test nuclear charges and ammunition, for large-scale testing of strategic-purpose missile troops facilities (silo launchers including missiles, unified command posts), fragments of state administration posts, nuclear ammunition storage facilities and their technological equipment, to try out the industrial explosion technologies. Peaceful underground nuclear tests were also conducted at the test site. Technologies were developed to use underground nuclear tests for the purpose of creating artificial water bodies, channels, seismic earth crust probing in searching for minerals, creating underground reservoirs, extinguishing burning gas flares and oil gushers and others. Over the period of STS activity, there were 7 underground nuclear explosions conducted for industrial purposes.

A. V. Panitskiy (✉) · S. N. Lukashenko · E. G. Batyrbekov · S. A. Berezin · M. A. Umarov
Y. G. Strilchuk · V. V. Yakovlev
National Nuclear Center of Kazakhstan, Kurchatov, Kazakhstan
e-mail: Panitskiy@nnc.kz; lukashenko@nnc.kz; batyrbekov@nnc.kz; berezin@nnc.kz; umarov@nnc.kz; strilchuk@nnc.kz; yakovlev@nnc.kz

D. K. Gupta, A. Voronina (eds.), *Remediation Measures for Radioactively Contaminated Areas*, https://doi.org/10.1007/978-3-319-73398-2_13

One more line of research at STS is experiments of incomplete chain reaction. Besides nuclear explosions, there were tests conducted at the test site using radiological warfare agents (sites «4» and «4a») and 175 explosions performed using chemical explosives. As a result of testing, areas were formed at STS of high values of radionuclide activity concentrations which require remediation activities to be performed.

In recent years, a great work has been done to specify proliferation boundaries of radionuclide contamination at these radiation hazardous STS sites (Lyakhova et al. 2012; Aidarkhanov et al. 2013; Turchenko et al. 2014; Panitskiy and Lukashenko 2015). The applied up-to-date techniques of a large-scale pedestrian gamma-spectrometry resulted in clearly enough localized spots posing hazard to the population and requiring measures to limit the effect of such spots on the environment and man. Remediation of radiation hazardous STS lands can be implemented by different mays the choice of which, first of all, depends on natural conditions in a specific area and levels of radionuclide contamination. In view of sufficiently large areas of high radionuclide content levels in soil, complete removal of the amount of contaminated soil does not seem possible as it is not always cost-effective. Besides, another problem arises– storage of removed soil. Therefore, remediation is possible–deep plowing by soil overturning on STS lands («Experimental Field", plumes of excavation nuclear explosions («Atomic lake», «Telkem»), where soil contamination is spread in insignificant areas. Such method of soil rehabilitation is also applicable to these areas because these represent sloping, leveled sites of covering loose sediments thicknesses up to 2–3 m and thicker. Therefore, despite crushed stone level in soils, plowing these territories is feasible.

Specialists of the National Nuclear Center of the Republic of Kazakhstan have estimated the remediation efficiency by soil plowing through the experimental work in the STS territory followed by applying the technology studied in practice. Surface contamination in individual spots was also reduced by means of removal of radioactively contaminated soil.

2 Remediation in Radioactively Contaminated STS Areas by Plowing

2.1 Estimating the Remediation Efficiency in the Radioactively Contaminated Territory by Soil Plowing

2.1.1 Materials and Methods

2.1.1.1 Selection of Research Areas

An area of light chestnut soils has been selected for the research in the region of «Experimental Field» ground in the territory of «P-2» experimental site where

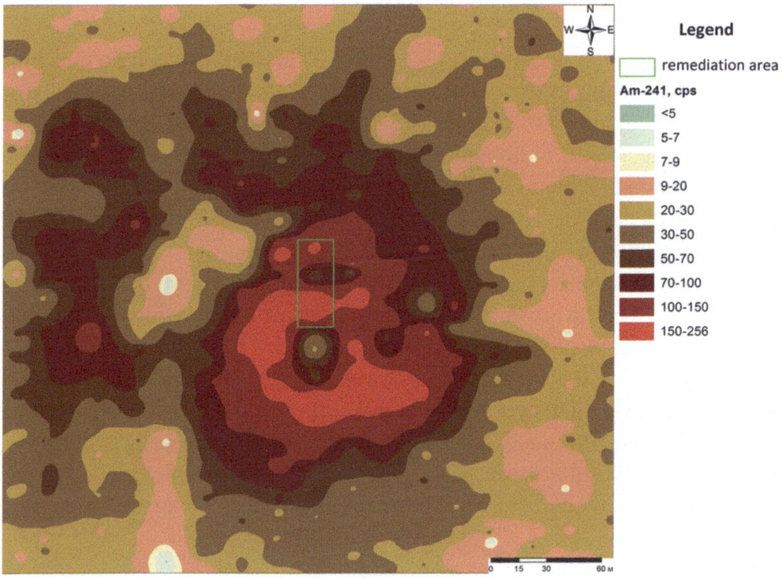

Fig. 1 Area of marked research sites (counts per second (cps))

low-yield nuclear tests were conducted (Fig. 1). This area is characterized by high content of technogenic radionuclides.

2.1.1.2 Key Spot Remediation

2.1.1.2.1 The Choice of Remediation Technology

Based on soil characteristics of the research area and the data on technogenic radionuclide distribution in soil profile of the investigated region, a complete turn of at least 30 cm is recommended. Taking into account meteorological characteristics of «Experimental Field» site territory, where the speed of wind gusts can reach 17 m s^{-1}, it is possible to expect intensification of soil cover deflation after plowing. Therefore, it was decided to evaluate the effectiveness of three possible remediation technologies:

1. Soil plowing to a depth of 30 cm with a turnover of 180° followed by soil backfilling with crushed stone;
2. Soil plowing to a depth of 30 cm with a 180° turnover followed by sowing with steppe vegetation;
3. Soil plowing to a depth of 30 cm with a turnover of 180°.

Crested wheatgrass (*Agropyron cristatum*) was selected to sow in plots. It is a cereal growing from root every year and is considered to be the most frost-resistant and drought-resistant plant that sustains soil salinity.

2.1.1.2.2 Research Areas Remediation

There are 3 spots of 16 × 15 m (U-1, U-2 and U-3) marked out in light chestnut soils of the P-2 site according to remediation technologies chosen (Fig. 1). Pedestrian γ-, β-radiometric surveying with a grid of 2 × 2 m was carried out to evaluate the effectiveness of decreased radiometric indices at spots marked out. A description of the relief and vegetation cover was made at each spot; 1 sample of the aboveground plant parts was selected for radionuclide analysis; one soil profile cut was marked out by soil sampling layer wise (5 cm each) down to 50 cm.

After carrying out these measures, the soil was plowed on all the plots to a depth of 30 cm with an 180^0 turnover. DT-75 tractor performed plowing, equipped with a suspended three-hulled plough PLN-3, with a heavy tooth harrow BZT-1. After plowing, to determine the degree of radionuclides penetration, layer-by-layer soil sampling was made at each spot from central soil profile cuts every 5 cm to a depth of 30 cm for radionuclide analyses. To evaluate the effectiveness of various technologies to prevent wind erosion, U-2 spot was backfilled with crushed stone. U-3 spot was sown with seeds of crested wheatgrass on the 14 kg of seeds basis per 1 hectare.

2.1.1.3 Evaluating Radionuclide Carry-Over by Plants

For 3 years after spots were remediated, vegetation was sampled at each of them to determine a potential radionuclide carry-over. The middle plant sample from each spot was formed by 5 points (Fig. 2). 1 m^2 of vegetation was mown at each point. 2 plant species were additionally sampled having different root system types in order to evaluate the impact of a root system type on a potential radionuclide carry-over from soil at U-1 spot on the 2 and 3 years after remediation. 2 species were considered– crested wheatgrass (*Agropyron cristatum*), having fibrous root system and sericeous sagebrush (*Artemisia sericea*), having taproot system. Both plants were mown at the same spot of 1 m^2.

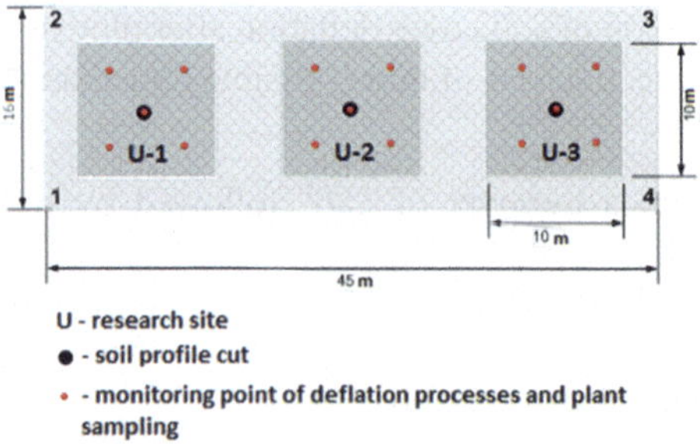

a)

b)

Fig. 2 Research sites. (**a**) Diagram to mark out sites. (**b**) U-2 site after backfilling with broken rock

2.1.1.4 Estimating Deflation Processes

To observe deflation processes, right after remediation, 5 points were laid at each spot. Soil of 5 cm was sampled at each point at a depth of 15 cm for physico-chemical research (mechanical composition and humus content). 2 years after remediation, soil was resampled 5 cm at a time at a depth of 15 cm at monitoring points of deflation processes for physico-chemical research.

2.1.1.5 Estimating Processes in Remediated Areas Overgrowing with Vegetation

The naturally overgrowing steppe and halophytic vegetation was monitored in three experimental variants in order to study zonal steppe and intrazonal halophytic revegetation in remediated areas of «Experimental Field» ground throughout 2012–2016.

2.1.2 Research Findings

2.1.2.1 Estimating the Efficiency of Decreased Values of Radiometric Parameters in the Remediation Area

Results of the radiometric γ-, β-survey conducted before and after plowing key remediation spots, are shown in the diagram of integral indicator distribution (Figs. 3 and 4).

Analyzed findings of equivalent dose rate (EDR) measurements showed that before plowing, EDR values varied in a relatively wide range of values– from 0.3–3.7 $\mu Sv\ h^{-1}$. After plowing, EDR values decreased on average by 56% and the value range narrowed significantly– 0.3-1.4 $\mu Sv\ h^{-1}$ (Fig. 4). Analyzed findings of measured β- particle flux density as in the case of EDR showed that before remediation activities, these radiometric indicators varied relatively widely from 26 particles/(min $\times$ cm^2) to 980 particles/(min $\times$ cm^2). After spots were plowed, values of β- particle flux density decreased on average by 87%. The value range also narrowed significantly– from <10 particles/(min $\times$ cm^2) to (160 particles/(min $\times$ cm^2) (Fig. 4).

2.1.2.2 Estimating the Efficiency of Reduced Radionuclide Activity Concentration in Top (Contact) Soil

Figure 5 demonstrate the nature of radionuclide distribution in soil before and after plowing down to 30 cm with an 180° overturn. Graphs reflect well variations in the nature of soil cover contamination before and after plowing spots. Analyzed findings showed that before remediation activities, radionuclides of interest were in upper soil horizons of 0–5, 5–10 cm (Fig. 5).

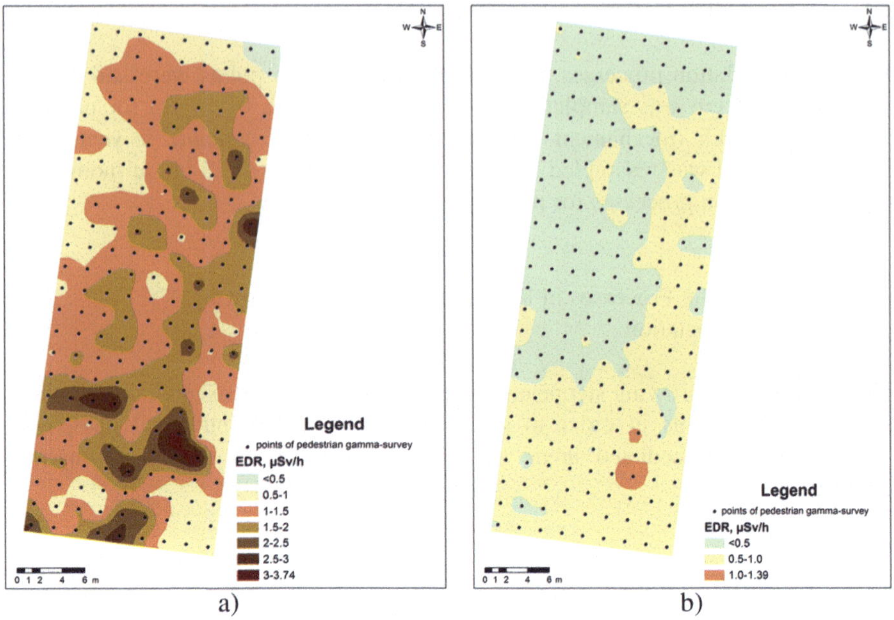

Fig. 3 Nature of EDR distribution in the territory by soil plowing before (**a**) and after (**b**) remediation activities

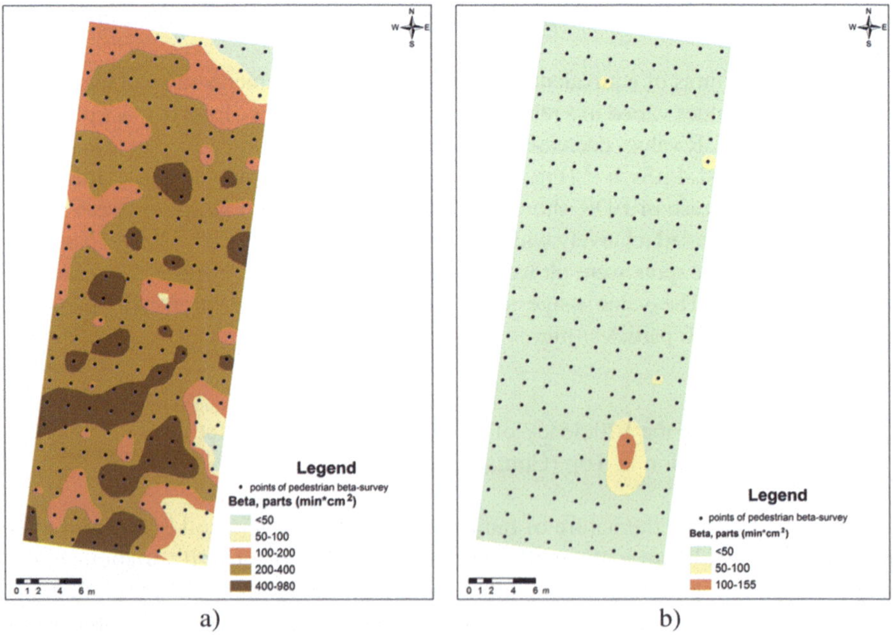

Fig. 4 Distribution nature of beta particle flux density in the territory before (**a**) and after (**b**) remediation activities by soil plowing

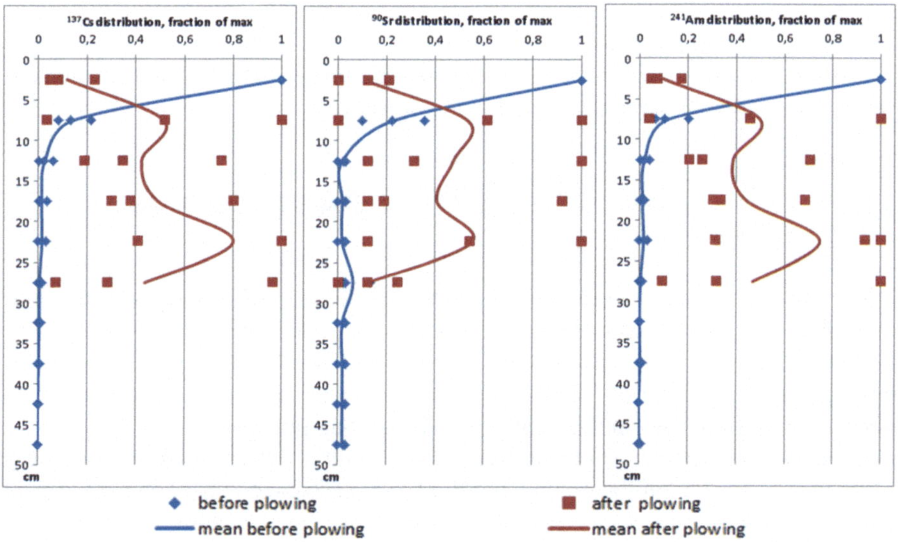

Fig. 5 Radionuclide distribution nature in soil before and after plowing at a depth of 30 cm by overturning, the fraction of a maximum value (mean based on the data on each of the 3 plots)

After plowing at a depth of 30 cm, radionuclide distribution nature in the 30 centimeter soil layer greatly changed. On the whole, a sharp reduction in activity concentration in top soil of 0–5 cm can be observed for all the radionuclides. At spot 1, the maximum for ^{137}Cs and ^{241}Am shifted to the 5–10 cm layer. For ^{90}Sr, the maximum value of activity concentration can be observed in 10–15 cm. A similar nature of radionuclide distribution in soil is also observed in other remediated spots of high activity concentrations. However, most of ^{137}Cs and ^{241}Am radionuclide content at U-2 spot fall on the two layers of 20–25 and 25–30 cm. For ^{90}Sr, the maximum value of activity concentration can be observed in the 5–10 cm layer. Maximum of all the radionuclides of interest at U-3 spot falls on the 20–25 cm layer.

2.1.2.3 Estimating Deflation Processes in Soil Cover of Key Spots

The extent of soil deflation is determined by changes in physical and chemical soil properties, and first of all, by the amount of humus and physical clay lost (Dzhanpeisov et al. 1990). Analysis results showed that during the observed period of time (2 years after remediation), an increase in sand soil fraction and a decrease in silt and pulverescent soil fractions occurred in all of the key spots. Loss of humus can be observed in the upper horizon of U-1, U-2 and U-3 spots. However, it is not time to speak of processes caused by blowing wind, because changes are observed not only in upper horizons but also in lower horizons. Perhaps, this is due to gravitational travel of silt fine fractions into the depth of soil cover with precipitation. The selected research technique did not enable to trace the variation in the mechanical composition of soil horizons below 15 cm.

A visual examination of spots showed signs of a forming crushed stone pavement which will further assist in preventing wind erosion.

2.1.2.4 Estimating Radionuclide Carry-Over by Plants into the Daylight Surface

Table 1 lists activity concentrations in plants of different root system, selected from one spot of 1 m^2. 2 species were considered- crested wheatgrass (*Agropyron cristatum*), having fibrous root system and sericeous sagebrush (*Artemisia sericea*), having taproot system. Both plant species were mown on the second and third years after plowing from one spot of 1 m^2 at U-1 site. Sampling points at the site were marked out at the spot with both species available within 1 m^2. Therefore, a sampling place may mismatch on the second and third year. Thus, a comparison can only be made here between these species without considering the dynamics.

The obtained data on radionuclide activity concentration in vegetation of researched spots before plowing, after 1 year, after 2 and 3 years shows that, on average, radionuclide activity concentration in plants of all spots located at P-2 site, decreased in the first year of plowing but the intake by plants continues (Table 2).

On the whole, there is an increase of radionuclide activity concentration at all spots on the 3rd year after remediating. The exception is U-2 spot backfilled with crushed stone where a rise in the activity concentration of ^{137}Cs and ^{241}Am radionuclides hardly occurs. The increase of $^{239+240}$Pu and ^{90}Sr is particularly noticeable at all spots. Such a sharp increase in radionuclide activity concentration on the third year after spots plowing is due to the fact that the plant root system achieved deeper buried radionuclides during plowing.

It should be noted that radionuclide activity concentration in plants on the 3rd year of research is far higher than before remediation activities. This may be due to the fact that before plowing, radionuclides were in the 0–5 cm top soil, and the bulk of roots are in the 0–25 layer. After plowing, radionuclides migrated into the soil depth, and, respectively, the area of the contacting root surface with contaminated soil increased.

When comparing plants with stalky and fibrous root systems, one can see that the type of root system has an ambiguous effect on radionuclide transfers (Table 2). Thus, ^{137}Cs and ^{90}Sr activity concentration in vegetation of fibrous root system is higher than in plants of taproot system. If for ^{90}Sr activity concentrations differ insignificantly, then values for ^{137}Cs are virtually 2 times different. $^{239+240}$Pu and ^{241}Am activity concentrations in vegetation of fibrous root system, on the contrary, are much lower than in plants of taproot system. At that, this tendency was traced on the 3rd and 4th year of research Thus, for $^{239+240}$Pu in 2014 and 2015 the difference was 2.6 and 3.7 times respectively. For ^{241}Am in 2014 and 2015–1.6 and 3.4 times respectively. Considering that the research territory is contaminated mainly with transuranic radionuclides, vegetation cover of fibrous root system is more preferable here.

Table 1 Radionuclide activity concentration in aboveground parts of the plants at key spots of different overgrowing stages

Sampling point	Radionuclide activity concentration, Bq kg^{-1}							
	[a]2012	2013	2014	2015	2012	2013	2014	2015
	^{241}Am				^{137}Cs			
P-2, U-1	13 ± 3	3.5 ± 0.7	3.2 ± 0.6	27 ± 5	3.5 ± 0.6	1.0 ± 0.2	0.9 ± 0.2	110 ± 20
P-2, U-2	4.3 ± 0.9	3.9 ± 0.8	3.3 ± 0.7	3.1 ± 0.6	9.4 ± 1.9	1.6 ± 0.3	1.4 ± 0.3	0.3 ± 0.06
P-2, U-3	7.1 ± 1.4	3.9 ± 0.8	26 ± 5	6.1 ± 1.2	2.4 ± 0.5	0.4 ± 0.08	2.1 ± 0.4	30 ± 6
	$^{239+240}$Pu				^{90}Sr			
P-2, U-1	160 ± 20	65 ± 6.3	110 ± 10	800 ± 40	7.6 ± 1.1	6.2 ± 0.9	8.9 ± 2	980 ± 150
P-2, U-2	180 ± 20	12 ± 1.1	32 ± 4	51 ± 6	11 ± 2	25 ± 4	8.5 ± 2	540 ± 80
P-2, U-3	26 ± 2	120 ± 11	94 ± 19	200 ± 20	5.1 ± 0.8	45 ± 7	13 ± 2	400 ± 60

[a]in 2012 plants were samples before plowing

Table 2 Radionuclide activity concentration in aboveground parts of the plants with different root systems

Plant species	Radionuclide activity concentration, Bq kg^{-1}			
	Point 1	Point 2	Point 1	Point 2
	^{241}Am		^{137}Cs	
Crested wheatgrass (*Agropyron cristatum*)	1.5 ± 0.3	7 ± 1,4	1.1 ± 0.2	1.5 ± 0.3
Sericeous sagebrush (*Artemisia sericea*)	2.4 ± 0.5	24 ± 5	1.2 ± 0.2	0.6 ± 0.12
	$^{239+240}$Pu		^{90}Sr	
Crested wheatgrass (*Agropyron cristatum*)	27 ± 3	130 ± 10	9.4 ± 2	1200 ± 200
Sericeous sagebrush (*Artemisia sericea*)	74 ± 16	480 ± 30	8.5 ± 2	710 ± 110

2.1.2.5 Estimating Processes of Remediated Spots Overgrowing with Vegetation

During natural overgrowing, in each of the three experiments, species of three florocenotypes are involved in succession– steppe, ruderal and halophytic. In revegetation, variation in total projective cover is the most noticeable in vegetation.

Thus, at U-1 site sown with crested wheatgrass, the projective cover on the 3rd year of overgrowing was 40–50%, on the 5th year 60–70%. At U-2 site backfilled with crushed stone, the projective cover was 30–40% on the 3rd year of overgrowing and 40–50% on the 5th. At U-3 site, with no additional cultivation, the projective cover was 20–20% on the 3rd year of overgrowing and 40–42% on the 5th.

At the spot of sown crested wheatgrass (U-3), phytocoenosis with one or several dominating sod grasses are formed faster. Thus, at the spot of no additional cultivation, 13 steppe plant species can be observed, at the spot backfilled with crushed stone – 8 steppe plant species and at the spot sown with crested wheatgrass, 22 steppe plant species are now observed.

Thus, proceeding from the data given, arguably, the most effective technology for vegetation to form fast and to create a sustained phytogenic medium at remediated spots, is a tehcnology that provide for sowing steppe grasses.

2.2 Remediating Radioactively Contaminated Areas of «Experimental Field» Ground by Plowing Soil

In 2014 activities began to remediate territories exposed to significant radioactive contamination. An area was selected for remediation in the plume of radioactive fallout on «Experimental Field» ground.[1] The most effective method is chosen to reduce and limit the surface contamination– a method of plowing contaminated areas

[1]Technical ground «Experimental Field» at STS – between 1949 and 1963, 30 aboveground and 86 atmospheric nuclear tests were conducted.

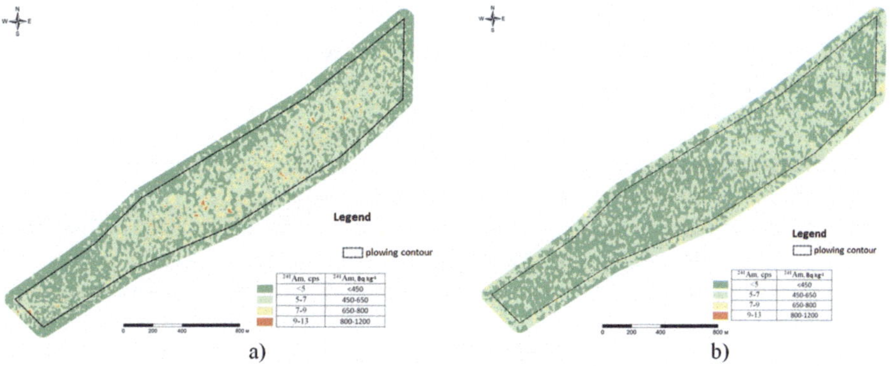

Fig. 6 Schematic maps of ^{241}Am distribution in the radioactive fallout plume area from a test at spot 14 of P-7 site before (**a**) and after (**b**) remediation activities

followed by harrowing. Plowing was made at a depth of 30–40 cm. The total area of a plowing area selected was 4 km^2.

Before plowing started, levels of surface contamination were estimated. Areas were selected based on available information on the content of the major dose-forming radionuclides like ^{137}Cs, ^{241}Am and $^{239+240}$Pu. Based on the data on the ^{241}Am and $^{239+240}$Pu content (by calculation) in the topsoil, remediation area was delineated (Fig. 6a) and a statement of work was prepared to perform remediation activities and the work on limitation of contamination spread in the area revealed.

$^{239+240}$Pu activity concentration in the topsoil of the selected area exceeded 3.8×10^3 Bq kg^{-1}.

In order to estimate the plowing efficiency, the following work was performed:

- Pedestrian γ-spectrometric survey which results are presented in figures (Fig. 6b);
- Soil sampling from analyses in laboratory to determine γ-, β- and α-emitting technogenic radionuclides.

Compared γ-survey results (Fig. 6b) allow a definite conclusion to be made that remediation activities performed significantly reduced the level of the major dose-forming radionuclide content in the surface soil and thereby reduced radiation hazard level for these territories and improved the environmental situation.

Based on results of field and in vitro measurements, it can be concluded that, on the whole, the remediation work performed in the area of 4 km^2 has significantly reduced surface areal contamination in the radioactive fallout plume. Mean values of radionuclide activity concentration in the top soil fell as low as 10 times which will diminish the probability of radionuclide redistribution owing to wind transfer as well as decrease levels of radionuclide intake by dominating plants in this region and thereby reduce the level of radiation hazard in the given territories, which will improve the environmental situation.

3 Remediation at Radioactively Contaminated Areas of «Experimental Field» Ground by Removing Residues of Nuclear Activity

One effective remediation method of territories exposed to radioactive contamination includes the removal of radioactive material.

The National Nuclear Center of the Republic of Kazakhstan in 2014 and in 2016 carried out the work to remove radioactively contaminated soil from radioactive «spots» detected on «Experimental Field» ground. «Spots» were relatively small in size, as large as 300 m^2, but with a high content of radionuclides. In view of small sizes of radioactive «spots», radioactively contaminated soil was manually removed.

The technology of the extraction process was adopted as follows.

At first, a «spot» was delineated, i.e. the boundaries of the site were determined, where soil removal was supposed to be. To do this, stakes were placed along the contour of the site.

Then soil was removed layer wise by placing it in the tractor bucket (Fig. 7).

After the first layer was removed, at spade's depth, measurements were performed with a field gamma spectrometer and, in case certain values of the total activity were exceeded, soil was removed at a deeper depth and in some cases beyond the originally defined contours of the site.

The removed soil was subjected to special tillage, which was supposed to prevent or at least significantly hamper the extraction of radioactive material from the ground. To do this, a special component was added to the soil followed by mixing and cementing the resulting mass. Once tilled, soil was placed in transport packaging containers (TPC) to be transported to a temporary storage area (Fig. 8). In total, by now, about 100 m^3 of radioactively contaminated soil has been removed. Figure 9 shows the radiation characteristics of plot before and after removing contaminated soil layer.

Fig. 7 Removal of radioactively contaminated soil

Fig. 8 Packing of contaminated soil into transport containers

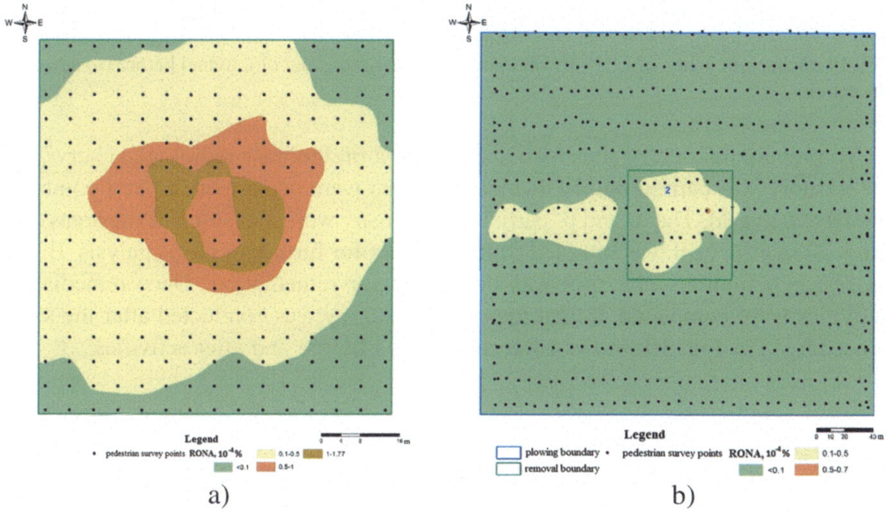

Fig. 9 Residues of nuclear activities (RONA) Plot 1 before (**a**) and after (**b**) removal, $10^{-4}\%$

4 Remediation in Radioactively Contaminated Areas by Filling with Clean Soil

In order to limit the spread of radioactive contamination at «Degelen» site in the period from 2011 to 2013, remediation activities were performed at six near-entry sites of tunnels with maximum parameters of radioactive contamination in soil at spots of 0.065 km^2 total areas. The work was carried out by layer-by-layer filling the contaminated spot with clean soil to decrease the dose rate on the surface as low as 10 μSv h^{-1}. The thickness of uncontaminated soil layer reached 20 cm.

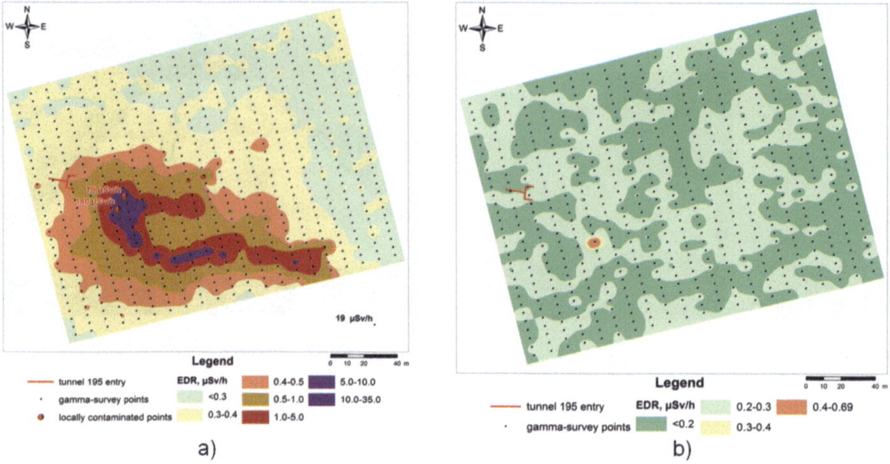

Fig. 10 Schematic maps of EDR distribution at the near-entry site of a tunnel before (**a**) and after (**b**) activities

To prepare near-entry sites for remediation, a preliminary radiological survey of the territory was carried out. EDR measurements were performed at fixed points on the 10×5 m grid. The α- and β- particle flux density were measured in site areas on a 20×20 m grid. Soil was sampled from site areas for further in vitro analysis. Based on research findings, the spot was delineated for remediation. Once remediation activities were completed, a final radiological survey was conducted after the same procedure. The Fig. 10, shows EDR distribution before and after activities.

5 Remediation Prospects in the Territory of Semipalatinsk Test Site

Rehabilitation of radioactively contaminated land at the test site will include extraction of radioactive waste followed by placing in a special storage area, deep plowing by overturning soil, creating physical barriers in the form of trenches and fencing with barbed wire, placing warning signs and creating radiation monitoring systems.

Aboveground nuclear tests conducted on «Experimental field» ground at STS (30 tests) resulted in contamination of both the site and the test site area, and in some cases outside STS at a distance of hundreds and thousands of kilometers. The territory of Experimental Field, in accordance with the Land Code, as a venue of nuclear weapon testing, cannot be transferred into ownership or land use. Therefore, remediation measures will be as follows. At epicenters of nuclear explosions, in the most contaminated areas, radioactive waste (with a volume of 1000 m^3) will be removed followed by transporting and dumping radioactive waste at the disposal site. Traces of radioactive fallout within the «Experimental Field» will be plowed (2500 ha area). At the same time, a significant part of the radioactive contamination will remain in the territory of «Experimental Field». To limit access to radioactively contaminated areas of «Experimental Field», physical barriers will built as trenches, fences with barbed wire and warning signs.

Other nuclear tests, which led to a significant contamination of the STS territory, are underground tests with soil outburst. As a result of these tests, millions of m^3 of radioactively contaminated soil was ejected; craters and traces of radioactive fallout were formed at a distance of several to tens of kilometers. The most contaminated areas are the «Atomic Lake» and 2 craters at the «Telkem» site, several craters (3) at «Sary-Uzen» site. Remediation at these sites does not seem possible due to the large volume of radioactively contaminated soil ejected, so rehabilitation will consist in plowing the territory adjacent to these sites, making trenches and placing a fence.

Underground nuclear tests at STS were conducted in tunnels of «Degelen» site (209 tests) and in boreholes at «Balapan», «Sary-Uzen» and «Telkem» sites (131 tests). Mostly, as a result of these tests, the products of nuclear explosions remained buried deeply underground. Nevertheless, there were situations when, after nuclear tests, radioactive substances were released into the atmosphere, which resulted in contamination of near-entry and near-mouth spots of tunnels and boreholes. Radioactive contamination of soil and vegetation cover is mainly localized in areas of hundreds of meters, in rare cases- several kilometers. Contamination levels at near-mouth spots commensurate with levels of radioactive waste.

At «Degelen» site, activities were previously performed to eliminate the infrastructure of nuclear tests, and work was also done to reduce the level of radioactive contamination. Across the perimeter of «Degelen» site, a physical protection system was created. This system proved itself to be effective as it completely restricts unauthorized access to the site area. Remediation within this site will be to backfill radioactively contaminated spots with clean soil at near-entry sites of tunnels. In future, security operations will be related to maintenance of the operability of the existing physical protection system.

It is planned to remove and dump about 1.5 thousand m^3 of radioactive waste in the storage facility at sites of underground nuclear tests «Balapan» and «Sary-Uzen» from the most contaminated spots. Near-mouth site areas will be remediated by plowing. Since the site area of «Balapan» and «Sary-Uzen» occupies a significant area, the access control system will only be created in contaminated near-mouth site areas. To this end, trenches and fences with barbed wire will be constructed.

Besides nuclear tests, there were radiological warfare agent tests at Semipalatinsk test site. Radioactive waste containing mostly ^{90}Sr, was brought to STS and dispersed (pulverized by different methods) within «4» and «4 a» sites. Such testing resulted in ground contamination at more than 30 spots within these sites. Level of radioactive contamination at some spot is commensurate with medium active radioactive waste. Radionuclide speciation in soil, their migration capacities and high levels of radioactive contamination make these spots the most hazardous to the natural environment in the STS territory and population who will later be engaged in economic activities on the test site land transferred to economic use. Therefore, the highest amount of remediation activities is planned at «4» and «4a» sites. AS a result of remediation, over 45,000 m^3 of radioactive waste (RAW) will be removed and dumped. Once RAW is removed, contaminated spots will be remediated by deep plowing. The access to contaminated spots will be restricted by physical barriers (trenches and fences of barbed wire). One mote site that poses a radiation hazard in a «mortuary» located 20 km away from Kurchatov city. As a result of earlier activities, 5 m^3 of radioactive waste were removed and transported followed by putting into storage at «Baikal-1». Estimated radioactive contamination of the «mortuary» made it possible to calculate the amount of radioactive waste left. The site remediation will consist in removing 1000 m^3 of radioactive waste and dumping it in a storage facility. The physical protection system in this area was created earlier. There are currently no radioactive waste repositories in the Republic of Kazakhstan that allows, dumping STS waste. Therefore, creation of a radioactive waste storage facility is planned directly in the test site territory, namely, on experimental field ground. Map construction to store radioactive waste of 50 m^3 will enable to dump all of the radioactive waste removed at the test site.

Thus, rehabilitation measures at STS will allow it to remove radioactively contaminated soil within the most radiation hazardous sites, make repositories and dump this waste there. In the rest of the territory– to reduce levels of radioactive contamination thereby mitigating radiation dose on the natural environment, create a system of physical barriers to restrict the access to radiation hazardous sites by the population as well as by wild and domestic animals. After rehabilitation activities, economic operations in the territory of the former test site beyond testing areas and fenced territories will reliably allow radiation safety assurance for the population and personnel. Radionuclide content in the natural environment and radiation burdens on population beyond controlled territories will not exceed RK regulatory standards. In the future, STS work will consist in maintaining the integrity of access restriction systems created (physical barriers) and creating a water and air radiation monitoring system which will allow radiation situation control at Semipalatinsk test site.

6 Conclusions

Estimation of plowing efficiency at spots contaminated with radionuclides showed that after the soil overturn to a depth of 30 cm, due to the movement of radionuclides from top soil to underlying horizons, a fairly great decrease in EDR levels and β-particle flux density is observed. Thus, the selected plowing technology is fairly effective in terms of reduced surface soil contamination in conditions of chestnut soils at Semipalatinsk test site.

Studies have shown that, despite the reduction in surface contamination of soil, radionuclides are carried out with vegetation into the daylight soil surface. This factor should be taken into account when remediating territories. In the course of studies, there was no univocal influence by the type of the root system on radionuclide transfer to plants. The technology including wheat cowing is the most preferable since in this case it is steppe vegetation that forms quite quickly, which prevents overgrowing with ruderal species that have predominantly taproot system capable of penetrating into the soil layer with deeper buried radionuclides.

Thus, based on the conducted studies, it was found that the most preferable remediation technology for spots contaminated with technogenic radionuclides in conditions of the steppe zone at Semipalatinsk test site is soil plowing to a depth of 30 cm with a 180° turnover followed by sowing with steppe vegetation.

Overall, a comprehensive environmental survey of the STS territory with the area of 7860 km^2 has been carried out. In order to eliminate testing consequences, activities began on remediation of most heavily contaminated spots on «Experimental Field» and «Degelen» sites. Contaminated soil with a volume of ~70 m^3 was removed followed by putting into long-term storage in «Baikal-1» repository. In the territories adjacent to these spots with an area of 4.05 km^2 the level of radioactive contamination was decreased by deep plowing. Remediation at «Degelen» site was carried out in the territory of the most contaminated near-entry sites of 40 tunnels in the area of 0.065 km^2. The remediation method is filling with clean soil. To limit access to venues of both nuclear tests on «Experimental Field» and «Degelen» sites and radiological warfare agents (RWA), access control systems were created at «4» and «4A» sites.

Further rehabilitation of radioactively contaminated land at the test site will include the extraction of radioactive waste followed by dumping in a special storage facility, deep plowing by overturning soil, creating physical barriers in the form of trenches and fencing with barbed wire, installing warning signs and developing radiation monitoring systems.

References

Aidarkhanov AO, Lukashenko SN, Lyakhova ON, Subbotin SB, Yakovenko YY, Genova SV, Aidarkhanova AK (2013) Mechanisms for surface contamination of soils and bottom sediments in the Shagan river zone within former Semipalatinsk nuclear test site. J Environ Radioact 124:163–170

Dzhanpeisov RD, Zonov GV, Smagulov T (1990) Effect of erosion and deflation on soil cover structure of semi-deserts and deserts of Kazakhstan. Nauka Publishing House, Alma-Ata, pp 28

Lyakhova ON, Lukashenko SN, Larionova NV, Tur YS (2012) Contamination mechanisms of air basin with tritium in venues of underground nuclear explosions at the former Semipalatinsk test site. J Environ Radioact 113:98–107

Panitskiy AV, Lukashenko SN (2015) Nature of radioactive contamination of components of ecosystems of streamflows from tunnels of Degelen massif. J Environ Radioact 144:32–40

Report on scientific and technical activities of the Institute of Radiation Safety and Ecology (2011) NNC RK made within the framework of measure 0346 «development of nuclear energy in the Republic of Kazakhstan« for 2010–2011: IRSE NNC RK; headed by Lukashenko SN, Report by STP – Kurchatov, p 98

Turchenko DV, Lukashenko SN, Aidarkhanov AO, Lyakhova ON (2014) Studying of tritium content in snowpack of Degelen mountain range. J Environ Radioact 132:1–6

Basic Radiation Protection Related to the Assessment of Remediation Measures in Radioactively Contaminated Areas

Jozef Sabol

1 Introduction

Current requirements for releases of radioactive substances into the environment are so strict that under normal circumstances, when regulations are observed, any excessive radioactive contamination is expected and thus the contribution to the total average exposure of the population due to these discharges is trivial and much lower than the natural background exposure. Just for a comparison and a realistic perception aimed at the assessment how much is too much, Fig. 1 illustrates the typical contributions of specific components to the total average exposure of individual members of the public worldwide (UNSCEAR 2011) and in the USA (NCRP 2009; UNSCEAR 2011). Similar situations can also be found in some other developed countries, and it is presumed that many other countries all over the word will soon be in a similar position. The exposure is given in terms of effective dose in mSv, which is directly related to radiation stochastic health effects. In any case, there has obviously been an enormous recent increase in exposure from the medical field.

This, however, does not mean that we do not need to pay appropriate attention to controlling exposure from man-made activities, where the radioactive releases and contamination of the environment should be minimized and optimized.

Currently, excessive radioactive contamination of the environment may occur as a result of incidents and accidents, causing uncontrollable contamination of the air, water and soil or the surface contamination of the terrain and other surfaces surrounding the place of radioactive releases, including those far away from the source of contamination since the air containing radioactive dust may transport radioactivity to very remote locations. In fact, the major portion of any radioactive contamination of the air will sooner or later be deposited on the surface of the earth somewhere. In all these cases the assessment of the magnitude of the exposure of persons affected

J. Sabol (✉)
Department of Crisis Management, PACR, Prague, Czech Republic

D. K. Gupta, A. Voronina (eds.), *Remediation Measures for Radioactively Contaminated Areas*, https://doi.org/10.1007/978-3-319-73398-2_14

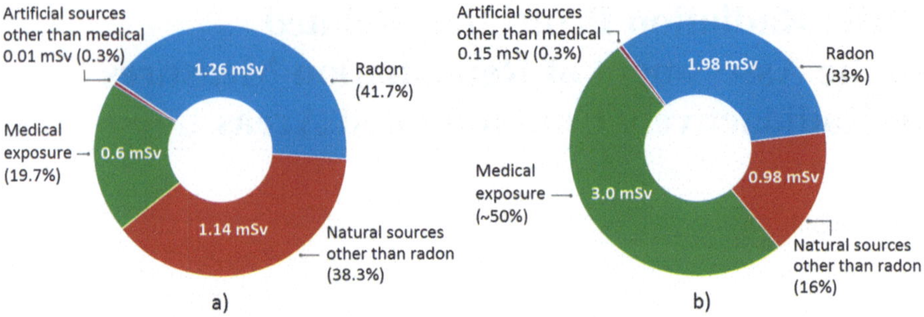

Fig. 1 Individual contributions to the overall average annual population exposure worldwide (UNSCEAR 2010) and in the USA (NCRP 2009)

requires the use of internationally accepted quantification which should be preferably consistent with quantities and units recommended by the ICRP (International Commission on Radiological Protection) (ICRP 2007) and ICRU (International Commission on Radiation Units and Measurements) (ICRU 2011).

Radiological compliance criteria for determining acceptable remediation actions at radioactively contaminated sites should also be expressed in the internationally recognized quantities and units. Any remediation should be justified taking into account that such intervention should result in the radiation exposure situation of the contaminated territory doing more good than harm. In another words, the remediation actions have to be optimized so that they would result in the likelihood of incurring exposures, the number of people exposed and the magnitude of their individual doses all being kept as low as reasonably achievable, taking into account economic and social factors (Voigt and Fesenko 2009).

Site characterisation based on the monitoring of radiation level and radioactive contamination, as the first necessary stage in the remediation of any contaminated environment, as well as all criteria in relevant regulations should be expressed in quantities and units which are unambiguously understood so that all interpret them in the same way and the data could be compared for the reliable evaluation of the results and the effectiveness of remediation actions. This is why a sound introduction of an appropriate system of quantities and units applicable to the evaluation of radioactive contamination and its impact on the exposure of persons is so important.

2 Quantification of Radioactive Sources and Radiation Fields

2.1 *Radioactive Sources*

All radioactive sources consist of one or more radionuclides, each of which is characterized by some unique properties such as the type of decay, type and energy of particles emitted, activity and half-life. It is well-known that a radioactive atom contains an unstable nucleus which is spontaneously disintegrating (decaying) into other atoms, releasing energy by the emission of ionizing radiation.

Although the amount of radioactive material can be expressed in its mass, a more appropriate measure of its ability to emit ionizing radiation is the ***activity (A)*** of radionuclides present. The activity of a radionuclide is related to the rate at which it decays i.e.

$$A = -\frac{dN}{dt} = \lambda\, N \cong \frac{\Delta N}{\Delta t} \tag{1}$$

where dN is the expectation value of the number of spontaneous nuclear transformations from the given energy state in the time interval dt and λ is the decay constant. The negative sign is needed here because N decreases as the time t increases. In practical use, the definition of this quantity, as well as in the definition of other similar quantities, the infinitesimal values dN and dt can be approximated by the finite delta values Δt (sufficiently small) and corresponding ΔN. A more detailed treatment of the activity and other radiation protection quantities and units can be found in a number of relevant publications, e.g. (Sabol and Weng 1995; Martin 2000; Martin et al. 2012).

The activity SI unit is *Bq* (becquerel) corresponding to one decay per second (s^{-1}). The old unit is Ci (curie) equals 37 GBq (1 GBq = 10^{12} Bq).

Radioactive sources may be represented formally in the form of point sources (where their size is very small so that they can be essentially considered negligible in relation to their surroundings) or the radioactive material may be distributed throughout a certain volume of the material or on its surface. In these cases, it may be useful to introduce such additional quantities as an activity concentration corresponding to the activity per unit of mass or volume containing certain radionuclides, or the activity may be related to the unit of an area on the surface of the contaminated material or ground.

Consequently, for the ***volume activity concentration (a_v), mass activity concentration (a_m)*** and ***surface activity concentration (a_s)*** the following relationships can be written

$$a_v = \frac{dA}{dV}, a_v = \frac{dA}{dV}, a_v = \frac{dA}{dV}, \tag{2}$$

where dA is the activity in the volume element of dV, in the mass element dm and on the surface of dS, respectively. The basic units of these quantities are Bq/m^3 (or Bq l $^{-1}$), Bq kg^{-1} and Bq/m^2.

The activity as a function of time can be expressed in the form

$$A(t) = A_0\, e^{-\lambda\, t} = A_0\, e^{-T_{1/2}} \tag{3}$$

where $A(t)$ is the activity A at the time t, A_0 is the activity at the beginning (at the time t_0) and $T_{1/2}$ is the half-life, i.e. the time during which the original activity falls to

Table 1 The half-life $T_{1/2}$ of some of the radionuclides

Radionuclide	Half-life $T_{1/2}$
U-238	$4{,}5 \times 10^9$ year
Cs-137	30.1 year
Co-60	5.3 year
P-32	14.3 day
I-131	8.0 day
Rn-222	3.8 day
Tc-99 m	6 h
F-18	110 min

one-half of its value. The half-life of some commonly used radionuclides is shown in Table 1.

Radioactive decay is a random process, and therefore the ***measurement of radioactivity*** must be treated with statistical methods. We have to observe a sufficient number of events (counts) with detectors to achieve an acceptable level of fluctuations (standard deviation) of the mean rate (counts/time). If the activity in the sample is low, we have to measure it for a longer time, increase the size of the sample, or use a more sensitive detector system to achieve satisfactory ***count statistics***.

The emission of radiation from a radionuclide is isotropic, which means that particles or photons are emitted in all directions with the same probability. However, this may not always be the case for all radiation sources, where sometimes the quantity ***emission rate*** (N_r) is more appropriate. This quantity is defined as

$$N_r = \frac{dN_e}{dt} \tag{4}$$

where dN_e is a number of particles or photons emitted by the source during the interval dt in all directions (into the 4π space). The unit of emission rate is s^{-1} equivalent to the number of particles emitted by the source 1 s^{-1}. Sometimes, it is useful to relate the emission to a selected direction, which is important for the characterization of radiation sources the emission of which is not homogenous.

2.2 Radiation Field

Around any radiation source, a radiation field is formed. A complete description of a radiation field requires the (particle) fluence distribution as a function of particle type, e.g. electrons, photons, neutrons, and their energy. The field can be characterized by such quantities as the fluence, fluence rate, energy fluence and energy fluence rate.

The ***fluence*** (Φ) is defined as the quotient of the number of particles dN incident on a sphere of cross-sectional area da

$$\Phi = \frac{dN}{da} \tag{5}$$

where *dN* is the number of particles incident on a sphere of cross-sectional area *da*. The use of a sphere expresses the fact that one considers the area perpendicular to the direction of each particle. By using a sphere, the area perpendicular to the direction of each particle is accounted for so that all particles passing through this volume of space are included. The basic unit of the fluence is m^{-2} (number of particle per square meter).

The ***energy fluence (Ψ)*** is defined as the energy *dE* of particles incident on a sphere of cross-sectional area *da*. If you have a fluence Φ of particles all of energy *E*, then the energy fluence is $\Psi = \Phi E$. The SI unit of energy fluence is J/m^2 or eV/m^2 (keV/m^2, MeV/m^2 etc.).

Both (particle) fluence and energy fluence can be related to the unit of time, thus defining two useful quantities, namely the ***fluence rate (φ)*** and ***energy fluence rate (ψ)*** introduced as

$$\varphi = \frac{d\Phi}{da}, \psi = \frac{d\Phi}{da} \tag{6}$$

The SI units of these field quantities are (m^{-2} s^{-1}) and J/(m^2 s^1), respectively. In practice, instead of J, a conventional energy unit eV (and its multiples) is often used, where the approximate conversion between eV and J is 1 eV $= 1.6 \times 10^{-19}$ J.

3 Interaction of Radiation with Matter

The interaction of radiation with matter is the most essential part of radiation physics, which is particularly ***important*** for the understanding of dosimetry, radiation protection, the biological effects of radiation, and also the interpretation of results obtained from detectors and monitors in terms of relevant quantities.

Radiation emitted by any radionuclide or radioactively contaminated material always forms a radiation field around such a source. If the nearby space is filled with a substance consisting of specific nuclides present, radiation will interact with their atoms and molecules. The processes will result in the absorption of radiation energy, causing ionization of the medium.

There is a substantial difference between the interaction of ***uncharged particles*** (indirectly ionizing radiation) and ***charged particles*** (directly ionizing radiation). While the fluence of indirectly ionizing radiation lies with the thickness of the absorber going down exponentially, on the other hand, the charged particles are losing their energy in small portions along their path and, eventually, at the end of the ***range*** they lose their entire energy. Because of these characteristic phenomena, in the case of indirectly ionizing radiation one can only refer to its ***attenuation***, or the reduction in terms of number of individual uncharged particles or photons depending

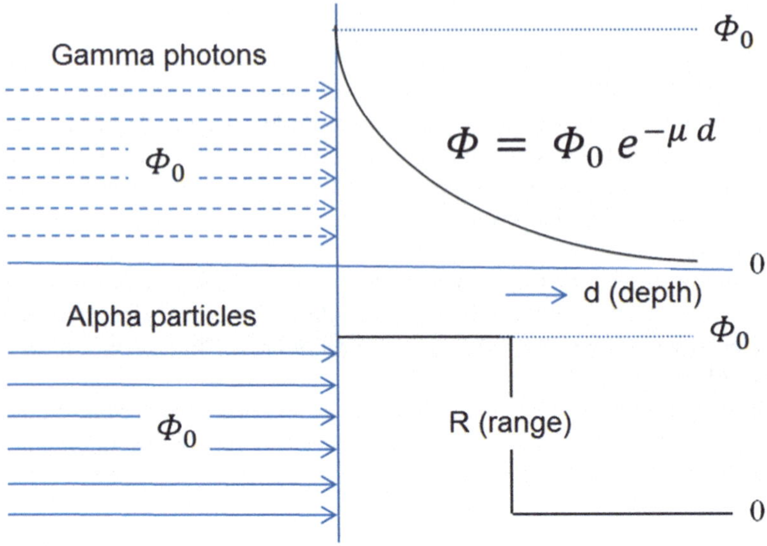

Fig. 2 Principal difference between the interaction of gamma photons as a typical representative of indirectly ionizing radiation (exponential decrease of their fluence) and alpha particles characterized by their definite range

on the thickness of the absorber. This is schematically illustrated in Fig. 2, where μ is the linear attenuation coefficient for gamma radiation depending on the energy of the photons and the composition of the absorbing material.

Any single uncharged particle or photon may interact with atoms and nuclei of the matter by several different major processes, each of which is characterized by a certain probability. Generally, on a certain distance a particle may not interact at all or it can be absorbed or scattered. In this case, one cannot predict the behavior of any individual particle which, in principle, may be absorbed at a very short distance in the material or it may travel some long distance without any interactions. This is why in the case of uncharged particles we cannot refer to their definite ***range*** but we can only specify the ***attenuation*** of a sufficient number of such particles.

X-ray and gamma radiations interact with matter through a variety of alternative mechanisms, the three most important of which are the ***photoelectric effect***, ***Compton scattering*** and ***pair production***.

In the ***photoelectric effect***, all the energy of a photon is transferred to an atomic electron located on the one of the atomic shells. The energy of the ejected electron, which then begins to pass through the surrounding matter, is sooner or later absorbed in the medium. Since the interaction creates a vacancy in one of the electron shells, typically the K or L, an electron moves down to fill it. The drop in energy of the filling electron produces a characteristic x-ray photon.

Conversely, when ***Compton scattering*** occurs, only part of the energy of the photon is transferred to an atomic electron, which then ionizes the medium around it. The scattered photon then continues with reduced energy. This photon leaves the site of the interaction in a direction different from that of the original photon.

If the energy of the incident photons is higher than the so-called threshold energy of 1.02 MeV, a ***positron-electron pair*** is produced in the intense electric field close to a nucleus. Any energy above the threshold energy is, according to the conservation law, split between the kinetic energy of these two particles. The positron is the anti-particle of the electron, so when a positron comes to rest, it interacts with another electron, resulting in the annihilation of both particles; this process is associated with the emission of two oppositely directed 0.511 MeV photons. The probability of pair production increases approximately with the square of atomic number Z of the absorber and increases with photon energy.

The result of these interactions is the transfer of the entire or a substantial portion of the energy of the incident photon to an electron and the scatter of the original photons. In other words, one cannot completely stop all photons in the beam; one can only reduce their number and this depends on the energy of the photons as well as on the type and thickness of the absorbing materials where a specific quantity, the ***half-value layer (HVL)***, is a useful characteristic parameter of the properties of the material in terms of its ability to attenuate a photon beam. The half-value layer corresponds to such thickness of the material which will reduce the number of original photons to the half of their number. It can be expressed as

$$HVL = \frac{\lg 2}{\mu} \cong \frac{0.693}{\mu} \tag{7}$$

where HVL is in m (meter) when the linear attenuation coefficient μ is given in m^{-1}.

A similar situation is also found with other commonly encountered indirectly ionizing radiation. Neutrons, which essentially interact only with the atomic nucleus, may undergo a variety of processes; the probability (cross-section) of each process occurring depends on the neutron energy and the composition of the material. In one of these interactions a new radioactive nucleus is produced by the absorption of a neutron by a stable nucleus of the material. Many man-made radionuclides are produced by the ***neutron activation*** of a suitable sample of relevant material.

Interaction coefficients for ***indirectly ionizing radiation***, in addition to the linear attenuation coefficient μ, include also the ***mass attenuation coefficient (μ/ρ)***, ***mass energy transfer coefficient (μ_{tr}/ρ)***, **mass energy transfer coefficients (μ_{tr}/ρ)** and **mass energy absorption coefficients (μ_{en}/ρ)** which are mainly used for photons, while for neutrons the ***cross section (σ)*** is usually used. The values of these coefficients depend on the type and energy of radiation, and on the composition of the material in which the interactions take place.

The above mentioned interaction coefficients are given by the following relationships

$$\frac{\mu}{\rho} = \frac{1}{\rho\, dl}\frac{dN}{N}, \quad \frac{\mu_{tr}}{\rho} = \frac{1}{\rho}\frac{dR_{tr}}{R}, \quad \frac{\mu_{en}}{\rho} = \frac{1}{\rho}\frac{dR_{tr}}{R}(1-g), \sigma\frac{P}{\Phi} \tag{8}$$

where dN/N is the fraction of particles that experience interactions in traversing a distance dl in the material of density ρ, dR_{tr}/R is the fraction of incident radiant energy that is transferred as kinetic energy of charged particles by interactions in traversing a distance dl in a material of density ρ, g is the fraction of the energy of liberated charged particles that is lost in radiative processes in the material, and P is the probability of a particular interaction for a single target entity when subjected to the particle fluence Φ.

The SI unit of the first three interaction coefficients stated above is $m^2\, kg^{-1}$, while the cross section is given in the unit m^2.

High-energy photons transfer their energy to matter in complex interactions with atoms, nuclei, and electrons. For practical purposes, however, these interactions can be viewed as simple collisions between a photon and a target atom, nucleus, or electron.

Charged particles, as ***directly ionizing radiation*** (electrons, positrons, protons, heavy ions), ionize the matter directly by removing outer electrons from originally neutral atoms, which then become positive ions. In contrast to charged particles, uncharged particles or photons can ionize the matter only indirectly through the ***secondary charged particles*** formed as a result of the interaction processes of this radiation with matter.

In the case of ***directly ionizing*** radiation, the main ***interaction coefficients*** are ***linear stopping power*** **(S_l)**, ***mass stopping power*** **(S_m)**, ***linear energy transfer*** **(L)** and ***mean energy expended in a gas per ion pair production*** **(W_i)**. These quantities are defined by the following equations

$$S_l = -\frac{dE_k}{dl}, S_m = \frac{S_l}{\rho}, \quad W_i = \frac{E_c}{N_i} \tag{9}$$

where dE_k is the kinetic energy lost by the charged particle on the distance dl (i.e. S_l is the energy loss per unit path length of the charged particle), ρ is the density of the material, E_c is the initial energy of the particle and N_i is the number of ion pairs produced by this particle.

In the assessment of the biological effects of ionizing radiation, the quantity ***linear energy*** **transfer** ***(LET)*** of charged particles in a medium plays an important role. This quantity corresponds to the quotient dE/dl, where dE is the average energy ***locally*** imparted to the medium by a charged particle of specified energy in traversing a distance of dl. One has to distinguish between the LET and the linear stopping power (formally defined in a similar way): Stopping power is closely related to LET except that LET does not include radiative losses of energy (which are lost to the medium and so not absorbed) since the LET is related to biological damage. The severities of biological changes are directly related to the local rate of energy deposition along the particle track.

Charged particles (alphas, betas, positrons and heavy ions) interact with the electrons of atoms by transferring gradually some small portion of their kinetic energy to these electrons. If charged particles possess sufficient energy they may also transfer enough energy to an electron (generally one in an outer shell) to eject the electron from the atom, which is in this way ionized. When radiation causes the ejection of an electron from an atom of an absorber, the resulting positively charged atom and free negatively charged electron are called an ion pair. The process is known as ***ionization***. The amount of energy transferred per ion pair created, W_i, is characteristic of the materials in the absorber. For example, approximately 33 eV (25–40 eV) is transferred to the absorber for each ion pair created in air or water.

It is often convenient to refer to the number of ion pairs created per unit distance the radiation travels as its ***specific ionization***. Heavier particles with more charge (alpha particles) have a higher specific ionization than lighter particles with less charge (electrons). Linear energy transfer is related to specific ionization: alpha particles are classified as high-LET radiation, and beta particles and photons as low-LET radiation.

In traversing an absorber, an electron loses its energy at each interaction with the atoms of the absorber. The energy loss per interaction is variable. Therefore, the total distance travelled by electrons of the same energy can vary by as much as 3–4%. This variation in range is called the straggling of the range. The heavier alpha particles are not affected to a significant degree and demonstrate very little straggling of range.

Both alpha and beta particles lose energy mainly through interactions with atomic electrons in the absorbing medium. The energy transferred to the electrons causes them either to be excited to a higher energy level (excitation) or separated entirely from the parent atom (ionization). Another important effect is that when charged particles are slowed down very rapidly, they emit energy in the form of x-rays. This is known as ***bremsstrahlung*** (braking radiation) and is of practical importance only in the case of beta radiation.

As has already been mentioned, the ***range*** of a charged particle is the distance it travels before coming to rest. It is assumed that the particle loses its energy in a continuous way and at a rate equal to the linear stopping power S_l. Since the stopping power is the energy loss of the particle per unit path, the range can be calculated by

$$R(E_i) = \int_{E_0}^{E_i} \frac{dE}{S_l(E)} \tag{10}$$

where E_i is the initial energy of the charged particle, E_0 is the energy where the particle is effectively absorbed. The range is the path length travelled by the particle. Since the particle's path is usually not straight due to scattering, its actual range is always longer than the projected range, which is the distance between the point where the particle enters the stopping medium and the point where the particle is absorbed (or comes to rest), projected onto the original direction of travel. The

Fig. 3 Ranges in cm of electrons, protons and alpha particles in air

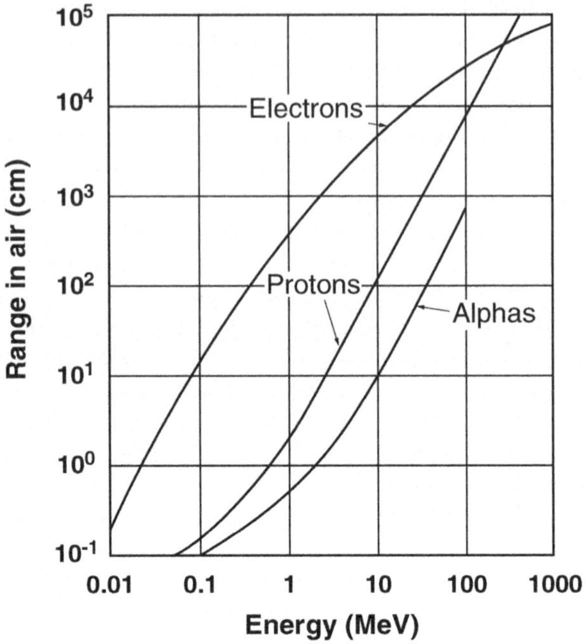

ranges of electrons, positrons and alpha particles in air under standard temperature and pressure are illustrated in Fig. 3.

4 Dosimetric Quantities and Units

At the beginning of the radiation era, for the assessment of biological consequences of radiation exposure, some simple physical quantities were used. These quantities were successfully evolved into more sophisticated physical ***dosimetric quantities*** which were gradually modified using various appropriate factors and converted into what we now know as ***radiation protection quantities.*** These current quantities more accurately approximate the biological effects of ionizing radiation at lower levels, where only stochastic effects occur.

4.1 Exposure

The unit *roentgen* (*R*) was the first unit introduced in 1928. The original definition has changed several times before coming to the present concept of the quantity *exposure*, the unit of which is now C kg^{-1} instead of R (roentgen).

The current definition of the ***exposure*** (***X***), the ***exposure rate*** (***dX/dt***) and the relationship of the exposure with the quantities characterizing the photon radiation field (Ψ) and the interaction of photon radiation (including x-radiation, γ-radiation or annihilation radiation) with the air is as follows

$$X = \frac{dQ}{dm}, \quad \dot{X} = \frac{dX}{dm}, \quad X = \Psi \left(\frac{\mu_{en}}{\rho}\right)_{air} \frac{e}{W_{i,air}} \tag{11}$$

where dQ is the absolute value of the total charge of the ions of one sign produced in air when all the electrons and positrons liberated or created by photons in air of mass dm are completely stopped in air, Ψ is the energy fluence, μ_{en}/ρ is the mass energy absorption coefficient, $W_{i,air}$ is the average energy required to produce an ion pair in air and e is the electronic charge.

The SI unit of exposure and exposure rate is C kg^{-1} and W kg^{-1}, respectively. The old (traditional) unit of the exposure is related to the SI unit as 1 R = 2.54 × 10^{-4} C kg^{-1}.

4.2 Absorbed Dose

The term ***dose of radiation*** was initially used in a pharmacological sense that means it is analogous to its meaning when used in prescribing a ***dose of medicine***. Radiation dosimetry is a now a pure ***physical science.***

The quantity of the ***absorbed dose*** (***D***), often only dose, is the most universal dosimetry quantity; it can be used for any radiation and any material or substance. The quantities *absorbed dose* (or sometimes referred to as only the dose), the ***dose rate*** (***dD/dt***) and the expression of the dose at the point of interest for indirectly ionizing radiation (D_i) and directly ionizing radiation (D_d) can be written in the following forms

$$D = \frac{d\bar{\varepsilon}}{dm}, \quad \dot{D} = \frac{d\bar{\varepsilon}}{dm}, \quad D_i = \Psi \left(\frac{\mu_{en}}{\rho}\right)_m, \quad D_d = \Phi E \left(\frac{S}{\rho}\right)_m \tag{12}$$

where $d\bar{\varepsilon}$ is the mean energy imparted to the matter of mass dm. The SI units of the dose and the dose rate are gray (Gy) and gray per hour (Gy h^{-1}). In practice, units as mGy, μGy and mGy h^{-1} or μGy h^{-1} are also used quite often. The relationship of the main SI unit to the old unit of the dose is 1 Gy = 100 rad.

It should be emphasized that in the definition of the dose the energy imparted is not necessarily only the energy absorbed since the energy ε is the sum of all energy deposits in the volume considered. The energy imparted and energy deposit can be expressed in the form

$$\varepsilon = \sum_i \varepsilon_i, \quad \varepsilon_i = \varepsilon_{in} - \varepsilon_{out} + Q_r \tag{13}$$

where ε_i is the energy deposit in a single interaction i, ε_i is the energy of the incident ionizing particle (excluding rest energy), ε_{out} is the sum of the energies of all ionizing particles leaving the interaction (excluding rest energy), Q_r is the change in the rest energies of the nucleus and of all particles involved in the interaction ($Q_r > 0$, decrease of rest energy, and $Q_r < 0$, increase of rest energy).

For a differential fluence $\Phi(E)$ of identical charged particles the dose D is given by

$$D = \int \frac{S_l(E)}{\rho} \, \Phi(E) \, dE \tag{14}$$

where $S_l(E)/\rho)$ is the tabulated mass stopping power of the charged particle as a function of energy E for the considered material.

4.3 Kerma

This quantity kerma, which is the acronym for Kinetic Energy Released per unit Mass, can only be used for indirectly ionizing radiation in any matter. The ***kerma (K)*** is defined as the following quotient

$$K = \frac{dE_{tr}}{dm} \tag{15}$$

where dE_{tr} is the sum of the initial kinetic energies of all the charged particles liberated by uncharged particles in a mass dm of material. The medium should always be specified. There are various primary standards to realise K for various particle types and energies.

The special name for the unit of kerma is gray (Gy); the unit for the kerma and dose is the same, namely gray (Gy). The older unit which is no longer supposed to be used was rad, where 1 Gy = 100 rad.

For a fluence $\Phi(E)$ of uncharged particles of energy E, the kerma (K) in a specified material and the relationship between the kerma in air (K_{air}) and the exposure (X) are given by

$$K = \int \Phi(E) \, E \, \frac{\mu_{tr}(E)}{\rho} \, dE, \quad K_{air} = \frac{W_{i,air}}{e\,(1-g)} \, X \tag{16}$$

The average fraction of the energy transferred to electrons that is lost through radiative processes is represented by a factor referred to as the radiative fraction g. Analogically one can derive a formula expressing the dose D_m in a given material m at a point where the air kerma is K_{air}.

The ***air kerma rate constant*** is often used to characterize gamma emitting radioactive sources in terms of their ability to produce the kerma rate at a distance of 1 m from a point source related to its unit activity.

The kerma is usually expressed in terms of the distribution $\Phi(E)$ of the uncharged particle fluence with respect to energy. The kerma K is then given by

$$K = \int \left(\frac{\mu_{tr}}{\rho}\right) \Phi(\mathrm{E})\mathrm{EdE} \tag{17}$$

where (μ_{tr}/ρ) is the tabulated mass energy transfer coefficient of the material for uncharged particles of energy E.

As the mass of a sample decreases in general, the energy per unit mass will become more random (stochastic). The energy imparted per unit mass can still be defined in region z, but the definition of absorbed dose implies an averaging to give D (a non-stochastic quantity).

All above-mentioned quantities can be related to the unit of time, dX/dt, dK/dt and dD/dt, which are then called the exposure rate, kerma rate and dose rate with the units of $A.kg^{-1}$ and $Gy.s^{-1}$, respectively.

5 Radiation Protection Quantities

The main aim of radiation protection is to ensure adequate protection of persons including radiation workers, patients and members of the public as well as satisfactory protection of the environment. Recently, due attention is also being paid to the security of high activity radioactive sources which may be misused for radiological terrorism.

The ICRP in its latest general recommendations (ICRP 2007) redefined or modified some quantities which had been previously recommended for specification in terms of the limits on exposure to external radiation and to intakes of radionuclides. In addition, the ICRU (ICRU 2011) has defined a set of operational dose equivalent quantities to be used in radiation protection measurements of external radiation.

5.1 Dose Equivalent and Dose Equivalent Rate

The ***dose equivalent*** (***H***) represents one of the initial quantities in radiation protection. It is the point quantity obtained by multiplying the ***dose*** (***D***) at a point in tissue by the ***quality factor*** (Q). The modifying factor Q depends on the unrestricted linear energy transfer (L) in water of the charged particles responsible for the dose. The dose equivalent and dose equivalent rate at a point are defined by the relationships

$$H = \int_L Q(L)D(L)dL, \quad \dot{H} = \frac{dH}{dt} \tag{18}$$

where $Q(L)$ and $D(L)$ are the quality factor and the dose as a function of L, respectively. The unit of the dose equivalent is sievert (Sv) which corresponds to J kg^{-1} (multiplied by Q). In a simplified manner the definition of the dose equivalent can be written as

$$H = D\,Q \tag{19}$$

where D is the absorbed dose and Q is the mean value of the quality factor for the specific radiation at this point.

5.2 Equivalent Dose and Committed Equivalent Dose

The ***equivalent dose*** $(\boldsymbol{H_T})$ is specified as the summation of ***doses*** $(\boldsymbol{D_{T,R}})$ averaged over a tissue or organ T due to radiations of type R incident on the body or emitted by radionuclides in the body and weighted by ***radiation weighting factors*** $(\boldsymbol{w_R})$.

The ***committed equivalent dose*** $\boldsymbol{H_T(\tau)}$ is the time integral of the equivalent dose rate in a particular tissue or organ that will be received by an individual following intake of radioactive material into the body by a Reference Person (an idealized person for whom the organ or tissue equivalent doses are calculated), where τ is the integration time in years (50 years for adults or 70 years in the case of children). This quantity reflects the contribution of the internal exposure to the total equivalent dose.

The equivalent dose H_T, mean dose D_T and committed equivalent dose $H_T(\tau)$ are given by

$$H_T = \sum_R w_R\, D_{T,R}, \quad D_T = \frac{1}{m_T}\int_{m_T} D\, dm, \quad H_T(\tau) = \int_{t_0}^{t_0+\tau} \frac{d\,H_T}{dt}\, dt \tag{20}$$

where m_T is the mass of the tissue or organ and D is the dose in the mass element dm. In fact, the dose D_T equals the ratio of the energy imparted ε_T to the tissue or organ and the mass of this tissue or organ m_T. The integration time τ follows the intake at time t_0.

Since the radiation weighting factor is considered to be a dimensionless factor, the unit of both the equivalent dose and committed equivalent dose is Sv (provided the dose is in Gy).

Table 2 Recommended radiation weighting factors (w_R) and tissue weighting factors (w_T)

Type of radiation	w_R	Tissue or organ	w_T
Photons, all energies	1	Bone marrow (red), colon, lung, stomach, breast, remainder tissues[1]	0.12
Electrons, all energies	1	Gonads	0.08
Protons	2	Bladder, oesophagus, liver, thyroid	0.04
Alpha particles, fission fragments, heavy ions	20	Bone surface, brain, salivary glands, skin	0.01

[1]Note: Adrenals, brain, small intestine, kidney, muscle, pancreas, spleen, thymus, uterus.

5.3 *Effective Dose and Committed Effective Dose*

The *effective dose* (*E*) is the main quantity in radiation protection for the assessment of biological effects at low doses. It has been defined only for stochastic effects. The effective dose is defined as:

$$E = \sum_T w_T \sum_R w_R D_{T,R} \tag{21}$$

The contribution to the total effective dose due to intakes of radionuclides is quantified by the *committed effective dose* ($E(\tau)$) which is defined as the sum of the products of the committed equivalent doses and the appropriate tissue weighting factors w_T. The basic unit for both the above-mentioned quantities is Sv. The values of weighting factors w_R and w_T are given in the latest recommendations of the ICRP (Table 2).

5.4 *Assessment of External Exposure*

The body-related radiation protection quantities such as the equivalent dose and effective dose are, due to their nature, not directly measurable and so they cannot be used directly as quantities in radiation monitoring. This is why so-called ***operational quantities*** have been introduced by the ICRU.

The purpose of these quantities is to provide an estimate or upper limit for the value of the protection quantities related to an exposure, or potential exposure of persons under most irradiation conditions. Operational quantities are often used in practical regulations or guidance aimed at controlling external exposures. Individual operational quantities (which are point quantities) can be defined as follows:

The ***ambient dose equivalent $H^*(10)$***, at a point in a radiation field, is numerically equal to the dose equivalent that would be produced by the corresponding expanded and aligned field a 30 cm tissue-equivalent ICRU sphere at a depth of 10 mm on a radius vector opposing the direction of the aligned field. It should be understood/ noted that the ambient dose equivalent is now the only quantity to be applied for the measurement of external radiation.

The ***directional dose equivalent*** $H'(0.07, \Omega)$, which is intended for the assessment of low-penetrating radiation, is the dose equivalent that would be produced by the corresponding radiation field in the ICRU sphere at a depth 0.07 mm on a radius in a specified direction Ω.

The ***personal dose equivalent*** $H_p(d)$ is the dose equivalent in soft tissue at an appropriate depth d below a specified point on the body. For strongly penetrating radiation a depth of 10 mm is usually used while for weakly penetrating radiation a depth of 0.07 mm is normally employed.

The basic unit of all operational quantities is Sv, which also serves as a unit for some other radiation protection quantities.

5.5 Assessment of Internal Exposure

While the contribution of external radiation to the total effective dose is related to the period during which a person is exposed, in case of internal exposure its contribution may be spread over a much longer period of time following intakes by inhalation or ingestion. Moreover, there are no operational quantities for the direct assessment of internal exposure. Any assessment due to the intake of radionuclides relies first on the evaluation of the intake of a radionuclide, based typically either on direct measurement of the radioactivity of the body (e.g., by a whole-body counter or external detectors measuring the radioactivity of specific organs and tissues) or indirect measurements (e.g., measuring the activity of radionuclides in urine, faeces, air or other environmental samples).

Once the intake is known, the effective dose is calculated using reference dose coefficients recommended by the ICRP; these coefficients have been adopted by the EU (Basic Safety Standards Directive) and IAEA (International Basic Safety Standards).

5.6 Radiation Source as a Cause of the Exposure of Persons

Many attempts have been made to define a universal quantity to quantify the exposure to a person in order to assess the probability of detriment regardless of the type of radiation (or irradiation conditions), including nonhomogeneous exposure of individual organs. The quantity of the effective dose has been adopted for this noble purpose and extensive effort has been made to associate its value with whole-body stochastic effects. In principle, this has been achieved by the introduction of relevant weighting factors – the radiation weighting factor and the tissue weighting factor, which convert a physical quantity - the absorbed dose - into a bio-physical quantity expressed in Sv. These factors, based largely on epidemiological studies, have often been changed (and presumably this process will continue also in the future) in order to match the value of the effective dose with the impact of the

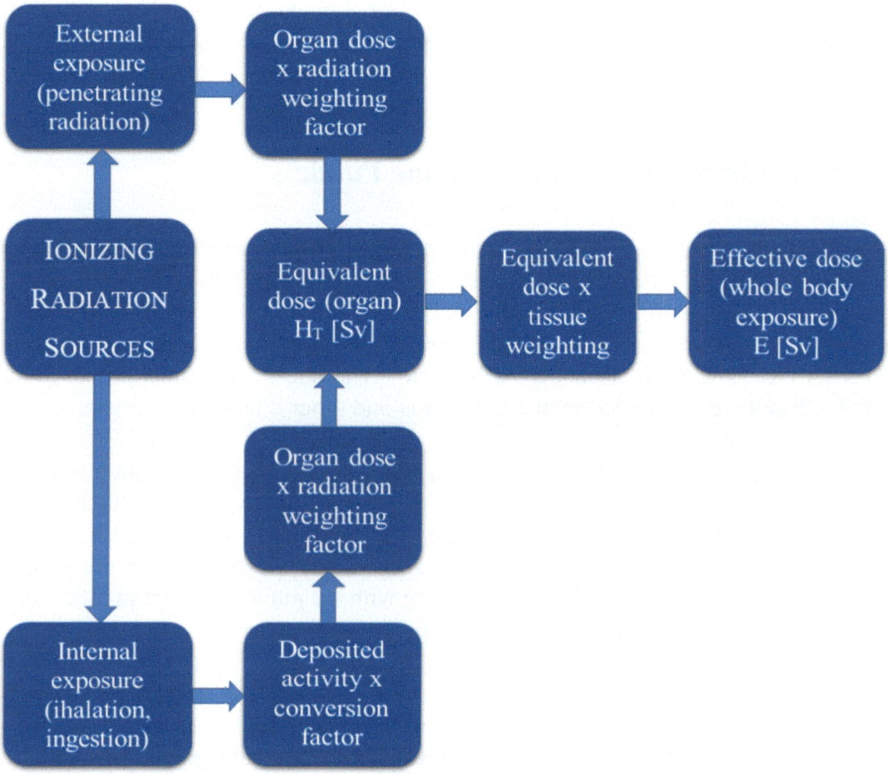

Fig. 4 Illustration of ionizing radiation sources and the resulting exposure in terms of the effective dose reflecting the total stochastic effects due to external and internal exposure

exposure in terms of stochastic effects based on the latest scientific findings. A similar concept, relying on the quantity of equivalent dose, has also been adopted for the stochastic effects occurring in an individual organ or tissue exposed to ionizing radiation.

In principle, the stochastic dose has to consider both components, namely from external and internal exposure (Fig. 4).

The present system has been developed through many phases reflecting available information and scientific knowledge at that time. The primary aim of the system was based on the effort to fully assess harmful stochastic effects. Obviously one had to have considered some important biological factors related to the harm caused by different types of ionizing radiations in individual tissues or organs. Consequently, the quantity, which would adequately represent stochastic effects, would no longer be a physical quantity, but rather a bio-physical quantity, which apparently could not be assessed using only pure physical principles. This is why the effective dose as well as the equivalent dose cannot be measured directly. These quantities can only be approximated using specifically defined operational quantities. Even here one experiences some problems in measuring or monitoring operational quantities directly.

6 Dose Limits for Workers and the Public

The adequate protection of radiation workers (professionally exposed) and members of the public is ensured by keeping radiation exposures not only below recommended dose limits but at the same time reducing the exposure to a very minimum in accordance with one of basic principles of radiation protection – ***ALARA*** (As Low As Reasonably Achievable). The latest dose limits proposed by the ICRP and recommended by the IAEA, EU and other international organizations are given in Table 3.

All professionals in radiation protection, including the users of radiation sources, should be familiar with the average values of the exposure to natural sources, dose limits, and at least be approximately aware of the exposure due to most important man-made sources and their applications. They are supposed to make use of this information when dealing and communicating with the public. In order to establish a good communication atmosphere, no complex technical or scientific terminology should be used. Under normal conditions, in accordance with the basic radiation protection requirements, the exposure is not supposed to exceed the above-mentioned dose limits. On the contrary, everything possible has to be done to keep actual exposures at the lowest achievable level well below those limits. However, in the case of an accident, some guidance levels have been recommended in order to restrict the exposure of emergency workers. In such situations the guidance values for restricting exposure to external penetrating radiation of emergency workers, who are engaged in life saving actions or actions to prevent severe deterministic effects and actions to prevent the development of catastrophic conditions that could significantly affect people and the environment, can reach up to 500 mSv.

Table 3 Recommended dose limits in planned exposure situations

Type of limit	Annual occupational exposure	Annual public exposure
Effective dose	20 mSv[a]	1 mSv[b]
Equivalent dose to lens of the eye	20 mSv	15 mSv
Equivalent dose to the skin	500 mSv	50 mSv
Equivalent dose to hands and feet	500 mSv	–

Notes:
[a]Averaged over 5 years, with no more than 50 mSv in any one year
[b]Exceptionally, a higher value of effective dose could be allowed in a year provided that the average over 5 years does not exceed 1 mSv in a year

7 Biological Effects of Radiation Exposure

It is well known that when radiation undergoes an interaction processes, it can deposit some energy in the material either directly (by charged particles) or indirectly (by secondary charged particles produced by uncharged particles). This energy usually results in ionization and excitation of atoms and molecules. If living organs or tissues are exposed to radiation, some specific biological effects may appear which, at low exposure levels, may induce cancer (with a certain probability proportional to the amount of exposure), or some biological damage as long as the exposure is above a certain threshold level. The first type of biological effects is referred to as ***stochastic effects*** (they may appear in an exposed individual or not), while the second type of effects are called ***deterministic effects*** (they will surely develop in an individual who has received high exposure).

At low exposure levels, no visible or clinical effects can be recognized following the exposure of a person. However, deterministic effects (known now as harmful tissue reactions) occur when sufficient cells are killed or changed following high exposure above a certain dose, the so-called threshold level, i.e. a level below which a visible health effect is absent, but above which the effect is expected with certainty. This means that a person exposed to such high doses will be directly affected and the severity will increase with increasing doses.

It has been recognized that high-LET radiations, e.g., alpha particles, heavy ions and also neutrons, produce greater damage per unit of absorbed dose than radiations characterized by low LET, namely photons and electrons or positrons. For the quantification of low-LET radiations the dose is completely adequate but for high-LET radiations an RBE-weighted dose has been proposed. This quantity, proposed by the ICRP and adopted by the IAEA for the evaluation of emergency exposure, is defined as the product of the averaged absorbed dose $D_{T,R}$ due to the radiation of type R in an organ or tissue T and the relative biological effectiveness $RBE_{T,R}$ related to the tissue or organ T and radiation R. In order to differentiate the unit of the dose (Gy) from that associated with the RBE-weighted dose, a unit equivalent/gray (Eq-Gy) has been attributed to this quantity.

The IAEA has developed an International Nuclear and Radiological Event Scale (INES) (IAEA 2009) for ease in comparing various nuclear accidents, taking into account many factors reflecting the impact of such accidents (Fig. 5). As with most things, it is not perfect but it serves the purpose at least partially of meeting the needs of public awareness in a broad sense. The difficulties in using this scale became apparent following the Fukushima accident (Thielen 2012), where there was some hesitation concerning classifying it in accordance with the INES. Formally, it should fall into category 7, but everything suggests that the Fukushima impact was much lower than that of Chernobyl (Balonov 2012, 2013), which was classified as an event at the same level 7 (no higher ranking).

The public would welcome something as simple and illustrative as the Richter scale, which reflects the magnitude of earthquakes with its non-linear representation of the massiveness of the event rather than its consequences. In the case of nuclear

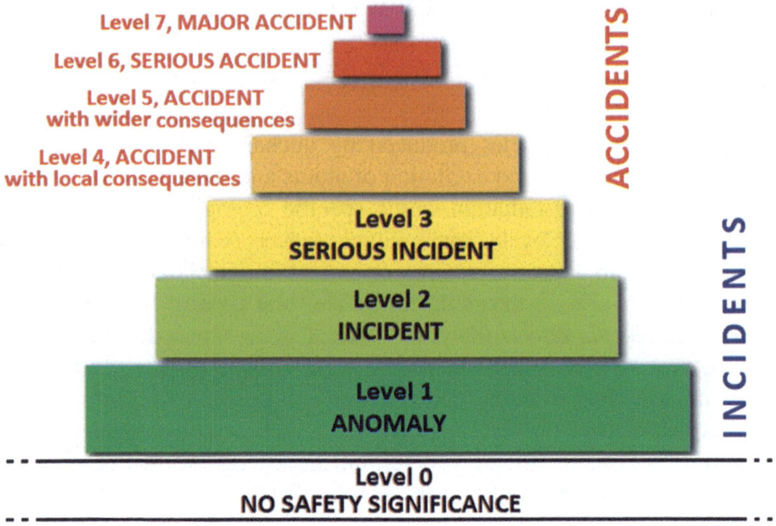

Fig. 5 An illustration of individual levels of the IAEA categorization of nuclear and radiation incidents and accidents in accordance with the INES

accidents, the public would prefer a scale reflecting the number of immediate and potential casualties as well as the impact on the environment and thus the influence on future generations.

8 Radiation Measurements and Monitoring

The interactions of various types of ionizing radiation with matter form a basis for measuring the amount of radiation present and emitted by a source or absorbed in a particular material. Radiation instruments include portable survey instruments that are designed to detect radiation and measure exposure or absorbed dose and laboratory instruments that allow precise quantitation and identification of the radiation source. Various detectors are used in them, and these can be roughly divided into two categories: gas-filled chambers and crystalline materials. The operation of each of these various devices is based on the liberation of electrons in a medium and the collection and processing of the ions by electronic means.

In addition to the detection and monitoring instruments using an active detector (sensor) which produces an electrical signal and must be powered by a source of electricity, there are other types of radiation sensors – passive dosimeters. These sensors accumulate the information about the measured quantity (usually the dose) and after the measurement they have to be read using specific processes and a reader. An example of these types of radiation monitors are film badges or thermoluminescent dosimeters (TLDs) to measure the radiation dose received by persons over a

Table 4 Main features of the survey instruments relying on specific detectors as sensors of radiation

					Detectable energy range	
Instrument	Alpha	Beta	Photon	Neutron	Low-end boundary	High-end boundary
Ionization chamber	NA	Fair	Good	NA	40–60 keV	1.3–3 MeV
Gas-flow proportional counter	Good	Good	Poor	Poor	5–50 keV	6–9 MeV
Geiger-Muller detector	Fair	Good	Poor	Poor	30–60 keV	1–2 MeV
ZnS(Ag) scintillation detector	Good	Poor	NA	NA	30–50 keV	8–9 MeV
NaI(Tl) scintillation detector	NA	Fair	Good	NA	10–60 keV	0.1–3 MeV
Cs(Tl) scintillation detector	NA	Poor	Good	NA	40–60 keV	1.3–3 MeV
Plastic scintillation detector	NA	Fair	Good	NA	40–60 keV	1.3–3 MeV
BF_3 proportional counter	NA	NA	NA	Good	0.025 eV	100 MeV
^3He proportional detector	NA	NA	Poor	Good	0.025 eV	200 MeV

period of time, usually a month or a quarter. For short-term monitoring of work, a pocket ion chamber dosimeter is worn or an electronic dosimeter. Electronic dosimeters can be useful for this purpose because they can be set to indicate an alert or warning level of exposure.

For field measurements and surveys a variety of instruments based on specific detectors are widely used. However, one has to be aware of their limitations since none of the monitors can be universally used in all cases. Table 4 (Voigt and Fesenko 2009) shows an overview of survey instruments and their properties from which we can see the potential for their applications.

9 Main Principles of Radiation Protection

The main task of radiation protection is aimed at avoiding the occurrence of any deterministic effects and minimizing the risk due to stochastic effects to acceptable levels. Consequently, the principal objectives of radiation protection may be formulated as follows:

(a) ***To prevent deterministic effects*** by keeping the dose below levels approaching the threshold level, and
(b) To ensure adequate protection under normal or planned situations by implementing effective protective actions and other measures to reasonably reduce the risk of stochastic effects by ***keeping the dose below levels***

approaching the generic criteria, namely the dose limits and dose constraints set by regulatory authorities.

The precept of ***justification*** postulates that any decision which may alter the radiation exposure situation should do more good than harm. The ***optimization*** of protection consists in adopting such measures that the likelihood of incurring exposures, the number of people exposed, and the magnitude of their individual doses should all be kept as low as reasonably achievable, taking into account economic and societal factors. The ***limitation*** means that the total dose to any individual from authorized sources in planned exposure situations other than the medical exposure of patients is not supposed to exceed the dose limits set by the regulatory authority.

The regulatory body is responsible for the planning, implementation and verification of remedial actions including

- approval of the remedial action plan and granting of any necessary authorisation;
- establishment of criteria and methods for assessing safety;
- review of work procedures, monitoring programmes and records;
- review and approval of significant changes in procedures or equipment that may have an environmental impact or may alter the exposure conditions of remediation workers or of members of the public;
- receipt and assessment of reports of abnormal occurrences;
- performance of regular inspections and, where necessary, any enforcement actions;
- verification of compliance with the legal and regulatory requirements, including criteria for waste management and discharges established for the remediation programme; and
- where necessary, establishment of regulatory requirements for post-remediation control measures (Voigt and Fesenko 2009; IAEA 2007, 2016).

The overall aims and objectives of implementing remedial measures in a radioactively contaminated ecosystem is to improve the radiological situation for the affected populations where compliance with the criteria related to the exposure of people and impact on the environment should be unambiguously defined using the relevant quantities and units. In achieving this, the main goal consists in reducing the adverse radiological impact imposed by the remaining radioactive contamination.

In accordance with the international standards, the main objectives of remediation actions should be aimed at

- the reduction of the doses to individuals or groups of individuals who may be exposed;
- the averting of doses to individuals or groups of individuals who are likely to get exposed in the future; and
- the prevention or reduction of environmental consequences due to the radionuclides present in the contaminated area.

The reductions in the exposure to individuals as well as environmental impacts should be to be achieved by means of interventions focusing on removing the existing sources of contamination, modifying the pathways of exposure, and/ or reducing the numbers of individuals exposed to radiation from the source.

10 Conclusion

Nuclear science and technology offer many beneficial peaceful uses, including the generation of energy and the production of radionuclides for use in medical examinations and cancer treatment. All radioactive sources and nuclear materials as well as facilities housing them have to be carefully disposed at the end of their useful lives. At present, before embarking on any new programmes involving the use of radioactive material, preliminary plans for the eventual decommissioning, the safe disposal of radioactive material and necessary remediation are developed. However, this was not always the case since at the time when many radiation, nuclear and mining facilities were built and operated, there were no such strict safety and security requirements as we have today. Many countries are now implementing or devising plans for decommissioning such facilities and for remediation of radioactively contaminated sites. The IAEA helps them to do so, bringing its international expertise (Amano 2016). An essential component of planning for decommissioning and environmental remediation is knowledge sharing. This requires using of unified terminology where the correct understanding and interpretation of use of quantities and units in radiation protection plays an important role.

References

Amano Y (2016) Decommissioning and remediation: enhancing safety of the public and the environment. IAEA Bull:1 www.iaea.org/bulletin

Balonov MI (2012) On protecting the inexperienced reader from Chernobyl myths. J Radiol Prot 32:181–189

Balonov MI (2013) The Chernobyl accident as a source of new radiological knowledge: implications for Fukushima rehabilitation and research programmes. J Radiol Prot 33:27–32

IAEA (2007) Remediation process for areas affected by past activities and accidents. IAEA safety standards series no. WS-G-3.1. IAEA, Vienna

IAEA (2009) The international nuclear and radiological event scale user's manual. IAEA, Vienna

IAEA (2016) Advancing implementation of decommissioning and environmental remediation programmes. IAEA nuclear energy series no. NW-T-1.10

ICRP (2007) The 2007 Recommendation of the International Commission on Radiological Protection. ICRP Publication 1003. Annual report ICRP 37:(2–4)

ICRU (2011) Fundamental quantities and units for ionizing radiation (Revised). ICRU report no. 85. J ICRU 11:1

Martin JE (2000) Physics for radiation protection. Wiley, New York

Martin A, Harbison S, Beach K, Cole P (2012) An introduction to radiation protection. Hodder Arnold, London

NCRP (2009) Ionizing radiation exposure of the population of the United States. NCRP report no. 160. NCRP, Bethesda
Sabol J, Weng PS (1995) Introduction to radiation protection dosimetry. Scientific World Publisher, Singapore
Thielen H (2012) The Fukushima Daichi nuclear accident – an overview. Health Phys 103:169–170
UNSCEAR (2011) Report of the United Nations Committee on the effects of atomic radiation sources and effects of ionizing radiation (UNSCEAR) 2010. United Nations, New York
Voigt G, Fesenko S (2009) Remediation of contaminated environments. Elsevier Science, Oxford

Index

A
Absorbed dose, 301–302
Absorption, 94
Adsorption, 94
Air dose rates, remediation effect, 264–268
Air kerma rate constant, 303
Algae, 96, 97
Aluminosilicates, 226, 227
Ambient dose equivalent, 305
Ameliorants, 113
  application, 151–154
  biological mobility, 152–154
  mineral and organic fertilizers, 150
  physicochemical properties, 151–152
Aminopolycarboxylic acid, 188
Anthropogenic radionuclides, IWF, 236–238
Arbuscular mycorrhizal fungi (AMF), 190, 191
As low as reasonably achievable (ALARA), 308
Atmospheric tests, 273
Atomic power engineering, 65

B
Bacteria, 98–99
Bioaccumulation, 178, 179, 187, 202
Bioaccumulation coefficient (BC), 15, 206
Bioaugmentation, 176
Biochar
  accumulation, 127
  agriculture, 116–117
  agrochemical analysis, 119
  agrochemical properties, 116
  ameliorative properties, 134
  bergmeal, 118, 126, 130, 131
  bokashi, 118, 125, 128, 129
  cesium-selective sorbents, 114
  characteristics
    carbon, 114
    graphite, 115
    macro-elements, 115
    magnesium, 115
    microelements, 115
    microstructure, 115
    organic compounds, 115
    organic heterocycles, 115
    phenoxides, 115
    phosphorus, sulfur and nitrogen, 115
  Chernobyl accident, 114
  composts, 117
  experimental soil mixture, 118
  fossils, 113
  fuel and energy resources, 113
  mangold, 120, 124
  measurement activity concentration, ^{137}Cs, 119–120
  mechanisms, 134
  pollutants, 114
  pyrolysis processing, 114
  root absorption, 113
  soil ameliorants
    agrochemical indicators, 122–124
    mangold, 124–126
    peat soil, 127–131
    physical and chemical forms, ^{137}Cs, 131–134
  soil-improving additives, 119
  valuable soil ameliorant, 113
  vegetative experiment, 118
Bioconcentration factors (BCF), 183
Biodiesel, 113

D. K. Gupta, A. Voronina (eds.), *Remediation Measures for Radioactively Contaminated Areas*, https://doi.org/10.1007/978-3-319-73398-2

Biogas, 113
Biological effects, radiation exposure, 309–310
Biological treatment methods, 3
Biological wastewater treatments, 95
Biomineralisation, 178, 179, 181
Bioremediation, 164–170, 178–179, 201–203
AMF, 190, 191
anthropogenic environmental contamination, 163
applications, 182–191
artificial/technogenic radionuclides, 163
bio- and phytoremediation, 164
Chernobyl nuclear power plant disaster, 163
contaminated soils, 174–182
conventional strategies, 192
cost effective process, 175, 176
depleted uranium, 173
genetics and genome-level characteristics, 192
IAEA, 193
in situ *vs.* ex situ, 176
laboratorial conditions, 192
laboratory experiments, 185
in medicine, 174
mesocosm systems, 192
natural attenuation, 177
nuclear disasters, 170–171
nuclear fuel cycle (*see* Nuclear fuel cycle)
nuclear test sites, 171–172
nuclear weapons, 163
physicochemical characteristics, 164
vs. physico-chemical remediation, 177–178
phytoextraction, 188–190
and phytoremediation, 201–203
poly-metal decontamination, 182
radionuclides (*see* Radionuclides)
uranium mines, 166
Biosorbents, *see* Biosorption
Biosorption, 7
algae, 96
bacteria, 98–99
biomass/adsorbents, 95
factors, 95–96
fungi and yeast, 96–97
plants, 96
polymeric substances, 95
solid and liquid phases, 95
Biostimulation, 176
Bremsstrahlung, 299
Brookhaven National Laboratory (BNL), 211

C
Carbamide, 227
Carbon-encapsulated magnetic nanoparticles (CEMNPs), 102
Carbon nanomaterials (CNM), 103
Carbon nanotubes (CNTs), 103
Cartridge filters, 4
Catastrophic radiation accidents, 223
Cation exchange capacity (CEC), 143
Cesium and strontium, 53
Cesium-134 (^{134}Cs), 215, 260–262
Cesium-137 (^{137}Cs), 260–262
ability of plant species, 208
Chernobyl fallouts, 207
hydroponic experiments, 207
Kochia scoparia, 214
phytoremediation, 209
rhizospheric mobilization, 207
soil contamination, 207
and ^{90}Sr, 210
Cesium sorption, 54
Cesium transfer decreasing factors, 62
Chemical pollutants, 95
Chemical vapour deposition (CVD), 103
Chemisorbents, *see* Chemisorption
Chemisorption
chemical-conjugate absorbent, 100
graphite based chemical adsorbents, 100
natural and industrial effluents, 100
Chernobyl disaster, 51, 114, 144, 165, 170, 180, 204, 224, 234
Chernobyl's nuclear power plant accident, 183
Clay aquifuges, 248
Coal ash slag (CAS), 186
Collective water filters (CWF), 230, 232, 233
Colorado River, 8
Committed effective dose, 305
Committed equivalent dose, 304
Comparative tests, 230
life test of IWF, 236–238
natural radionuclides, 238–239
Competitive sorption
a priori estimation, 66
aqueous electrolyte solution, 86
chromatographic equipment, 87
countermeasures, 66
decontamination, 65, 67
diffusion-kinetic mode, 87
extractant, 86
flow-type reactor, 86
heterogeneous equilibria, 67
ion-exchange desorption, 66
kinetics
analytic solution, 77
degree of reversibility, 80
diffusion-controlled sorption, 77
direct and inverse reaction constants, 79
equilibrium desorption, 75
equilibrium system, 77

exponential factors, 79
inflection point, 80
iron-potassium cyanoferrate, 75
linear adsorption mechanism, 77
mini-reactors, 76
non-equilibrium decontamination coefficient, 77–79
polyethylene test tubes, 80
relaxation time values, 77
silica gel powder, 80
ultra-diluted solution, 77
material, 69
microbe strains and phyto-sorbents, 66
mini-reactors, 86
multiphase and multicomponent system, 66
potassium fertilizers, 65
radionuclides, 65
relaxation time values, 88
RIP, 66
self-sustaining process, 88
soil deactivation technologies, 66
sorbents, 66, 86
sorption-active clay fraction, 65
statics
acceptable strategy, 73
complexing agent solutions, 72
convenience of analysis, 72
convenient phase separation method, 72
decontamination coefficient, 71, 72
desorption efficiency growth, 72
electrolyte, 69, 74
equilibrium competitive system, 69
exchange center types, 74
heterogeneous equilibrium, 70
homogeneous reaction, 70
hydromica microcrystals, 74
isotherm, 70, 73
liquid desorbing substance and sorbent, 73
microelement, 67, 69, 70
microelement sorption equilibrium, 72
nonselective exchange centers, 74
PB and soil solution, 74
polyantimonous acid, 75
potassium/ammonium ions, 74
soils, 74
Sr(II) ions, 75
thermodynamic equilibrium, 67
thermodynamic estimation, 69
Compton scattering, 296, 297
Constituents of concern (COC), 3
Crystallic FB-2 filters, 238

D

Daylight surface, radionuclide carry-over by plants, 280–282
Deactivation
characteristics, soils and grounds, 251
distribution coefficients of cesium, 252
drinking water
CWF, 230, 232, 233
filters life test, 233–236
IWF, 230, 232, 233
large-scale, 240
radionuclides, 230
liquid and dry methods, 247
LRW, 247–248
radioactive matter, 246
solid radioactive waste, 248–250
solutions, 246
Decontamination Pilot Project, 259
Deflation processes, 277
Dehydroascorbate reductase (DHAR), 191
Depleted uranium (DU), 173
Desorbing electrolytes, 248
Deterministic effects, 309
Diethylenetriaminepenataacetic acid (DTPA), 188
5-Dimethyl-phenylenedene (DDPD), 100
Directional dose equivalent, 306
Dose equivalent rate, 301, 303–304
Dose of medicine, 301
Dose of radiation, 301
Dosimetric quantities and units
absorbed dose, 301–302
exposure, 300–301
kerma, 302, 303
radiation protection quantities, 300
Drinking water decontamination, 230–240
activation products, 229
deactivation (*see* Deactivation, drinking water)
fissionable materials, 229
fission products, 229
maximal permissible activities, 230, 231
temporary permissible levels, 230

E

East Urals Radioactive Trace (EURT), 171
EDR distribution, 286
Effective dose, 305
Electric double layer, 34
Electrokinetic decontamination, 39
Electrokinetic fences, 42
Electrokinetic method of deactivation, 248

Electrokinetic remediation, 11–12, 32, 33
  acid and alkaline fronts, 35
  advantages, 33
  ammonium ion, 41
  contamination, 36
  electrode compartment, 40
  electrokinetic removal, 36–39
  electromigration, 40
  pilot scale and field applications, 41–42
  potassium nitrate, 41
  principle, 32
  radionuclides, 39–41
  scale and field applications, 41–42
  soil flushing/washing, 39
  transport mechanisms, 33
Electrokinetic removal, 36–39
Electromigration, 34, 35
Electroosmosis, 33
Electro-osmotic flow, 35
Elementary competitive system, 88
Eluent deactivation, 247
Energy fluence rate, 295
Environmental decision support system (EDSS), 18
Equivalent dose, 304
Equivalent dose rate (EDR), 277, 278
Estimating deflation processes, soil cover, 279–280
Estimating processes, overgrowing with vegetation, 277
Ethylenediamine disuccinate (EDDS), 188
Ethylene-diamine-tetraacetic acid (EDTA), 189
Exposure, 300, 301
  radiation source, 306–308
  rate, 301, 303
Ex-situ deactivation, 247, 248
Extracellular polymeric substances (EPS), 180

F
Ferrocyanide ions, 58
Ferrocyanide sorbents, 54, 58
Ferrocyanides leaching, 59
Filling with clean soil, 285–286
Filters
  comparative tests, 236–239
  life test, 233–236
  sanitary qualifying, 239, 240
  and spent sorbent, 253–254
Fluence, 294
Fluence rate, 295
Frayed edge sites (FES), 141
Fresh water
  clinoptilolite, 228
  equilibrium distribution coefficients, 227–229
  sorption properties, 227
FSUE "Radon", 254
Fukushima Daiichi accident, 144, 168, 224
Fukushima Daiichi nuclear power plant (FDNPP), 10, 183, 259
Fukushima-1 nuclear power plant, 248
Fungi, 96–98

G
Geobacter, 8
GeoMelt-ISV project, 13, 14
Groundwater, 8

H
Half-value layer (HVL), 297
Harmful tissue reactions, 309
Henry's law, 67, 70, 77, 81
Heteropolyacids (HPAs), 100
Hexafluoride, 166
High-energy photons transfer, 298
High-level radioactive waste (HLW), 223
High Purity Germanium (HPGe) gamma spectroscopy, 260
Humic acids (*HA*)
  acid-base properties, 82
  adsorption regularities, 81
  ANOVA techniques, 83
  energy inhomogeneity, 83
  geochemical systems, 81
  humate solution, 85
  Langmuir isotherm, 82
  logarithmical form, 84
  strontium ion sorption, 84, 85
  thermal/chemical treatment, 82
Hydrosol™ solution, 210
Hydrous ferric oxide (HFO), 102
Hydrous manganese oxide (HMO), 102
Hygienic norms, 57
Hyperaccumulators, 184, 202, 217

I
Indirectly ionizing radiation, 297
Individual water filters (IWF), 227, 230, 232, 233, 239
In-situ deactivation, 247
In situ leaching (ISL), 164

In situ remediation's technologies, 11
In situ soil flushing, 43
In situ vitrification (ISV), 12
Intensive Contamination Survey Areas (ICSAs), 259
Interaction of radiation with matter
  absorption of radiation energy, 295
  attenuation, 295, 296
  biological effects, 295
  bremsstrahlung, 299
  charged particles, 295, 298, 299
  Compton scattering, 296, 297
  directly ionizing radiation, 298
  electron loses, 299
  energy loss, 299
  high-energy photons transfer, 298
  HVL, 297
  interaction coefficients, 297, 298
  ionization, 299
  LET, 298
  neutrons, 297
  photoelectric effect, 296
  positron-electron pair, 297
  range, 295, 296
  secondary charged particles, 298
  specific ionization, 299
  threshold energy, 297
  uncharged particles, 295, 296
  X-ray and gamma, 296
Intermediate Level Radioactive Waste (ILRW), 44
International Atomic Energy Agency (IAEA), 1, 168, 246
International Commission on Radiation Units and Measurements (ICRU), 292
International Commission on Radiological Protection (ICRP), 292
International Energy Agency (IEA), 166
International Nuclear and Radiological Event Scale (INES), 309
Ion pair, 299
Isomorphous substitution, 141
ISV equipment, 13
IWF-1, 232
IWF-2, 232
IWF-3, 232
IWF-4, 232
IWF-5, 232
IWF-6, 232
IWF-7, 233
IWF-8, 233
IWF-9, 233
IWF-10, 233

J
Japan Atomic Energy Agency (JAEA), 259

K
K-channel, 122
Kerma, 302, 303
Kinetic Energy Released per unit Mass, 302
Kyshtym accident, 143

L
Laboratory-scale experiments, 11
Langmuir equation, 67
Langmuir theory, 66
Large-scale deactivation, drinking water, 240
Least square method, 85
Life test of IWF
  anthropogenic radionuclides, 236–238
  count rate, 235
  decontamination factors, 236
  loading spiked and non-spiked water, 234, 235
  measurement, 233
  NPF-C and KU-2, 235
  permissible activities of radionuclides, 234
  principles of radioactive tracer's method, 234
  protection shield, 234
  before treatment, 234
  value of retention volume, 233
  values of counting coefficients, 235
Liming, 149, 150
Linear energy transfer (LET), 298
Linear stopping power (S_l), 298
Liquid radioactive waste (LRW), 247–249, 251, 253, 255
Logarithmic Langmuir equation, 81
Low-level LRW, 248
Low Level Nuclear Waste (LLNW), 186

M
Maghemite, 102
Magnetically assisted chemical separation (MACS), 102
Manganese oxides (MnOs), 101
Mangold, 120, 121, 124–126
Mass attenuation coefficient (μ/ρ), 297
Mass energy absorption coefficients (μ_{en}/ρ), 297
Mass energy transfer coefficients (μ_{tr}/ρ), 297
Mass stopping power (S_m), 298

Mathcad 14 program, 78
Mathematical modelling approaches
  acetate biostimulation, 19
  EDSS, 18
  EXPURT model, 21
  factors, 17
  hydrogeochemical modelling, 20
  kinetic and equilibrium models, 21
  software based models, 16
  STRATEGY project, 17
Mean mass depth, 262
Mean residual dose rate factors, 269
Measured activity depth profiles, 268–270
Mechanical remediation of radiocesium
  air dose rates, 264–268
  costs and benefits, 260
  ^{134}Cs, 260–262
  ^{137}Cs, 260–262
  farms, schoolyards, gardens, parks, 263
  FDNPP, 259
  JAEA, 259
  measured activity depth profiles, 268–270
  strategies, 263
  topsoil mixing, 263
  topsoil stripping, 260, 263, 264
  total cost, 263
Metal sulfide ion exchangers (MSIEs), 100
Metal wastes, 93–94, 104
Microbial bioreduction, 6–9
  anaerobic microbial communities, 6
  bacterial electron transport, 9
  decontamination test process, 10
Mill tailings, 5
Mineral composition, 60
Ministry of Education, Culture, Sports, Science and Technology (MEXT), 260
Ministry of the Environment (MoE), 259
Mixed Waste Management Facility (MWMF), 212
Monodehydroascorbate (MDHAR), 191
Monte Carlo radiation transport, 264
Monticello Mill, 5
Montmorillonite clays, 154
Multi-walled carbon nanotubes (MWCNTs), 103

N
Nanomaterial
  aqueous systems, 101
  bioremediation approaches, 101
  CNM, 103
  Fe and Fe^0, 102
  functions, 101
  maghemite, 102
  Mn and Zr, 101–102
  nano-sorbents, 101
  Ti, MgO and ZnO nanoparticles, 103
Nano-remediation, 104
Nanoscaled zero valent iron (NZVI), 102
Nanosorption, 101
Nanotechnology, 101
National Nuclear Center of the Republic of Kazakhstan, 284
Natural radionuclides, 238–239
Natural water bodies
  advantages, 227
  aluminosilicates, 226, 227
  fresh water, 227–228
  seawater, 229
  sorbents, 226
  sorption, 226
  surface-modified sorbents, 226
  thin-layer modification method, 227
  TLIS, 226, 227
Naturally occurred radioactive materials (NORMs), 163, 225
Nevada Test Site, 172
N-(hydroxyethyl)-ethylenediaminetriacetic acid (HEDTA) trisodium salt, 190
Nickel oxide, 103
Nickel-potassium ferrocyanides (NPF), 52
Nitrogen fertilizers, 146
N,N-disalicylidene-4, 100
Nonwoven filtering materials, 226
Novaya Zemlya test site, 172
Nuclear Energy Agency (NEA), 166
Nuclear explosions in atmosphere, 273
Nuclear fuel cycle
  nuclear fuel reprocessing, 169–170
  nuclear power plants, 167–169
  uranium fuel fabrication, 166–167
  uranium mining and milling, 164–166
  waste repositories, 169–170
Nuclear Non-Proliferation Treaty (NPT), 172
Nuclear power plants (NPPs), 167
Nuclear Regulation Authority (NRA), 260
Nuclear weapon tests, 223, 224, 248

O
Old Rifle processing site, 8
Operational quantities, 305
Organic RW, 249
Oxidize dehydroascorbate (DHA), 191

P
Pacific Northwest Laboratory (PNL), 12
Passive dosimeters, 310
Permeable treatment wall (PTW), 4
Personal dose equivalent, 306
Phosphate fertilizers, 148
Photoelectric effect, 296
Physicochemical methods
  electrokinetic remediation, 32
  electro-osmotic flow, 34
  on-site soil remediation technologies, 32
  soil washing and incineration, 32
  solid/liquid wastes, 31
  technologies, 32
Phytoaccumulation, 14
Phytoextraction, 14
  arbuscular mycorrhizal fungi, 188
  Brassica juncea, 189
  chelating agents, 189
  CO_2 levels, 190
  concentrations of Cs, 189
  description, 205
  fertilization, 189
  phytoaccumulation, 205–209
  radionuclides, 188, 189
  rizhosphere, 188
  schematic representation, 206
  soil microbial community, 188
  U-citric acids complexes, 190
Phytomining, 182
Phytoremediation, 14–16, 104
  applications, 204–213
  bioremediation, 201–203
  ^{137}Cs, 209
  Cs contaminated soils, 213–215
  ^{3}H, 212
  history, 203–204
  phase of, 218
  plants, 218
  radionuclides, 208, 217–218
  soil remediation techniques, 219
  Sr, Tc and Cl contaminated soils, 216–217
  U contaminated soils, 216
Phytostabilization, 182, 183, 190, 191
  description, 205
  radionuclides, 212, 213
  schematic representation, 206
Phytovolatilization, 182
  description, 205
  plant's ability, 211, 212
  schematic representation, 206
Polyethylene terephthalate (NTFM), 226
Polypropylene membranes (MP), 226
Portland cement, 45
Positron-electron pair, 297
Potassium fertilizers, 147
Prussian blue (PB), 74
PTW performance monitoring system, 5

R
Radiation accidents, 245, 248, 249
Radiation dose limits, 308, 312
Radiation exposure, 292, 300, 308–310, 312
Radiation field, 294, 295
Radiation measurements and monitoring, 310–311
Radiation protection, 303–308
  average annual population exposure, 291, 292
  biological effects of radiation exposure, 309–310
  dose limits, 308
  dosimetric quantities and units, 300–303
  interaction, 295–300
  measurements and monitoring, 310–311
  principals, 311–313
  quantities (*see* Radiation quantities)
  radiation field, 294, 295
  radioactive contamination, 291
  radioactive releases, 291
  radioactive sources, 292–294
  radiological compliance criteria, 292
  remediation actions, 292
  site characterisation, 292
Radiation quantities
  assessment
    external exposure, 305, 306
    internal exposure, 306
  committed effective dose, 305
  committed equivalent dose, 304
  dose equivalent rate, 303–304
  effective dose, 305
  equivalent dose, 304
  ICRP, 303
  radiation source, 306–308
Radiation sensors, 310
Radiation source, 306–308
Radiation weighting factors, 304–306
Radioactive biological waste, 254, 255
Radioactive contamination, 36, 201, 204
  caesium levels, 157
  ^{137}Cs, 139
  landscapes, 51
  sod-podzolic soils, 147
  territory, 157

Radioactive fallouts, 223
Radioactively contaminated lands, *see* Radioactive waste (RW)
Radioactively contaminated water, *see* Water decontamination
Radioactive sources, 303
activity, 293
amount of radioactive material, 293
characterization, 292
count statistics, 294
emission rate, 294
mass activity concentration (a_m), 293
measurement, 294
point sources, 293
surface activity concentration (a_s), 293
volume activity concentration (a_v), 293
Radioactive tracer's method, 234
Radioactive waste management, 312
Radioactive waste (RAW), 10, 93–94, 246, 288
deactivation (*see* Deactivation)
disposal, 255
IAEA, 246
rehabilitation method, 246
treatment (*see* Treatment of RW)
Radiocesium, 186, 213–215, 259
See also Mechanical remediation of radiocesium
Radiocesium gamma rays, 263
Radiocesium interception potential (RIP), 54, 66
Radiological warfare agents, 274
Radiometric parameters, 277, 278
Radionuclide carry-over by plants
daylight surface, 280–282
evaluating, 276
Radionuclides, 36, 229
activity concentration, 277–279
americium, 250
bacteria and fungi, 179
beta emitting, 255
catastrophic radiation accidents, 223
cesium, 252
concentrations, 225
from contaminated soils and ground, 250
dissolving and desorbing, 247
drinking water (*see* Drinking water decontamination)
durability of fixation, 246
elimination, 250
hydrolyzable, 251
ions, 248
leaching, 252, 254, 255
LRW, 255
mechanisms, 178
and metals, 178
microbial transformation and immobilization, 179–182
migration, 248
natural waters, 225
natural waters with salt content, 226–229
and radioactive aerosols, 224
separation, 251
short-lived, 223
storage of spent sorbents, 253
strontium, 254
volatile, 254
water migrating, 224
Radiostrontium, 206
Reagent deactivation, 250
Regular Exchange Sites (RES), 141
Rehabilitation, 246
activities, 245–246
measures, 288
radioactively contaminated land, 286
RW (*see* Radioactive waste (RW))
Remediation, 104, 273
overgrowing with vegetation, 282
STS (*see* Semipalatinsk test site (STS))
Residual dose rate factors, 264–267
Residues of nuclear activities (RONA), 285
Rhizofiltration
description, 205
plant roots, 209, 210
schematic representation, 206
Rifle site, 8

S

Salt composition, 54
Sanitary qualifying of filters, 239, 240
Savannah River Site (SRS), 211
Seawater decontamination, 229
Self-Assembled Monolayers on Mesoporous Supports (SAMMS), 101
Semipalatinsk test site (STS), 224, 249, 274–283
atmospheric tests, 273
nuclear testing, 273
plowing (*see* Soil plowing)
radiation hazardous, 274
research, 274
soil rehabilitation, 274
storage of removed soil, 274
territory, 286–288
underground nuclear tests, 273
Shubert method, 83

Siderophores, 178
Single-walled carbon nanotubes (SWCNTs), 103
Sod-podzolic soil
  cation exchange capacity, 152
  characteristics, 146
  dry meadow, 157
  field experiments, 149, 150
  humus, 149
  organic fertilizers, 150
  soil solution, 147
  ^{90}Sr, 152
Soil-based remedial options, 143
Soil cover, deflation processes, 279–280
Soil layer interchange, 263, 264, 266, 268, 270
Soil mixing, 264, 270
Soil plowing
  choice of remediation technology, 275
  deflation processes, 277
  estimating deflation processes, 279–280
  filling with clean soil, 285–286
  ground by plowing soil, 282–283
  ground by removing residues, nuclear activity, 284
  overgrowing with vegetation, 277
  radiometric parameters, 277, 278
  radionuclide activity concentration in top (contact) soil, 277–279
  radionuclide carry-over by plants, 276, 280–282
  remediated spots overgrowing with vegetation, 282
  selection of research areas, 274–276
Soil rehabilitation, 274
Soil remediation
  electrokinetic remediation process, 11
  ISV, 13
  phytoremediation, 14
  PNL, 12
  radioactive soil, 9
  radioecological models, 20
  and waste, 9
Soil un-remediated, 260–262
Solid radioactive waste, 248–250
  after deactivation
    organic RW, 249
    soil and ground, 248
    spent sorbents, 249–250
Sorbents, 226–230, 232–240
Sorption method
  aluminosilicates, 54
  cesium radionuclides, 56, 59–62
  cesium sorption, 54
  characteristics, 53
  clinoptilolite, 52
  ecological safety, 57–59
  ferrocyanide sorbents, 58
  ferrocyanides leaching, 58
  glauconite, 55
  materials, 51, 52
  nickel and ferrocyanide ions, 57
  NPF-Cl sorbent, 54, 56
  NPF-Gl sorbents, 54
  radionuclides, 52
  rehabilitation, 52–53
  Shivertooyskoye deposit, 55
  specificity and capacity, 53
  strontium radionuclides, 56
  surface-modified sorbents, 60
Special Decontamination Area (SDA), 259
Spent sorbents
  and filters, 253–254
  solid radioactive waste, 249–250
^{88}Sr isotope, 37
Stabilisation/solidification (S/S), 44
Stochastic dose, 307
Stochastic effects, 309
Straggling of the range, 299
Strontium (II), *see* Humic acids (*HA*)
Strontium-90 (^{90}Sr)
  action mechanism, 151–154
  agricultural production, 159
  agricultural techniques, 159
  agrochemical indicators, 151
  ameliorants, 150
  behaviour, 140–143
  bioavailability parameter, 154
  biological hazard, 139
  cereal crops, 150
  Chernobyl accident, 139
  distribution coefficients, 147
  grasslands, 157–158
  liming, 149, 150
  mechanical and physicochemical methods, 140
  mechanical methods, 143–145
  mineral fertilizers
    biological mobility, 152–154
    crops, 146
    nitrogen, 146–147
    NPK, 148–149
    phosphate, 148
    physicochemical properties, 151–152
    potassium, 147, 148
    types, 146
  minerals and montmorillonite clays, 155

Strontium-90 (^{90}Sr) (*cont.*)
mobility factors, 142
nuclear weapon tests, 139
organic fertilizers, 149
physicochemical remediation methods, 145–154
radiostrontium transfer, dry meadows, 158
reduction factors, 144
RSr soil-plant transfer, 158
sod-podzolic soil, 148, 156
soil-based remedial techniques, 140
soil-plant system, 153
sorbent physicochemical properties, 156
sorbents, 154, 156
Styrofoam™, 210
Surface-modified sorbents, 226

T
Technical Assessment Group (TAG), 13
Temporary permissible levels, 230
Territory STS, 286–288
Tetrafluoride, 166
Thermoluminescent dosimeters (TLDs), 310
Thin-layer inorganic sorbents (TLIS), 226, 227
Thin-layer modification method, 227
Threshold energy, 297
Threshold level, 309
Tissue weighting factors, 305, 306
Titanium dioxide, 103
Top (contact) soil, radionuclide activity concentration, 277–279
Topsoil mixing, 263, 264, 266, 268, 270
Topsoil stripping, 260, 263, 264, 266, 268, 270, 271
Transport packaging containers (TPC), 284
Treatment of RW, 250–253
decontamination
soils and grounds, 250, 251
solutions after deactivation, 251–253
radioactive biological waste, 254, 255
spent sorbent and filters, 253–254
Tsar Bomba test, 172

U
Underground nuclear explosions (tunnels and boreholes), 273
Underground nuclear tests, 273, 287
United Nations Environmental Program (UNEP) reports, 173
University of Science and Technology Houari Boumediene (USTHB), 37
Uranium
concentration, 225
drainage waters, 202
mine tailings, 212
mining and milling, 164–166
in shoots, 208

V
Vitrification, 45, 46

W
Waste disposal, 93
Waste management, 93
Water contamination, 225
Water decontamination, 223, 229
contamination of natural water, 225
drinking water (*see* Drinking water decontamination)
HLW, 223
natural water bodies, 224, 226–229
NORMs, 225
nuclear weapon tests, 224
radioactive fallouts, 223
radionuclides (*see* Radionuclides)
Water remediation
attenuation technologies, 6
COCs, 3
groundwater, 4
hydraulic performance, 5
mill tailings, 5
PTW project, 4
US Department of Energy's Hanford Site, 3
Water treatment plants, drinking water, 240
Waterproof clayey soils, 248
Weighting factors, 306
West Valley Demonstration Project treatment, 4
Wet gravity separation method, 250

Y
Yeast, 96–97

The manufacturer's authorised representative in the EU is Springer Nature Customer Service Centre GmbH, Europaplatz 3, 69115 Heidelberg, Germany. If you have any concerns regarding our products, please contact ProductSafety@springernature.com

Printed and bound by CPI Group (UK) Ltd, Croydon, CR0 4YY
05/10/2026
02228141-0009